ADVANCED TOPICS
IN SCIENCE AND TECHNOLOGY IN CHINA

ADVANCED TOPICS IN SCIENCE AND TECHNOLOGY IN CHINA

Zhejiang University is one of the leading universities in China. In *Advanced Topics in Science and Technology in China*, Zhejiang University Press and Springer jointly publish monographs by Chinese scholars and professors, as well as invited authors and editors from abroad who are outstanding experts and scholars in their fields. This series will be of interest to researchers, lecturers, and graduate students alike.

Advanced Topics in Science and Technology in China aims to present the latest and most cutting-edge theories, techniques, and methodologies in various research areas in China. It covers all disciplines in the fields of natural science and technology, including but not limited to, computer science, materials science, life sciences, engineering, environmental sciences, mathematics, and physics.

For further volumes:
http://www.springer.com/series/7887

Weiyan Pan · Kai Li

Propagation of SLF/ELF Electromagnetic Waves

ZHEJIANG UNIVERSITY PRESS
浙江大学出版社

Springer

Prof. Weiyan Pan
China Research Institute of Radiowave Propagation
Qingdao, China

Dr. Kai Li
Department of Information Science and Electronic Engineering, Yuquan Campus
Zhejiang University
Hangzhou, China

ISSN 1995-6819 ISSN 1995-6827 (electronic)
ISBN 978-3-662-52238-7 ISBN 978-3-642-39050-0 (eBook)
DOI 10.1007/978-3-642-39050-0
Springer Heidelberg New York Dordrecht London

Jointly published with Zhejiang University Press, Hangzhou
ISBN: 978-7-308-11639-8 Zhejiang University Press, Hangzhou

Softcover reprint of the hardcover 1st edition 2014

Printed on acid-free paper

Springer is part of Springer Science+Business Media (www.springer.com)

Preface

Super low frequency (SLF: 30–300 Hz) and extremely low frequency (ELF: below 30 Hz) are the lowest frequency ranges in the applied radio spectrum. In SLF/ELF ranges, the wavelength is over 1,000 km, and the wave propagation in the Earth's space will be effected inevitably by the ground and the ionosphere. It is known that the excitation and emission in the space between the ground and the lower boundary of the ionosphere are of extreme difficulty. Meanwhile, because of the properties of low carried frequency, narrow bandwidth, and low information capacity, the applications of SLF/ELF waves do not range widely. However, in the past century the subject *Propagation of SLF/ELF Electromagnetic Waves* has been intensively investigated because of its useful applications in communication with submarines, geophysical prospecting and diagnostics, and seismic electromagnetic precursors monitoring. In the pioneering works of Wait, Budden, and Galejs, analytical solutions on SLF/ELF wave propagation are carried out. The subject has been investigated widely and the findings have been summarized in the classic book *Terrestrial Propagation of Long Electromagnetic Waves* by Galejs (1972). And some important findings are included in the book *Electromagnetic Waves in Stratified Media* by Wait (1970). In the past decades, some important progress has also been made. In the book *Resonances in the Earth–Ionosphere Cavity* by Nickolaenko and Hayakawa (2002), the fundamentals of ULF/ELF wave propagation and lightning problems are well summarized.

In this book it is investigated with an emphasis placed on the solution for SLF/ELF wave propagating in different regions, especially including Earth–ionosphere cavity or waveguide, anisotropic ionosphere, sea water, and layered Earth or sea floor. It is concerned with the approximated analytical solutions of the electromagnetic field radiated by a vertical or horizontal dipole. Usually, a simplified or idealized physical model is meant for solving a practical problem. From Maxwell's equations, and subject to the boundary conditions, the formulas of the electromagnetic field are always represented in the exact form of general integrals or in terms of special functions. Obviously, it is necessary to treat these by using mathematical techniques. The corresponding computations are also carried out, and some new conclusions are obtained.

In Chap. 1, the historical and technical overview is addressed for SLF/ELF wave propagation in the past century. Chapter 2 presents the fundamental theory of SLF/ELF wave propagation in Earth–ionosphere waveguide or cavity. Especially, the new algorithm is addressed to ELF range and the lower end of SLF range specifically. In Chap. 3, we treat the spherical harmonic series solutions for SLF/ELF field in the non-ideal Earth–ionosphere cavity and the speed-up numerical convergence algorithm. In Chaps. 4 and 5, we are concerned with SLF/ELF wave propagation in the regions consisting of the Earth, air, and ionosphere. Chapter 6 deals with SLF/ELF wave propagation along the boundary between sea water and ocean floor and the marine controlled-source electromagnetics (mCSEM) method. Chapter 7 addresses SLF field on the sea surface generated by a space borne transmitter. In Chap. 8, the atmosphere's noise in SLF/ELF ranges is summarized.

The authors would like to express their gratitude to Professor Qingliang Li and Professor Hongqi Zhang of the China Research Institute of Radiowave Propagation (CRIRP). Many thanks are also due to Professor Kangsheng Chen, Professor Xianmin Zhang, and Professor Erping Li of Zhejiang University for their helpful support and encouragement. The authors are also grateful to their graduate students, especially including Huaiyun Peng, Yu Chen, Hong Lu, Yuanxin Wang, Xiqian Wang, and Xiaofei Zhao at China Research Institute of Radiowave Propagation, Yin Lin Wang, Yun Long Lu, Peng Fei Yu, and Bing Jun Xia at Zhejiang University. Finally, the authors would like to acknowledge the support and guidance of the editorial staffs of Zhejiang University Press and Springer.

The authors sincerely appreciate the financial support in part of the National Science Foundation of China under Grants 61271086 and 60971057, and the National Key Lab of Electromagnetic Environment, China Research Institute of Radiowave Propagation. The authors hope that the book will help stimulate new ideas and innovative approaches to radio wave propagation and applications in the years to come.

Qingdao, China
Hangzhou, China
July 2013

Weiyan Pan
Kai Li

Contents

Chapter 1
Historical and Technical Overview of SLF/ELF Electromagnetic Wave Propagation

In general, very low frequency (VLF) refers to electromagnetic waves with frequencies from 3 kHz to 30 kHz; ultra low frequency (ULF) refers to frequency range between 300 Hz to 3 kHz; the frequency range of 30 Hz to 300 Hz is defined as super low frequency (SLF); and the frequency below 30 Hz is referred to as extremely low frequency (ELF). In some early studies, alternative definitions may have been used with frequency range of 3 Hz to 3 kHz generally referred to ELF. In this book, the emphasis is on SLF (30–300 Hz) and ELF (below 30 Hz) ranges, using the MKS system of units and the time dependency $\mathrm{e}^{-\mathrm{i}\omega t}$.

1.1 Medium Characteristics of SLF/ELF Wave Propagation

Due to the long wavelength, SLF/ELF waves, when excited and propagated on and near the Earth's surface, will cover lithosphere, atmosphere and ionosphere, whose electromagnetic characteristics differ significantly in their propagation paths. The permeability of the atmosphere and ionosphere is approximately μ_0, the permeability in free space, as well as that of the lithosphere, except in the regions that are rich in iron, nickel, cobalt, etc. Therefore, the permeability of the lithosphere can be also treated as μ_0, if neither the transmitter nor the receiver is located in mineral-rich areas. The atmosphere is a non-conductive medium with negligible conductivity, whose permittivity is close to ε_0, the permittivity of free space. While the lithosphere is conductive, the conductivity of sea water σ_{sea} is in the range of 2.5–5.5 S/m, and the conductivity σ_g of rock and soil in the range of 10^{-2}–10^{-4} S/m. The dielectric constant ε_r of the lithosphere does not have large effects on the propagation of SLF/ELF waves.

The ionosphere is composed of partially ionized gas, which includes electrons, ions, and electrically neutral particles, above 70 km or so from the sea level. The electrons and ions make the ionosphere conductive, thus the majority of the energy of the incident VLF/SLF/ELF waves will be reflected by the ionosphere. When both the transmitter and the receiver are located in the space between the ground and the

W. Pan, K. Li, *Propagation of SLF/ELF Electromagnetic Waves*,
Advanced Topics in Science and Technology in China,
DOI 10.1007/978-3-642-39050-0_1,

lower boundary of the ionosphere, the excited wave will be reflected back and forth between the ground and the ionosphere. Due to the conductivity of the lithosphere and that of the ionosphere, the electromagnetic waves in the VLF range and below will have the reflection coefficients being close to 1 with low loss, the waves will thus be guided in the space between the ground and the lower boundary of the ionosphere, forming the Earth–ionosphere waveguide or cavity.

1.2 The VLF Waveguide Propagation Theory and Its Applications in Submarine Communication

In a conductive medium, the attenuation rate of the electromagnetic wave is proportional to the square root of its operating frequency, thus the low frequency wave is capable of penetrating a certain depth of sea water and can be picked up by a submerged submarine. The developments of the VLF submarine communication and navigation system began in both the United States and the Soviet Union in the 1950s. Driven by the defense requirements, massive effort was made and many resources were devoted to study the theory and application of VLF wave propagation, and numerous achievements were accomplished (Sommerfeld 1949; Bremmer 1949; Budden 1961a; Wait 1957, 1960, 1962b, 1968; Wait and Spies 1965; Galejs 1964a, 1967, 1968, 1972b). Especially, the two classic books *The Wave-Guide Mode Theory of Wave Propagation* (Budden 1961a) and *Electromagnetic Waves in Stratified Media* (Wait 1962a) laid the foundation of the VLF waveguide propagation theory.

In the works by Wait, the space between the ground and the lower boundary of the ionosphere was modeled as a waveguide with two concentric spherical reflective walls, and the ground-based VLF antenna as VED, as shown in Fig. 1.1. For this model, the VLF field components in the Earth–ionosphere waveguide can be expressed in Eq. (1.1) (Wait 1962a):

$$\begin{bmatrix} E_r \\ E_\theta \\ \eta H_\phi \end{bmatrix} = E_0(\pi x)^{\frac{1}{2}} \exp\left(-\mathrm{i}\frac{\pi}{4}\right)\frac{2}{y_0} \times \sum_{n=0}^{\infty} \begin{bmatrix} \Lambda_n G_n(z) \\ \Lambda_n G_n'(z)\cdot(-\mathrm{i}\gamma_n) \\ \Lambda_n G_n(z)\cdot(-\gamma_n) \end{bmatrix} G_n(z_0)\exp(-\mathrm{i}t_n x), \tag{1.1}$$

where

$$\gamma_n = \frac{1}{k[1+\frac{t_n}{2}(\frac{2}{ka})^{\frac{2}{3}}]}, \tag{1.2}$$

$$E_0 = \frac{\mathrm{i}I\mathrm{d}l\eta\times 10^3}{\lambda}\cdot\frac{\exp(-\mathrm{i}ka\theta)}{a(\theta\sin\theta)^{\frac{1}{2}}} = \mathrm{i}300\times(P_{\mathrm{kW}})^{\frac{1}{2}}\cdot\frac{\exp(-\mathrm{i}ka\theta)}{a(\theta\sin\theta)^{\frac{1}{2}}}. \tag{1.3}$$

Here E_0 could be understood as the electric field in mV/m on the ideal ground radiated by a VDM. The angle θ is the circle angular distance between transmitter

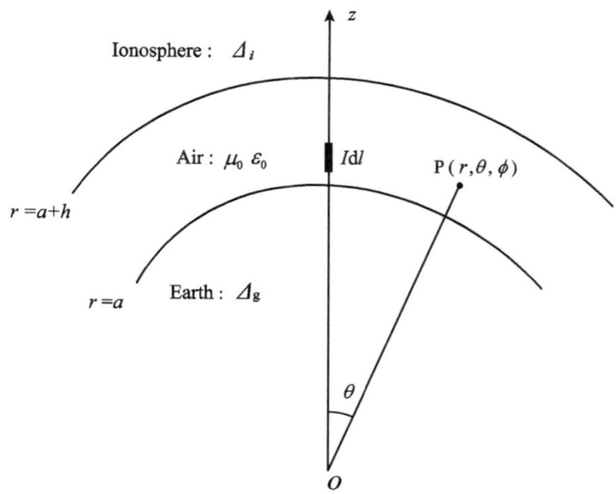

Fig. 1.1 A VED in the Earth–ionosphere waveguide

and receiver in rad; P_{kW} is the transmitting power in kW, and a is the Earth's radius, which is 6,370 km. The parameter x in Eq. (1.1) is determined by the angular distance θ from the transmitting source to the receiving point. It is

$$x = \left(\frac{ka}{2}\right)^{\frac{1}{3}} \cdot \theta. \tag{1.4}$$

Here G_n is the height-gain function, and $z = r - a$ is the height of the observation point. t_n is the nth root of the modal equation, which determines the relative phase velocity and the attenuation rate of the nth mode. Λ_n is the excitation factor of nth mode in the Earth–ionosphere waveguide excited by a VED.

From Eq. (1.1), the VLF wave excited by a vertical electric dipole in the Earth–ionosphere waveguide can be treated as the summation of each mode, which has only the three components E_r, E_θ, and H_ϕ traveling along the waveguide with the same phase velocity and attenuation rate. The relative phase velocity and the attenuation rate for the nth mode are expressed by

$$\frac{v_n}{c} - 1 = \mathrm{Re}\left[\frac{-t_n}{2^{\frac{1}{3}}(ka)^{\frac{2}{3}}}\right], \tag{1.5}$$

$$\alpha_n = 8.68 \times \mathrm{Im}\left[\frac{t_n \pi (\frac{2}{ka})^{\frac{2}{3}}}{\lambda}\right] (\mathrm{dB/km}), \tag{1.6}$$

where λ in km is the wavelength in free space.

It is noted that multiple modes exist in the Earth–ionosphere waveguide in the VLF range and the high-order modes decay significantly as the mode number increases. Thus the TM_1 mode is dominant in the far-field region, which is the most important one in engineering.

In 1967, Watt reviewed the study of US researchers in VLF communication and navigation, and published the monograph *VLF Radio Engineering*. The monograph was a summary of the contemporary accomplishments in VLF propagation researching, and discussed VLF communication and navigation system design, antenna design, and atmospheric noise in the VLF range as well.

1.3 SLF Communication System and SLF/ELF Propagation Theory

The operating frequency of SLF wave is even lower than that of VLF wave, thus SLF waves decay slower in sea water. The loss in sea water for the wave at 100 Hz is about 0.3 dB/m. With the development of modern submarine-detection technology, VLF receiving depth is no longer safe for a submarine. Project *Sanguine*, a plan to develop the submarine communication system in SLF range, thus started in the 1960s (Bannister 1975; Wait 1977). At that time, the properties of SLF/ELF wave propagation were investigated extensively. An excellent summary was given in the monograph *Terrestrial Propagation of Long Electromagnetic Waves* by Galejs (1972a).

Considering long wavelengths, no antenna can be put vertically on the ground in SLF range. Galejs therefore concentrated on analyzing electromagnetic waves excited by a HED in the Earth–ionosphere waveguide (Galejs 1972a). In SLF range, the wavelength in the atmosphere exceeds significantly the attitude between the ground surface and the lower boundary of the ionosphere, thus there only exists one propagating mode, which is a quasi-TEM mode, while all high-order modes are evanescent. In the case that both the transmitter and the receiver are located on or near the ground, ignoring high-order modes, SLF components radiated by an HED can be expressed as follows (Galejs 1972a):

$$E_r^{\mathrm{he}}(a,\theta,\phi) = -\frac{I\mathrm{d}s\eta\Delta_{gs}}{2ha}\cos\phi\Lambda_0^e\frac{\partial P_{\nu_0}(\cos(\pi-\theta))}{\sin\nu_0\pi\partial\theta}, \tag{1.7}$$

$$E_\theta^{\mathrm{he}}(a,\theta,\phi) = \frac{\mathrm{i}I\mathrm{d}s\eta\Delta_{gs}}{2ha}\Delta_{gr}\cos\phi\frac{\Lambda_0^e}{ka}S_0^{-2}\frac{\partial^2 P_{\nu_0}(\cos(\pi-\theta))}{\sin\nu_0\pi\partial^2\theta}, \tag{1.8}$$

$$E_\phi^{\mathrm{he}}(a,\theta,\phi) = -\frac{\mathrm{i}I\mathrm{d}s\eta}{2ha}\sin\phi\Delta_{gr}\Delta_{gs}\frac{\Lambda_0^e}{ka\sin\theta}S_0^{-2}\frac{\partial P_{\nu_0}(\cos(\pi-\theta))}{\sin\nu_0\pi\partial\theta}, \tag{1.9}$$

$$H_\theta^{\mathrm{he}}(a,\theta,\phi) = -\frac{\mathrm{i}I\mathrm{d}s}{2ha}\Delta_{gs}\frac{\sin\phi}{ka\sin\theta}\Lambda_0^e S_0^{-2}\frac{\partial P_{\nu_0}(\cos(\pi-\theta))}{\sin\nu_0\pi\partial\theta}, \tag{1.10}$$

$$H_\phi^{\mathrm{he}}(a,\theta,\phi) = -\frac{\mathrm{i}I\mathrm{d}s}{2ha}\Delta_{gs}\frac{\cos\phi}{ka}\Lambda_0^e S_0^{-2}\frac{\partial^2 P_{\nu_0}(\cos(\pi-\theta))}{\sin\nu_0\pi\partial^2\theta}. \tag{1.11}$$

In these formulas, Δ_{gs} and Δ_{gr} are the normalized impedances of the ground in the transmitting and receiving points, respectively. ν_0 is the eigenvalue of quasi-TEM mode, which is determined by the modal equation. For the first-order approximation, we have

$$\nu_0(\nu_0+1) \approx (ka)^2\left[1+\frac{\mathrm{i}(\Delta_i+\Delta_g)}{kh}\right], \tag{1.12}$$

where Δ_g and Δ_i are the normalized surface impedance of the ground and that of the ionosphere, respectively.

In the operating range of 65 to 140 Hz for SLF communication system, the condition $\nu_0 \gg 1$ is satisfied, thus the Legendre function is expressed in the asymptotic form (Galejs 1972a). It is

$$\begin{aligned} P_\nu(\cos(\pi-\theta)) &\approx \left(\frac{2}{\pi\nu_0\sin\theta}\right)^{\frac{1}{2}}\cos\left[\left(\nu_0+\frac{1}{2}\right)(\pi-\theta)-\frac{\pi}{4}\right] \\ &= \left(\frac{1}{2\pi\nu_0\sin\theta}\right)^{\frac{1}{2}}\left\{e^{-i[(\nu_0+\frac{1}{4})\pi-(\nu_0+\frac{1}{2})\theta]}\right. \\ &\quad \left.+\,e^{i[(\nu_0+\frac{1}{4})\pi-(\nu_0+\frac{1}{2})\theta]}\right\}. \end{aligned} \tag{1.13}$$

When the receiving point is located neither near the source point nor its antipodal point, the electromagnetic wave propagating along the path of large arc can be neglected, and the components at the receiving point can be further approximated as follows (Galejs 1972a):

$$\begin{aligned} E_r^{\text{he}}(a,\theta,\phi) &= \frac{iIds\eta}{2ha}\Delta_{gs}\cos\phi\sqrt{\frac{2}{\pi\sin\theta}} \\ &\quad\times(kaS_0)^{\frac{1}{2}}\Lambda_0^e\exp\left(ikaS_0\theta+i\frac{\pi}{4}\right), \end{aligned} \tag{1.14}$$

$$\begin{aligned} E_\theta^{\text{he}}(a,\theta,\phi) &= \frac{iIds\eta}{2ha}\Delta_{gs}\Delta_{gr}\cos\phi\sqrt{\frac{2}{\pi\sin\theta}} \\ &\quad\times(ka)^{\frac{1}{2}}S_0^{-\frac{1}{2}}\Lambda_0^e\exp\left(ikaS_0\theta+i\frac{\pi}{4}\right), \end{aligned} \tag{1.15}$$

$$\begin{aligned} E_\phi^{\text{he}}(a,\theta,\phi) &= -\frac{Ids\eta}{2ha}\sin\phi\Delta_{gs}\Delta_{gr}\sqrt{\frac{2}{\pi\sin\theta}} \\ &\quad\times(ka)^{-\frac{1}{2}}S_0^{-\frac{3}{2}}\Lambda_0^e\frac{\exp(ikaS_0\theta+i\frac{\pi}{4})}{ka\sin\theta}, \end{aligned} \tag{1.16}$$

$$\begin{aligned} H_\theta^{\text{he}}(a,\theta,\phi) &= -\frac{Ids}{2ha}\Delta_{gs}\frac{\sin\phi}{ka\sin\theta}\sqrt{\frac{2}{\pi\sin\theta}} \\ &\quad\times(ka)^{-\frac{1}{2}}S_0^{-\frac{3}{2}}\Lambda_0^e\exp\left(ikaS_0\theta+i\frac{\pi}{4}\right), \end{aligned} \tag{1.17}$$

$$\begin{aligned} H_\phi^{\text{he}}(a,\theta,\phi) &= -\frac{iIds}{2ha}\Delta_{gs}\sqrt{\frac{2}{\pi\sin\theta}} \\ &\quad\times(ka)^{\frac{1}{2}}S_0^{-\frac{1}{2}}\Lambda_0^e\cos\phi\exp\left(ikaS_0\theta+i\frac{\pi}{4}\right), \end{aligned} \tag{1.18}$$

where S_0 is sine of incident angle to the ionosphere for the quasi-TEM mode, and $\nu_0(\nu_0+1)\approx(kaS_0)^2$.

From Eqs. (1.14)–(1.18), it is seen that the lower conductivity (thus higher impedance) of the ground can place a boost to the radiation efficiency of the ground-based SLF transmitter, thus making such region a good candidate for construction site of SLF horizontal antenna.

Wave propagation in SLF/ELF ranges has been studied widely in the 1990s and 1970s by many investigators (Schumann 1952a, 1952b; Galejs 1961, 1962, 1964b, 1965, 1972b; Bannister 1974, 1984a; Greifinger and Grifinger 1978; Behroozi-Toosi and Booker 1980; Yuan 2011), especially including Galejs and Wait. Galejs, who made outstanding contributions to the SLF communication engineering, passed away at an early age of 49 in 1972. In April 1974, a special issue on ELF/SLF propagation and communication was published by *IEEE Transactions on Communications*, especially including papers on SLF antenna design and site locating (Burrows 1974a, 1974b), propagation computation and measurement (Pappert and Moler 1974; Bannister 1974), signal and noise processing (Evans and Griffiths 1974; Ginsberg 1974), and environmental effect of a transmitting antenna (Valentino et al. 1974). In this issue, Wait wrote a mourning article "Dedication to Janis Galejs". Lately, the ground-based SLF communication systems were completed and put in use in both the United States and the former Soviet Union (Russia).

In the whole ELF range, even the lower end of SLF range (below 50 Hz), the wavelength is very long, which can be compared with the Earth's circumference. For instance, at $f = 10$ Hz, the wavelength is about $3/4$ times of the Earth's circumference. It is noted that the space between the Earth's surface and the ionosphere is usually regarded as a cavity. In this case, it is found that the condition of $\nu_0 \gg 1$ is not satisfied for the mode eigenvalue ν_0 in the cavity, therefore Eq. (1.13) is no longer valid for calculating electromagnetic field in the Earth–ionosphere cavity excited by HED. In 2012, Peng et al. showed that in the lower SLF and ELF regions, the Legendre function $P_\nu(\cos(\pi - \theta))$, the distance factor of the field, can be better calculated by using numerical integrated algorithm (Peng et al. 2012, 2013). This new algorithm will be addressed in Chap. 2.

In 1999, Berrick presented a spherical harmonic series expression for electromagnetic field in atmosphere, with the lithosphere and the ionosphere being idealized as perfect conductors, and SLF/ELF radiation source as a VED (Barrick 1999). In the research group of Pan, the first author of this book, further works are concerned with the convergence acceleration algorithm for the electromagnetic field by HEDs and VEDs in SLF/ELF ranges, under the condition that the ground and the ionosphere are finitely conductive (Wang et al. 2007a, 2007b, 2009b). SLF/ELF electromagnetic field in Earth–ionosphere waveguide or cavity and the speed-up numerical convergence algorithm will be addressed in Chap. 3.

It should be pointed out that FDTD modeling of SLF/ELF wave propagation has greatly advanced (Simpson and Taflove 2004, 2007; Simpson 2006, 2009). However, this book mainly deals with the analytical solutions of SLF/ELF wave propagation, which is not concerned with the content of the FDTD modeling.

1.4 SLF/ELF Emission as Earthquake Precursor and Field on and near the Ground by Underground SLF/ELF Source

Earthquake is one of the most severe natural disasters that human beings have to face, and over 600,000 lives were lost due to earthquakes in China alone in the past century. Earthquake prediction is one of the great world scientific problems, which needs to be solved urgently, there being no satisfying solution till today.

In 1964, earthquake-related abnormal electromagnetic radiation in lower frequency ranges was observed when Soviet scientists were analyzing the data recorded by the “Alouette” satellite. Since 1983, the pre-earthquake electromagnetic anomaly in the lower frequency ranges observed by satellite was reported (Gokhberg et al. 1983; Parrot 1994; Serebryakova et al. 1992; Molchanov et al. 2004), where the cited data are from the satellites of OGO-06, INTERKOSMOS-19, GEOS-2, and AUREOL-3.

Early in 1978, the Soviet Union started setting up the ground-based electromagnetic radiation monitoring site. It was reported by Russian and Japanese scientists that pre-earthquake electromagnetic radiation abnormal strengthening phenomena in the lower frequency ranges were observed by the ground-based sites (Nikiforova et al. 1989; Morgunov and Matveev 1990; Uyeda et al. 2002). Since 1991, Chinese scholars also joined this researching field and some achievements had been made (Guan and Liu 1995; Yu et al. 2010; Zhang et al. 2012; Zhima et al. 2012).

The actual causes and mechanisms of the pre-earthquake electromagnetic radiation anomaly still remain in the dark by now and the controversies concerning several hypotheses continued for a long time. But it is generally believed that the radiation source is located in the seismogenic zone and further study on the earthquake-related electromagnetic radiation may possibly lead to a new method for short-term earthquake prediction. Thus some scientists suggested that the global monitor network for seismic electromagnetic radiation with both ground- and satellite-based monitoring sites should be built for collecting further data. In June 2004, the DEMETER (Detection of Electro-Magnetic Emissions Transmitted from Earthquake Regions) satellite, which was dedicated to seismic electromagnetic study, was launched in France (Parrot 2002), and many excellent works on the DEMETER satellite have been carried out. It is reported that there are similar research programs for seismic electromagnetism satellites in other countries including the United States, Russia, Japan, Ukraine, Greece, and China.

In order to speculate on the possible zone where an earthquake may take place, in the data of the electromagnetic radiation anomaly it is necessary to analyze the SLF/ELF field distributions near the ground or low-orbited space excited by an underground source. As the simplest model, the underground source can be idealized as an electric dipole. Due to the absorption of the lithosphere, only SLF/ELF waves may penetrate through the lithosphere and be picked up by the ground- or satellite-based receivers.

In this research field, some new progresses were also achieved by our research group (Zhang et al. 2009; Chen et al. 2009). In Chaps. 4 and 5, we will address the SLF/ELF field in the ionosphere radiated by underground HED in three-layered planar and spherical models, while the SLF/ELF wave propagation in the anisotropic ionosphere is treated analytically and numerically.

1.5 ELF Wave Propagation Along the Ocean Floor and the Marine Controlled Source Electromagnetics (mCSEM) Method

Compared with the solid seabed, the sea water has much higher conductivity. That is to say, ELF wave decays much slower in the ocean floor than in the sea water. Therefore, when the ELF radiator is placed on or near the bottom of the ocean, while a receiver a distance away is also close to the ocean floor, ELF wave radiated from the source will travel in a lateral-wave path, which is in the ocean floor along the boundary, the less lossy side. Therefore the magnitude and phase of the ELF wave at the receiver will carry the information on the conductivity of the seabed. Based on the above principles, Chave and Cox in the early 1980s suggested the mCSEM method for rough geologic survey of the sea floor on a large scale (Chave and Cox 1982). In Fig. 1.2, the diagram of the mCSEM method for marine exploration is shown. Since the mCSEM method requires the energy of the electromagnetic wave to penetrate the floor as much as possible, the transmitting source is usually employed a horizontal electric antenna closed to the sea or ocean floor. Following the above studies, Dunn analyzed the lateral wave in a three-layered medium model with sediments under the ocean floor (Dunn 1984, 1986). Pan's theoretical work (Pan 1985) on the lateral wave along boundary between one-dimensional anisotropic rock and sea water explained the experimental results by Young and Cox (1981). For a more general case, the integrated formulas were derived for the electromagnetic field in the n-layered model of the sea or ocean floor (King et al. 1992). Additionally, many excellent works on ELF wave propagation along the ocean floor were carried out by some investigators (Bannister 1984a, 1984b; Edwards et al. 1985; Inan et al. 1986; Bubenik and Fraser-Smith 1978; Fraser-Smith and Bubenik 1979; Fraser-Smith et al. 1988; Chave et al. 1990).

Starting from the late 1990s, as petroleum on land gradually depleted, several petroleum companies began to make more efforts in exploration of marine oil and methane clathrate. Due to the significant differences between the conductivity of water and the conductivities of oil, gas, and methane clathrate, the mCSEM method is capable of distinguishing the oil- or gas-rich layer apart from the layer of the water, and estimating the oil/water ratio. The mCSEM method is being developed towards the accurate investigations of the lithosphere's conductivity for geological studies of oil and gas resources. In the past decade, several companies specializing in marine exploration have been established in succession, especially including the British OHM, AGO of US, EMGS of Norway. Some exploration devices with the mCSEM method have been put in use today. In Fig. 1.2, the diagram for the mCSEM method is illustrated.

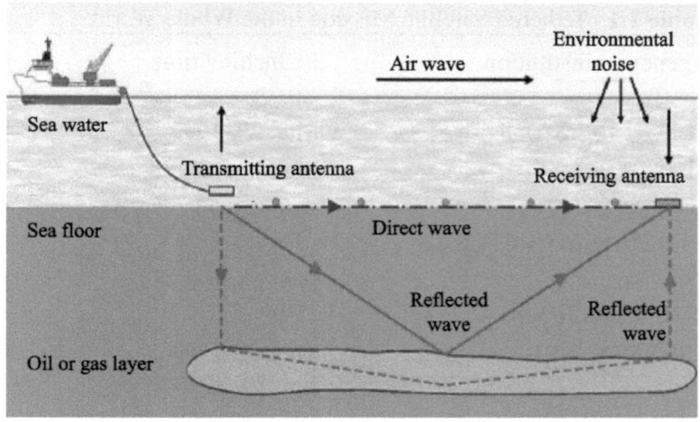

Fig. 1.2 The diagram for the mCSEM method

For practical exploration purposes, the theoretical investigations on the ELF propagation along seabed are evolving from idealized stratified plane models to inhomogeneous models, and some numerical methods for this problem are developed rapidly. Furthermore, based on the measured ELF field distributions, the inversion method for the conductivity and other parameters is also carried out. The available numerical methods for marine exploration can be found in the references (Gribenko and Zhdanov 2007; Plessix and van der Sman 2007; Zach et al. 2008a, 2008b).

In Chap. 6, we will attempt to address the ELF wave propagation along the boundary between sea water and sea floor and the mCSEM method.

1.6 VLF/SLF/ELF Field on Sea Surface Excited by Space Borne Transmitter

By now only VLF/SLF/ELF waves can be used for submarine communication. Building a ground-based VLF/SLF/ELF communication system requires the construction of gigantic transmitters that can often cover up to several square-kilometers, which is both expensive and vulnerable to enemy attacks, and it cannot be rebuilt overnight once destroyed during wartime. With the development of space technology, the US and Russia started testing to transmit VLF/SLF waves from small satellites on low Earth orbit (LEO) (Armand et al. 1988; Bannister et al. 1993). Additionally, VLF/SLF/ELF electromagnetic radiation can also be generated by high-frequency heating (Ferraro et al. 1984; Lunnen 1985; Barr et al. 1987, 1988).

In summer 1992, a joint experiment by NASA and Italian Space Agency (ASI) was carried out on board space shuttle "Atlantis" to deploy a tethered satellite, which includes a 20 km conducting tether and a 23 kg payload. The experiment ended in failure due to a jammed tether. The second attempt was made in February 1996, which was a partial success. Following these experiments, a new proposal made for the VLF/SLF transmitter is installed on a satellite with two long conductive tethers, where one tether is used as a transmitting antenna. The tether satellite missions by

Table 1.1 Tethered Satellite Mission in the Whole World

Agency or institution	Launching time	Orbit	Tether length
SEDS-1 (NASA)	Mar. 29, 1993	LEO	20 km
SEDS-2 (NASA)	Mar. 9, 1994	LEO	20 km
Oedipus-C (NRC/NASA, CRC, CSA)	Nov. 7, 1995	Suborbital	1 km
TSS-1R (STS-75)(ASI/NASA)	Feb. 22–Mar. 9, 1996	LEO	19.6 km
TiPS (NRO/NRL)	May 12, 1996 Jun. 20, 1996	LEO	4 km
YES (ESA and Delta-Utec)	Oct. 30, 1997	GTO	35 km
PICOSAT1.0 (Aerospace Corporation)	Jan. 27, 2000	LEO	30 m
PICOSAT1.1 (Aerospace Corporation)	Jul. 19, 2000	LEO	30 m
ProSEDS (NASA)	Canceled in 2003	LEO	25 km
YES2 (ESA and Delta-Utec SRC)	Sept. 14, 2007 Sept. 25, 2007	LEO	31.7 km

the US and Europe are listed in Table 1.1, and the specific materials can be found on the website: http://science.nasa.gov/missions/tss/.

The space borne VLF/SLF transmitting experiments were also carried out in Russia, and their design featured loop antenna as a transmitter. In 1987, two sets of loop antennas with the diameter 20 m, which were used to radiate a VLF signal, were deployed successfully on the spacecraft "Progress-28" (Armand et al. 1988).

A space borne VLF/SLF transmitter can benefit from the satellite's motion. As the transmitting source is much closer to a submarine, the propagation loss can be reduced significantly. Moreover, since the LEO satellite is within the ionosphere, so that the refractive index n is much larger than 1, the electrically length of antenna can be increased greatly without any change to antenna size, thus giving a boost to its radiation efficiency, and allowing the signal to be still detectable in sea water with a small transmitting power. Therefore, as a fundamental study of a new method of submarine communication, the investigation on the space borne VLF/SLF excitation and propagation may be useful.

In the analysis by Bannister et al. (1993), the ionosphere was treated as a homogeneous plasma with sharp boundary. Considering the effect on the refractive index of the ionosphere by the geomagnetic field, it is concluded that the whistle wave (VLF) and Alfvén wave (ELF) can be excited efficiently by an electric dipole and can propagate to the ground or sea surface. Additionally, the excellent works on the space borne VLF/SLF excitation and propagation were carried out in the past many years (Einaudi and Wait 1971; Tripathi et al. 1982; Armand et al. 1988; Serebryakova et al. 1992; Melnikov 1994), especially including the works by some Russian scientists.

Since 1995, the authors have been researching on this topic. The radiations of VLF/SLF electric and magnetic dipoles in infinite anisotropic plasma are treated analytically (Li and Pan 1997; Li 1998). The approximated formulas have been derived for the electromagnetic field on the ground or sea surface excited by a VLF source in the ionosphere (Pan 1996; Li et al. 2004). Additionally, the propagation of VLF/SLF/ELF waves in the lower ionosphere is investigated analytically (Li and Pan 1998, 1999; Li et al. 2011). The results obtained will be summarized in Chap. 7.

1.7 Atmospheric Noise in SLF/ELF Ranges

The seasonal, hourly, regional, and statistical distributions of the atmospheric noise in SLF/ELF ranges are worth studying for two reasons: First, they are important parameters for designing communication and detection system that works in SLF/ELF ranges; Second, they are also a major basis to identify whether an "electromagnetic anomaly", which can be an earthquake precursor, occurs or not (Ginsberg 1974; Fraser-Smith and Turtle 1993; Lee Boyce et al. 2003).

The main source of atmospheric noise in SLF/ELF ranges is the lightening discharge (Feldman 1972; Enge and Sawate 1988; Meloni et al. 1992; Chrissan and Fraser-Smith 1996a). In the industrial or urban area, the variety of human activity also creates significant local interference in SLF/ELF ranges, especially around power stations and power transmission lines. The human-related noise decays rapidly as the distance from the noise source increases, and it thus can be prevented and controlled in engineering. The atmospheric discharge, on the other hand, is an external background noise that all systems working in the frequency must encounter.

In the 1960s, Watt summarized and plotted a global distribution of atmospheric noise in VLF range and the noise data in VLF range accumulated by communication and navigation engineering in the United Station (Watt 1967). In this book, one year is divided into four seasons and one day into six time intervals. In the range of 1 kHz–100 kHz, the data on the spectral density of the atmospheric noise can be speculated on readily. However, this book leaves out information in the range below 1 kHz.

From the 1980s to the 1990s, Stanford University organized the monitoring of the atmospheric noise from 10 Hz to 32 kHz in the eight sites global wide that lasted several years continuously (Chrissan et al. 1997; Chrissan 1998). The observation data from four sites were published in the two reports (Chrissan and Fraser-Smith 1996a, 1996b).

Based on the observation data in the range of 10 kHz–32 kHz by Stanford University, the changes of the spectral density of the atmospheric noise versus the operating frequency in SLF/ELF ranges were investigated in our research group (Wang et al. 2009a). By using the available results on the global atmospheric noise distribution in VLF range by Watt (1967), the global distributions of atmospheric noise spectral density at the frequency of 80 Hz were deduced readily. In a similar manner, the corresponding data on the global distributions of the atmospheric noise spectral density at any frequency in the range from 10 Hz to 1 kHz can also be obtained readily (Wang et al. 2009a, 2009b, 2010). These studies will be included in Chap. 8.

References

Armand NA et al (1988) Experimental researches in the ionosphere of the Earth of the radiation of loop antenna in a range VLF waves, installed onboard the orbital complex “Progress-28”–“Souz TM-2”. Radiotech Electron 33:2225–2233. (In Russian)

Bannister PR (1974) Far-field extremely low frequency propagation measurements. IEEE Trans Commun 22(4):468–474

Bannister PR (1975) Project Sanguine quarterly technical summary, 1 July–30 September 1974. NUSC Technical Report, 4907

Bannister PR (1984a) ELF propagation update. IEEE J Ocean Eng 9(3):179–188

Bannister PR (1984b) New simplified formulas for ELF subsurface-to-subsurface propagation. IEEE J Ocean Eng OE-9(3):154–163

Bannister PR et al (1993) Orbiting transmitter and antenna for spaceborne communications at ELF/VLF to submerged submarines. AGARD Conf Proc 592:33.1–33.14

Barr R, Rietueld MT, Stubbe P, Kopka H (1987) Ionospheric heater beam scanning: a mobile source of ELF radiation. Radio Sci 22(6):1073–1083

Barr R, Rietueld MT, Stubbe P, Kopka H (1988) Ionospheric heater beam scanning: a realistic model of this mobile source of ELF/VLF radiation. Radio Sci 23(3):379–388

Barrick DE (1999) Exact ULF/ELF dipole field strengths in the Earth–ionosphere cavity over the Schumann resonance region: idealized boundaries. Radio Sci 34(1):209–227

Behroozi-Toosi AB, Booker HG (1980) Application of a simplified theory of ELF propagation to a simplified worldwide model of the ionosphere. J Atmos Terr Phys 42:943–974

Bremmer H (1949) Terrestrial radio waves. Elsevier, New York

Bubenik DM, Fraser-Smith AC (1978) ULF/ELF electromagnetic fields generated in a sea of finite depth by a submerged vertically directed harmonic magnetic dipole. Radio Sci 13:1011–1020

Budden KG (1961a) The wave-guide mode theory of wave propagation. Logos Press, London

Budden KG (1961b) Radio waves in the ionosphere. Cambridge University Press, Cambridge

Burrows ML (1974a) Surface impedance and the efficiency of horizontal-dipole extremely low frequency (ELF) antenna arrays. IEEE Trans Commun 22(4):399–401

Burrows ML (1974b) Optimizing the current distribution in a buried linear antenna. IEEE Trans Commun 22(4):409–411

Chave AD, Cox CX (1982) Controlled electromagnetic sources for measuring electric conductivity beneath the oceans. I: Forward problem and model study. J Geophys Res 87(B7):5327–5338

Chave AD, Flosadóttir AH, Cox CS (1990) Some comments on seabed propagation of ULF/ELF electromagnetic fields. Radio Sci 25:825–836

Chen Y, Wang YX, Pan WY (2009) The incidence, reflection, and transmission of SLF/ELF waves by the ionosphere. Technical Report, China Research Institute of Radiowave Propagation, Qingdao, China

Chrissan DA (1998) Statistical analysis and modeling of low-frequency radio noise and improvement of low-frequency communications. USA: ADA360417

Chrissan DA, Fraser-Smith AC (1996a) Seasonal variations of globally measured ELF/VLF radio noise. Radio Sci 31(5):1141–1152

Chrissan DA, Fraser-Smith AC (1996b) Seasonal variations of globally measured SLF/VLF radio noise. Technical Report D177–1, ADA358414/XAB

Chrissan DA et al (1997) Diurnal variations of globally measured SLF/VLF radio noise. USA: ADA358641

Dunn JM (1984) Electromagnetic lateral waves in layered media. PhD dissertation, Harvard University, Cambridge, MA, USA
Dunn JM (1986) Lateral wave propagation in a three-layered medium. Radio Sci 21(5):787–796
Edwards RN, Law LK, Wolfgram PA, Nobes DC, Bone MN, Trigg DF, Delaurier JM (1985) First results of the MOSES experiment; sea sediment conductivity and thickness determination, Bute Inlet, British Columbia, by magnetometric offshore electrical sounding. Geophysics 50(1):153–160
Einaudi F, Wait JR (1971) Analysis of the excitation of the Earth–ionosphere wave guide by a satellite-borne antenna. Can J Phys 49:447–457
Enge PK, Sawate DV (1988) Spread-spectrum multiple-access performance of orthogonal codes: impulsive noise. IEEE Trans Commun 36(1):98–106
Evans J, Griffiths A (1974) Design of a Sanguine noise processor based upon world-wide extremely low frequency (ELF) recordings. IEEE Trans Commun 22(4):528–539
Feldman DA (1972) An atmosphere noise model with applications to low frequency navigation systems. PhD thesis, Massachusetts Institute of Technology, Boston, USA
Ferraro AJ, Lee HS, Allshouse R, Carroll K, Lunnen R, Collins T (1984) Characteristics of ionospheric ELF radiation generated by HF heating. J Atmos Terr Phys 46:855–865
Fraser-Smith AC, Bubenik DM (1979) ULF/ELF electromagnetic fields generated above a sea of finite depth by a submerged vertically directed harmonic magnetic dipole. Radio Sci 14:59–74
Fraser-Smith AC, Turtle JP (1993) ELF/VLF radio noise measurements at high latitudes during solar particle events. AGARD Conf Proc 529:161–168
Fraser-Smith AC, Inan AS, Villard AG Jr, Joiner RG (1988) Seabed propagation of ULF/ELF electromagnetic fields from harmonic dipole sources location on the seafloor. Radio Sci 23:931–943
Galejs J (1961) Terrestrial extremely-low-frequency noise spectrum in the presence of exponential ionospheric conductivity profiles. J Geophys Res 66:2787–2793
Galejs J (1962) A further note on terrestrial extremely-low-frequency propagation in the presence of an isotropic ionosphere with an exponential conductivity-height profile. J Geophys Res 67:2715–2728
Galejs J (1964a) Propagation of VLF waves below a curved and stratified anisotropic ionosphere. J Geophys Res 69(17):3639–3650
Galejs J (1964b) Terrestrial extremely-low-frequency propagation. In: Bleil DF (ed) Natural electromagnetic phenomena below 30 kc/s. Plenum, New York, pp 205–258
Galejs J (1965) On the terrestrial propagation of ELF and VLF waves in the presence of a radial magnetic field. Radio Sci 69D(5):705–720
Galejs J (1967) Propagation of VLF waves below an anisotropic stratified ionosphere with a transverse static magnetic field. Radio Sci 2(6):557–574
Galejs J (1968) Propagation of ELF and VLF waves below an anisotropic ionosphere with a dipping static magnetic field. J Geophys Res 73(1):339–352
Galejs J (1972a) Terrestrial propagation of long electromagnetic waves. Pergamon, London
Galejs J (1972b) Stable solutions of ionospheric fields in the propagation of ELF and VLF waves. Radio Sci 7(5):549–561
Ginsberg LH (1974) Extremely low frequency (ELF) atmospheric noise level statistics for project sanguine. IEEE Trans Commun 22(4):555–561
Gokhberg MB, Pilipenko VA, Pokhotelov OA (1983) Observation from a satellite of electromagnetic radiation above the region of earthquake in preparation. Dokl Akad Nauk SSSR 268(1):56–58
Greifinger C, Grifinger P (1978) Approximate method for determining ELF eigenvalues in the Earth–ionosphere waveguide. Radio Sci 13:831–837
Gribenko A, Zhdanov M (2007) Rigorous 3D inversion of marine CSEM data based on the integral equation method. Geophysics 72:73–84
Guan HP, Liu GP (1995) The study on the relation between the electromagnetic anomaly of earthquake precursor and earthquake. ACtA Seismol Sin 17(2):237–246

Inan AS, Fraser-Smith AC, Villard OG Jr. (1986) ULF/ELF electromagnetic fields generated along the seafloor by a Straight current source of infinite length. Radio Sci 21(13):409–420
King RWP, Owens M, Wu TT (1992) Lateral electromagnetic waves: theory and applications to communications, geophysical exploration, and remote sensing. Springer, New York
Lee Boyce CO, Powell JD, Enge PK, Sherman CL (2003) A time domain atmospheric noise level analysis. In: Proceedings of the international Loran association 32rd annual meeting, Boulder, CO, USA
Li K (1998) Radiation of a magnetic dipole in an infinite anisotropic medium. Chin Phys Lett 26(12):898–900
Li K, Pan WY (1997) Radiation of an electric dipole in an infinite anisotropic medium. Indian J Radio Space Phys 26:340–345
Li K, Pan WY (1998) The VLF field on the sea surface generated by the transmitter antenna in the inhomogeneous ionosphere. Chin J Radio Sci 13(3):265–269. (In Chinese)
Li K, Pan WY (1999) Propagation of VLF electromagnetic waves penetrating the lower ionosphere. Indian J Radio Space Phys 28:87–94
Li K, Zhang HQ, Pan WY (2004) The VLF electromagnetic field on the sea surface generated by a space borne loop antenna. J Electromagn Waves Appl 18(1):121–135
Li K, Sun XY, Zhai HT (2011) Propagation of ELF electromagnetic waves in the lower ionosphere. IEEE Trans Antennas Propag 59(2):661–666
Lunnen RJ (1985) Detection of VLF and ELF long-path signals radiated from heated and modulated ionosphere current systems. PhD thesis, Penn State University, University Park
Melnikov VM (1994) Large space constructions formed by centrifugal forces. Space Bull 1:13
Meloni A, Palangio P, Fraser-Smith AC (1992) Some characteristics of the ELF/VLF radio noise measured near L'Aquila, Italy. IEEE Trans Antennas Propag 40:233–236
Molchanov O, Fedorov E, Schekotov A (2004) Lithosphere–atmosphere–ionosphere coupling as governing mechanism for preseismic short-term events in atmosphere and ionosphere. Nat Hazards Earth Syst Sci 4:757–767
Morgunov UA, Matveev MY (1990) Electromagnetic emission on Spitak earthquake. Geophysics 6:14–191
Nikiforova NN, Yudekhiu FN, Toktosopiev AM (1989) Studies of electromagnetic emission of seismotectonic origin in the Kirghiz SSR. Phys Earth Planet Inter 57(1–2):68–75
Pan WY (1985) Surface-wave propagation along the boundary between sea water and one-dimensionally anisotropic rock. J Appl Phys 58(11):3963–3974
Pan WY (1996) The VLF fields on the surface of the sea generated by the space borne transmitter. Chin J Space Sci 16(1):62–70. (In Chinese)
Pappert RA, Moler WF (1974) Propagation theory and calculation at lower extremely low frequency. IEEE Trans Commun 22(4):438–451
Parrot M (1994) Statistical study of ELF/VLF emission recorded by a low-altitude satellite during seismic events. J Geophys Res 99(A12):23339–23347
Parrot M (2002) The micro-satellite DEMETER. J Geodyn 33:535–541
Peng HY, Tao W, Pan WY, Guo LX (2012) Numerical integral method for ELF fields excited by vertical electric dipole in asymmetric Earth–ionosphere cavity. Chin J Radio Sci 22(2):333–338. (In Chinese)
Peng HY, Wu SH, Pan WY, Guo LX (2013) ELF fields excited by horizontal electric dipole in asymmetric Earth–ionosphere cavity. Submitted to Chin J Radio Sci. (In Chinese)
Plessix RE, van der Sman P (2007) 3D CSEM modeling and inversion in complex geologic settings. In: Society of exploration geophysicists, September 23–28
Schumann WO (1952a) On the radiation free self-oscillations of conducting sphere which is surrounded by an air layer and an ionospheric shell. Z Naturforsch 72:149–154. (In German)
Schumann WO (1952b) On the damping of electromagnetic self-oscillations of the system Earth–air–ionosphere. Z Naturforsch 72:250–252. (In German)
Serebryakova ON, Bdiehenko SV, Chmyrev VM (1992) Electromagnetic ELF radiation from earthquake regions as observed by low-altitude satellites. Geophys Res Lett 19(2):91–94

Simpson JJ (2006) FDTD modeling of a novel ELF radar for major oil deposits using a three-dimensional geodesic grid of the Earth–ionosphere waveguide. IEEE Trans Antennas Propag 54(6):1734–1741

Simpson JJ (2009) Current and future applications of 3D global Earth–ionosphere models based on the full-vector Maxwell's equations FDTD method. Surv Geophys 30(2):105–130

Simpson JJ, Taflove A (2004) Three-dimensional FDTD modeling of impulsive ELF antipodal propagation and Schumann resonance of the Earth-sphere. IEEE Trans Antennas Propag 52(2):443–451

Simpson JJ, Taflove A (2007) A review of progress in FDTD Maxwell's equations modeling of impulsive subionospheric propagation below 300 kHz. IEEE Trans Antennas Propag 55(6):1582–1590

Sommerfeld A (1949) Partial differential equations in physics. Academic Press, New York

Tripathi VK, Chang CL, Papadopoulos K (1982) Excitation of the Earth–ionosphere wave guide by an ELF source in the ionosphere. Radio Sci 17:1321–1326

Uyeda S et al (2002) Electric and magnetic phenomena observed before the volcano-seismic activity in 2000 in the Izu Island Region, Japan. Proc Natl Acad Sci 99(11):7352–7355

Valentino AR, Abromavage MM, Mclettan DW, Miller DA (1974) Project sanguine interference mitigation research. IEEE Trans Commun 22(4):562–569

Wait JR (1957) The mode theory of VLF ionosphere propagation for finite ground conductivity. Proc IRE 45(6):762–767

Wait JR (1960) Terrestrial propagation of VLF radio waves—a theoretical investigation. J Res Natl Bur Stand 64D:153–203

Wait JR (1962a) Electromagnetic wave in stratified media. Pergamon, New York

Wait JR (1962b) On the propagation of VLF and ELF radio waves when the ionosphere is not sharply bounded. J Res Natl Bur Stand 66D:53–61

Wait JR (1968) Recent theoretical advances in the terrestrial propagation of VLF electromagnetic waves. Adv Electron Electron Phys 25:145–209

Wait JR (1977) Propagation of ELF electromagnetic waves and project Sanguine/Seafarer. IEEE J Ocean Eng OE2:161–172

Wait JR, Spies KP (1965) Influence of finite ground conductivity on the propagation of VLF radio waves. Radio Sci J Res NBS/USNC-URSI 69D(10):1359–1373

Wang YX, Fan WS, Pan WY, Zhang HQ (2007a) Spherical harmonic series solution of fields excited by vertical electric dipole in Earth–ionosphere cavity. Chin J Radio Sci 22(2):204–211. (In Chinese)

Wang YX, Peng Q, Pan WY, Zhang HQ, Zhang ZW (2007b) The fields excited by SLF/ELF horizontal electric dipole in Earth–ionosphere cavity. Chin J Radio Sci 22(5):728–734. (In Chinese)

Wang YX, Pan WY, Jin RH, Zhang HQ (2009b) Electromagnetic fields in spherical Earth–ionosphere excited by SLF/ELF underground horizontal electric dipole. Chin J Radio Sci 24(6):1002–1008. (In Chinese)

Wang XQ, Pan WY, Zhang HQ, Ren Q (2009a) The global distribution of SLF atmospheric noise level and the comparisons to the measured data. Technical Report, China Research Institute of Radiowave Propagation, Qingdao, China. (In Chinese)

Wang YX, Pan WY, Jin RH, Zhang ST (2010) Antarctic SLF/ELF radio atmospheric radio noise measure and research in China. In: The 9th international symposium on antennas, propagation and EM theory, Guangzhou, China, pp 525–528

Watt AD (1967) VLF radio engineering. Pergamon, London

Young PD, Cox CS (1981) Electromagnetic active source sounding near the East Pacific Rise. Geophys Res Lett 8(10):1043–1046

Yu HY, Zhou HJ, Qiao XL (2010) Study on wave propagation of ELF emission anomaly before M_S8.0 Wenchuan earthquake. Acta Seismol Sin 32(6):641–648. (In Chinese)

Yuan Y (2011) Propagation and noise of SLF and ELF electromagnetic waves. National Defense Industry Press, Beijing. (In Chinese)

Zach JJ, Bjorke AK, Storen T, Maao F (2008a) 3D inversion of marine CSEM data using a fast finite-difference time-domain forward code and approximate hessian-based optimization. In: Society of exploration geophysicists, November 9–14
Zach JJ, Roth F, Yuan H (2008b) Preprocessing of marine CSEM data and model preparation for frequency-domain 3D inversion. In: PIERS proceedings, Cambridge, USA, July 2–6, pp 144–148
Zhang HQ, Chen Y, Pan WY (2009) Fields excited by earthquake ELF/SLF radiator on the ground and in the ionosphere. Chin J Radio Sci 24(3):432–439. (In Chinese)
Zhang XM, Shen XH, Miao YQ (2012) Electromagnetic anomalies around Wenchuan earthquake and their relationship with earthquake preparation. Proc Environ Sci 12A:693–701
Zhima ZR et al (2012) Possible ionospheric electromagnetic perturbations induced by the $M_S7.1$ Yushu earthquake. Earth Moon Planets 108:231–241

Chapter 2
Excitation and Propagation of SLF/ELF Electromagnetic Waves in the Earth–Ionosphere Waveguide/Cavity

In this chapter, the region of interest is a waveguide or cavity between the Earth's surface and an isotropic homogeneous ionosphere. The dipole (vertical electric dipole (VED), the vertical magnetic dipole (VMD), or the horizontal electric dipole (HED)) and the observation point are assumed to be located on or near the spherical surface of the Earth. The approximate all formulas are obtained for the electromagnetic field radiated by a VED and a VMD in the Earth–ionosphere waveguide or cavity. Based on the above results, the approximate formulas are derived readily for the electromagnetic field of an HED in the Earth–ionosphere waveguide or cavity by using the reciprocity theorem. Analyses and computations in SLF/ELF ranges are carried out specifically.

2.1 Introduction

The properties of the electromagnetic field generated by a VLF/ULF/SLF/ELF dipole source in the Earth–ionosphere waveguide or cavity have been investigated widely in the past 60 years because of its useful applications in submarine communication, navigation, geophysical prospecting and diagnostics, and earthquake electromagnetic detection (Bouwkamp and Casimir 1954; Budden 1961; Wait 1970; Galejs 1972a; Felsen and Marcuvitz 1973; Nickolaenko and Hayakawa 2002).

Remarkable progresses in VLF electromagnetic wave propagation in the Earth–ionosphere waveguide were made by many researchers, especially including several pioneers such as Budden (1961), Wait (1957, 1960, 1970), Wait and Spies (1965), and Galejs (1964, 1968, 1970, 1972a, 1972b). In early works by Wait and Galejs, the problem on VLF wave propagation in an Earth–ionosphere waveguide was treated analytically. In 1992, Wait examined analytically the problem on VLF radio wave propagation in an inhomogeneous Earth–ionosphere waveguide Wait (1992). Obviously, those works in VLF ranges were extended in the study on SLF/ELF radio wave propagation in the Earth–ionosphere waveguide or cavity (Wait 1957,

W. Pan, K. Li, *Propagation of SLF/ELF Electromagnetic Waves*,
Advanced Topics in Science and Technology in China,
DOI 10.1007/978-3-642-39050-0_2,

1960, 1970; Wait and Spies 1965; and Suchumann 1952a, 1952b; Galejs 1964, 1972a, 1972b). Based on the pioneering works by Budden, Wait, and Galejs, further works were also carried out by many researchers (Bannister et al. 1973; Bannister 1984; Tripathi et al. 1982; Carroll and Ferraro 1990; Fraser-Smith and Bannister 1998; Cummer 2000; Wang et al. 2005, 2008). In recent years, SLF/ELF wave propagation in the anisotropic Earth–ionosphere waveguide or cavity are treated analytically (Rybachek and Ponomariev 2007; Kirillov and Pronin 2007; Li 2012).

In the 1970s, some works in SLF/ELF wave propagation in the Earth–ionosphere waveguide or cavity were also carried out by Pan in China and summarized in a recent book (Pan 2004). In what follows, we will summarize some works on SLF/ELF wave propagation by Pan in the 1970s and some new research results in our research groups. In this chapter, the region of interest is a space between an electrically homogeneous Earth and a homogeneous ionosphere and both the dipole source and the observation point are assumed on or near the Earth's surface. First, the approximate formulas are obtained for the electromagnetic field of VEDs and VMDs in the Earth–ionosphere waveguide or cavity. Based on the results obtained, the approximate formulas are derived readily for the electromagnetic field of an HED in the Earth–ionosphere waveguide or cavity by using the reciprocity theorem. Finally, computations and analyses are carried out in SLF/ELF frequency ranges.

2.2 SLF/ELF Field of VED in the Earth–Ionosphere Waveguide/Cavity

In this section, we will summarize the derivations and analyses on the electromagnetic field of a VED in the Earth–ionosphere waveguide or cavity, which were carried out by Pan in the 1970s (Pan 2004). The results by Pan, in which the derivations are different from the available results, are similar to those by Galejs (1972a).

2.2.1 Formulations of the Problem

In SLF/ELF ranges, the effective waveguide height h, which is only 60–90 km, is less than the free-space wavelength λ, and only a zero-order mode can propagate. The geometry under consideration is the same as Fig. 2.1 in Chap. 1. We assume that a VED is represented by its current density, $\hat{z}I\,\mathrm{d}l\delta(x)\delta(y)\delta(z-b)$, where $b = z_s + a$, and $z_s > 0$ denotes the height of the dipole above the Earth's surface. The field components, for the TM waves radiated by a VED, are expressed as follows:

$$E_r = \left(\frac{\partial^2}{\partial r^2} + k^2\right)(Ur), \tag{2.1}$$

$$E_\theta = \frac{1}{r}\frac{\partial^2}{\partial\theta\partial r}(Ur), \tag{2.2}$$

$$H_\phi = \frac{\mathrm{i}\omega\varepsilon}{r}\frac{\partial}{\partial\theta}(Ur), \tag{2.3}$$

$$H_r = H_\theta = E_\phi = 0. \tag{2.4}$$

Here the potential functions U satisfy the scalar Helmholtz equation,

$$(\nabla^2 + k^2)U = 0. \tag{2.5}$$

It is noted that the normalized surface impedances at the boundaries can be represented as follows:

$$\left.\frac{E_\theta}{\eta_1 H_\phi}\right|_{r=a} = -\Delta_g; \qquad \left.\frac{E_\theta}{\eta_1 H_\phi}\right|_{r=a+h} = \Delta_i, \tag{2.6}$$

where η_1 is the wave impedance of the air, and Δ_g and Δ_i are the normalized Earth's surface impedance and the normalized ionospheric surface impedance, respectively.

With the substitution of Eqs. (2.2) and (2.3) into Eq. (2.6), the boundary conditions are written as follows:

$$\left.\frac{1}{rU}\frac{\mathrm{d}}{\mathrm{d}r}(rU)\right|_{r=a} = -\mathrm{i}k\Delta_g, \tag{2.7}$$

$$\left.\frac{1}{rU}\frac{\mathrm{d}}{\mathrm{d}r}(rU)\right|_{r=a+h} = \mathrm{i}k\Delta_i. \tag{2.8}$$

In a spherical coordinate system, because of symmetry, the expanded form of Eq. (2.5) can be written as follows:

$$\frac{1}{r^2}\frac{\partial}{\partial r}\left(r^2\frac{U}{r}\right) + \frac{1}{r^2\sin\theta}\frac{\partial}{\partial\theta}\left(\sin\theta\frac{\partial U}{\partial\theta}\right) + k^2U = 0. \tag{2.9}$$

The solution of the potential function U is represented in the form

$$U = \frac{1}{r}F(r)\Phi(\theta). \tag{2.10}$$

Then, we have

$$\frac{1}{\sin\theta}\frac{\mathrm{d}}{\mathrm{d}\theta}\left[\sin\theta\frac{\mathrm{d}\Phi(\theta)}{\mathrm{d}\theta}\right] + \nu(\nu+1)\Phi(\theta) = 0, \tag{2.11}$$

$$\frac{\mathrm{d}^2F(r)}{\mathrm{d}r^2} + k^2\left[1 - \frac{\nu(\nu+1)}{k^2r^2}\right]F(r) = 0. \tag{2.12}$$

With the substitutions of $x=\cos(\pi-\theta)$ and $\theta=\pi-\arccos x$, Eq. (2.11) can be transformed into the standard Legendre equation. We write

$$\left(1-x^2\right)\frac{\mathrm{d}^2\Phi}{\mathrm{d}x^2}-2x\frac{\mathrm{d}\Phi}{\mathrm{d}x}+\nu(\nu+1)\Phi=0. \tag{2.13}$$

The two solutions of Eq. (2.13) can be expressed by $P_\nu(\cos(\pi-\theta))$ and $Q_\nu(\cos(\pi-\theta))$, respectively. It is seen that the Legendre function of the first kind $P_\nu(\cos(\pi-\theta))$ has only one pole in the situation of the dipole source with $x=-1$ or $\theta=0$. The Legendre function of the second kind has two poles at $x=\pm1$. Obviously, the Legendre function of the second kind $Q_\nu(\cos(\pi-\theta))$ should not be included in the solutions of Eq. (2.11). We take

$$\Phi(\theta)=P_\nu\big(\cos(\pi-\theta)\big). \tag{2.14}$$

In the space between the Earth's surface and the lower boundary of the ionosphere, we have $a\le r\le a+h$, and h is very small compared with a. That is to say, $1-\frac{\nu(\nu+1)}{k^2r^2}$ changes very slowly in the region of $a\le r\le a+h$. Thus the WKB approximation can be taken in this case, and the two solutions of Eq. (2.12) can be obtained readily. We have

$$F_1(r)=\frac{1}{\sqrt[4]{1-\frac{\nu(\nu+1)}{k^2r^2}}}\exp\left(\mathrm{i}k\int_a^r\left[1-\frac{\nu(\nu+1)}{k^2t^2}\right]^{\frac{1}{2}}\mathrm{d}t\right), \tag{2.15}$$

and

$$F_2(r)=\frac{1}{\sqrt[4]{1-\frac{\nu(\nu+1)}{k^2r^2}}}\exp\left(-\mathrm{i}k\int_a^r\left[1-\frac{\nu(\nu+1)}{k^2t^2}\right]^{\frac{1}{2}}\mathrm{d}t\right), \tag{2.16}$$

where the first solution represents the wave propagating in the direction when r increases, while the second solution represents the wave propagating in the direction of decreasing r.

Letting $r=a+z_r$, $z_r>0$ denotes the height of the observation point above the Earth's surface. It is assumed that both the dipole and the observation point are on or close to the Earth's surface. Obviously, we have $z_s\ll a$ and $z_r\ll a$. Then, the following approximation can be taken as:

$$\frac{\nu(\nu+1)}{k_0^2r^2}\approx\frac{\nu(\nu+1)}{k_0^2a^2}\left(1-\frac{2z_r}{a}\right). \tag{2.17}$$

Thus, the solution of Eq. (2.12) in the Earth–ionosphere cavity becomes

$$F(z_r)=A\exp\left[-\mathrm{i}k\int_a^r\left(C^2+\frac{2z}{a}S^2\right)^{\frac{1}{2}}\mathrm{d}t\right]+B\exp\left[\mathrm{i}k\int_a^r\left(C^2+\frac{2z}{a}S^2\right)^{\frac{1}{2}}\mathrm{d}t\right], \tag{2.18}$$

where the parameters C and S are defined by

$$S^2 = \frac{\nu(\nu+1)}{k^2a^2}; \qquad C^2 = 1 - S^2. \tag{2.19}$$

From Eqs. (2.7) and (2.8), the boundary conditions are rewritten as follows:

$$\left[\frac{1}{F(z)}\frac{\mathrm{d}}{\mathrm{d}z}F(z)\right]\bigg|_{z=0} = -\mathrm{i}k\Delta_g, \tag{2.20}$$

$$\left[\frac{1}{F(z)}\frac{\mathrm{d}}{\mathrm{d}z}F(z)\right]\bigg|_{z=h} = \mathrm{i}k\Delta_i. \tag{2.21}$$

With Eq. (2.20), the reflection coefficient R_g of the lower boundary (Earth–air boundary) can be obtained readily. It is

$$R_g = \frac{B}{A} = \frac{C - \Delta_g}{C + \Delta_g}. \tag{2.22}$$

Applying Eq. (2.18) into Eq. (2.21), we get

$$\begin{aligned}&\left\{1 - R_g \exp\left[2\mathrm{i}k\int_0^h \left(C^2 + \frac{2z}{a}S^2\right)^{\frac{1}{2}}\mathrm{d}z\right]\right\}\cdot\left(C^2 + \frac{2h}{a}S^2\right)\\&= -\Delta_i\left\{1 + R_g \exp\left[2\mathrm{i}k\int_0^h \left(C^2 + \frac{2z}{a}S^2\right)^{\frac{1}{2}}\mathrm{d}z\right]\right\}.\end{aligned} \tag{2.23}$$

It is noted that

$$C' = \left(C^2 + \frac{2h}{a}S^2\right)^{\frac{1}{2}}; \qquad R_i = \frac{C' - \Delta_i}{C' + \Delta_i}, \tag{2.24}$$

where R_i is the reflection coefficient of the upper boundary (air–ionosphere boundary). Then, the modal equation can be obtained readily. We write

$$R_g R_i \exp\left[2\mathrm{i}k\int_0^h \left(C^2 + \frac{2z}{a}S^2\right)^{\frac{1}{2}}\mathrm{d}z\right] = 1. \tag{2.25}$$

If C_n is the nth root, the normalized height-gain function $F_n(z)$ is expressed in the following form:

$$\begin{aligned}F_n(z_r) = \frac{1}{1+R_g}&\left[\exp\left(-\mathrm{i}k\int_0^{z_r}\sqrt{C_n^2 + \frac{2z}{a}S_n^2}\,\mathrm{d}z\right)\right.\\&\left.+ R_g\exp\left(\mathrm{i}k\int_0^{z_r}\sqrt{C_n^2 + \frac{2z}{a}S_n^2}\,\mathrm{d}z\right)\right].\end{aligned} \tag{2.26}$$

From Eq. (2.19), we obtain

$$\nu(\nu+1) = k^2a^2S_n^2 = k^2a^2\left(1 - C_n^2\right). \tag{2.27}$$

In practical wave modes in the Earth–ionosphere waveguide or cavity, the parameter C_n is not close to 1. Considering that the Earth's radius is very large, in SLF range, $k^2a^2 \gg 1$, it is seen that the parameter $|\nu|$ is a large number. Then, we write

$$\nu = kaS_n - \frac{1}{2}. \tag{2.28}$$

When the parameter $|\nu|$ is very large, namely $|\nu| \gg 1$, while the angle θ is not close to 0 and π, the Legendre function of the first kind is approximated by

$$\begin{aligned} P_\nu\left(\cos\left(\pi - \theta\right)\right) &\approx \left(\frac{2}{\pi\nu\sin\theta}\right)^{\frac{1}{2}} \cos\left[\left(\nu + \frac{1}{2}\right)(\pi - \theta) - \frac{\pi}{4}\right] \\ &= \left(\frac{1}{2\pi\nu\sin\theta}\right)^{\frac{1}{2}} \mathrm{e}^{-\mathrm{i}kaS_n\pi + \frac{\mathrm{i}\pi}{4}} \cdot \left[\mathrm{e}^{\mathrm{i}kaS_n\theta} + \mathrm{e}^{\mathrm{i}kaS_n(2\pi-\theta) - \frac{\mathrm{i}\pi}{2}}\right]. \end{aligned} \tag{2.29}$$

It is noted that the asymptotic form in Eq. (2.29) is not valid in the vicinity of the source ($\theta \to 0$) or in the vicinity of the antipole ($\theta \to \pi$). From Eq. (2.29), the function $P_\nu(\cos(\pi - \theta))$, which varies with the angle θ, includes the wave $\mathrm{e}^{\mathrm{i}kaS_n\theta}$ propagating along the short circular propagation path and the wave $\mathrm{e}^{\mathrm{i}kaS_n(2\pi-\theta)-\frac{\mathrm{i}\pi}{2}}$ propagating along the long circular propagation path, which travels over the antipole. It is seen that when the observation point is not close to the antipole, the wave propagating along the long circular propagation path can be neglected. Thus, we write

$$P_\nu\left(\cos\left(\pi - \theta\right)\right) \approx \left(\frac{1}{2\pi\nu\sin\theta}\right)^{\frac{1}{2}} \mathrm{e}^{-\mathrm{i}kaS_n\pi + \frac{\mathrm{i}\pi}{4}} \mathrm{e}^{\mathrm{i}kaS_n\theta}. \tag{2.30}$$

Then, we get

$$\begin{aligned} P_\nu\left(\cos\left(\pi - \theta\right)\right) F_n(z) \propto &\left(\frac{1}{\sin\theta}\right)^{\frac{1}{2}} \mathrm{e}^{\mathrm{i}kaS_n\theta} \\ &\times \left\{\exp\left[-\mathrm{i}k\int_0^z \left(C_n^2 + \frac{2t}{a}S_n^2\right)^{\frac{1}{2}} \mathrm{d}t\right] \right. \\ &\left. + R_g \exp\left[\mathrm{i}k\int_0^z \left(C_n^2 + \frac{2t}{a}S_n^2\right)^{\frac{1}{2}} \mathrm{d}t\right]\right\}. \end{aligned} \tag{2.31}$$

Obviously, each propagation mode includes the wave traveling to the Earth's surface with the angle $\theta_n = \arccos C_n$ and the corresponding reflected wave. It is noted that the phase velocity v_n is inversely proportional to the real part of S_n, and the attenuation rate is proportional to the imaginary part of S_n. Then, we write

$$\frac{v_n}{c} = \frac{1}{\operatorname{Re} S_n} \approx \operatorname{Re}\left(1 + \frac{C_n^2}{2}\right), \tag{2.32}$$

$$\alpha_n = 8.68 \times \frac{2\pi \operatorname{Im} S_n}{\lambda} \approx 8.68 \times \frac{2\pi \operatorname{Im}(-\frac{C_n^2}{2})}{\lambda} \ (\mathrm{dB/km}). \tag{2.33}$$

In SLF/ELF ranges, both Δ_i and Δ_g are very small. Thus, the reflection coefficients R_i and R_g are approximated by

$$R_i \approx \exp(-2\Delta_i / C_n), \tag{2.34}$$

$$R_g \approx \exp(-2\Delta_g / C_n). \tag{2.35}$$

The approximated solutions of the mode equation (2.25) can be obtained readily. We write

$$C_n = \frac{n\pi}{2kh} + \sqrt{\left(\frac{n\pi}{2kh}\right)^2 - \frac{\mathrm{i}(\Delta_g + \Delta_i)}{kh}}, \tag{2.36}$$

$$S_n = \left\{1 - \left(\frac{n\pi}{2kh}\right)^2 \left(1 + \sqrt{1 - \frac{4\mathrm{i}(\Delta_g + \Delta_i)kh}{(n\pi)^2}}\right)\right\}^{1/2}. \tag{2.37}$$

For both Δ_i and Δ_g being very small, when $n\pi > 2kh$, the parameter S_n is a pure imaginary number, which represents an evanescent mode. For a typical daytime model with the ionospheric height $h = 70$ km, when $\lambda > 2h = 140$ km, $f < 2.2$ kHz, only the TM wave of zero order can propagate and the rest of the waves are evanescent. Similarly, for a typical nighttime model with the ionospheric height $h = 90$ km, when $\lambda > 2h = 180$ km, $f < 1.7$ kHz, only a zero-order TM mode can propagate and the rest, the high-order modes, are evanescent. Then, we have

$$C_0 \approx \mathrm{e}^{-\mathrm{i}\pi/4} \sqrt{\frac{\Delta_g + \Delta_i}{kh}}, \tag{2.38}$$

$$S_0 \approx \left[1 + \frac{\mathrm{i}(\Delta_g + \Delta_i)}{kh}\right]^{1/2}. \tag{2.39}$$

We examine the height-gain function $F_\mu(z)$, which is the solution of Eq. (2.12). At the Earth's surface $z = 0$, we have

$$F_\mu(0) = 1; \qquad F'_\mu(0) = -\mathrm{i}k\Delta_g. \tag{2.40}$$

It is noted that, for the parameter μ, its variations are regarded to be continuous. Both the height-gain functions $F_n(z)$ and $F_\mu(z)$ satisfy Eq. (2.12) and the boundary conditions at the Earth's surface. Then, we write

$$\int_0^h F_n(z) F_\mu(z) \mathrm{d}z = \frac{1}{k^2(S_n^2 - S_\mu^2)} \left[F_\mu(z) F'_n(z) - F_n(z) F'_\mu(z)\right]\Big|_0^h, \tag{2.41}$$

where

$$S_n^2 = \frac{\nu(\nu+1)}{k^2a^2}; \qquad S_\mu^2 = \frac{\mu(\mu+1)}{k^2a^2}. \tag{2.42}$$

Evidently, when $S_\mu \to S_n$, we have $F_\mu(z) \to F_n(z)$. Thus, we write

$$\begin{aligned} N_n &= \int_0^h F_n(z)F_n(z)\,\mathrm{d}z = \lim_{s_\mu \to s_n} \int_0^h F_n(z)F_\mu(z)\,\mathrm{d}z \\ &= \lim_{s_\mu \to s_n} \frac{1}{k(S_n^2 - S_\mu^2)}\left[F_n'(h)F_\mu(h) - F_n(h)F_\mu'(h)\right] \\ &= \frac{1}{2k^2S_n}\left[F_n(h)\frac{\mathrm{d}F_n'(h)}{\mathrm{d}S_n} - F_n'(h)\frac{\mathrm{d}F_n(h)}{\mathrm{d}S_n}\right]. \end{aligned} \tag{2.43}$$

Letting

$$H = \int_0^h \left(C_n^2 + \frac{2z}{a}S_n^2\right)^{\frac{1}{2}} \mathrm{d}z = \frac{a}{3S_n^2}\left(C_n'^{\,3} - C_n^3\right), \tag{2.44}$$

we have

$$\begin{aligned} N_n = {} & \frac{4R_g}{(1+R_g)^2}\frac{aC_n'}{3S_n^4}\left\{C_n'^{\,3} - C_n^3 + \frac{3S_n^2}{2}\left[C_n'\left(1 - \frac{2h}{a}\right) - C_n\right]\right\} \\ & - \frac{\mathrm{i}C_n'\Delta_g}{2C_n^3k} + \frac{\mathrm{i}(1-\frac{2h}{a})(\mathrm{e}^{-2\mathrm{i}kH} - R_g^2\mathrm{e}^{2\mathrm{i}kH})}{2kC_n'(1+R_g)^2}. \end{aligned} \tag{2.45}$$

In the case of the Earth being a perfect conducting sphere, $\Delta_g \approx 0,\ R_g \approx 1$. We obtain

$$\begin{aligned} N_n \approx {} & \frac{aC_n'}{3S_n^4}\left\{C_n'^3 - C_n^3 + \frac{3S_n^2}{2}\left[C'_n\left(1 - \frac{2h}{a}\right) - C_n\right]\right\} \\ & + \left(1 - \frac{2h}{a}\right)\frac{\sin 2kH}{4kC_n'}. \end{aligned} \tag{2.46}$$

When $C_n^2 \gg \frac{2h}{a}$, we have

$$C_n'^{\,3} - C_n^3 + \frac{3S_n^2}{2}\left[C_n'\left(1 - \frac{2h}{a}\right) - C_n\right] \approx \frac{3hS_n^4}{2aC_n}. \tag{2.47}$$

Then, a further simplification gives

$$N_n \approx \frac{h}{2}\left(1 + \frac{\sin 2kC_nh}{2kC_nh}\right). \tag{2.48}$$

By now, the potential function U for the electromagnetic field of a VED in the Earth–ionosphere waveguide or cavity can be represented in the following form:

$$U = \frac{1}{r} \sum_{s} A_s F_s(z_r) P_\nu\big(\cos(\pi - \theta)\big). \tag{2.49}$$

In the next step, it is necessary to determine the excitation coefficients A_s ($s = 0, 1, 2, \ldots$) in Eq. (2.49).

2.2.2 Determination of the Excitation Coefficients A_s

In the region of $|z_r - z_s| \le \delta$, where δ is a finite small value, when $\theta \to 0$, the influences of the electromagnetic field by both the Earth and the ionosphere can be neglected.

By multiplying the function $rF_n(z)$ on both sides of Eq. (2.49), and integrating from 0 to h for z, we have

$$A_n N_n P_\nu\big(\cos(\pi - \theta)\big) = \int_0^h rUF_n(z)\,\mathrm{d}z. \tag{2.50}$$

Obviously, the excitation coefficient A_n can be obtained by using the source singularity. Close to the source, the integral is written in the form

$$\int_0^h rUF_n(z)\,\mathrm{d}z = \int_0^{z_s-\delta} rF_n(z)U\,\mathrm{d}z + \int_{z_s-\delta}^{z_s+\delta} rF_n(z)U\,\mathrm{d}z + \int_{z_s+\delta}^{h} rF_n(z)U\,\mathrm{d}z, \tag{2.51}$$

where δ is very small. When the conditions $|z - z_s| \le \delta$ and $\theta \to 0$ are satisfied, the potential function U is approximated by

$$U \xrightarrow{\theta\to 0,\ |z-z_s|<\delta} U_0 = \frac{C_0 \mathrm{e}^{\mathrm{i}kR}}{r_0 R} \approx \frac{\mathrm{i}I\,\mathrm{d}l}{4\pi\omega\varepsilon_0 r_0 R}, \tag{2.52}$$

where

$$R = \left(r^2 + r_0^2 - 2rr_0\cos\theta\right)^{\frac{1}{2}}; \qquad C_0 = \frac{\mathrm{i}I\,\mathrm{d}l}{4\pi\omega\varepsilon_0}. \tag{2.53}$$

In the above formulas, $r_0 = a + z_s$ and $r = a + z_r$. In the intervals $(0,\ z_s - \delta)$ and $(z_s + \delta,\ h)$, both U and $F_n(z)$ are continuous functions, and the integration from 0 to $z_s - \delta$ and that from $z_s + \delta$ to h in Eq. (2.51) are finite. Thus, we get

$$\int_0^h rF_n(z)U\,\mathrm{d}z = r_0 F_n(z_s) \int_{z_s-\delta}^{z_s+\delta} U_0\,\mathrm{d}z + O(1) = \frac{\mathrm{i}F_n(z_s)I\,\mathrm{d}l}{4\pi\omega\varepsilon_0} \int_{z_s-\delta}^{z_s+\delta} \frac{\mathrm{d}z}{R} + O(1). \tag{2.54}$$

Close to the source, we have

$$\cos\theta \approx 1-\frac{\theta^2}{2}; \qquad R\approx r_0\left[\theta^2+\frac{(z-z_s)^2}{r^2}\right]^{\frac{1}{2}}. \tag{2.55}$$

Then, we obtain

$$\int_{z_s-\delta}^{z_s+\delta}\frac{\mathrm{d}z}{R}\approx \ln\left(\frac{\delta}{r_0}+\sqrt{\frac{\delta^2}{r_0^2}+\theta^2}\right)-\ln\left(\sqrt{\frac{\delta^2}{r_0^2}+\theta^2}-\frac{\delta}{r_0}\right)\xrightarrow{\theta\to 0}-\ln\theta^2. \tag{2.56}$$

The integral in the right side of Eq. (2.54) can be obtained readily:

$$\int_0^h rUF_n(z)\,\mathrm{d}z\xrightarrow{\theta\to 0}-\frac{\mathrm{i}I\,\mathrm{d}l\,F_n(z_s)}{4\pi\omega\varepsilon_0}\ln\theta^2. \tag{2.57}$$

When the observation point is close to the source, namely, $\theta\to 0$, the Legendre function of the first order $P_\nu(\cos(\pi-\theta))$ is approximated as

$$P_\nu\big(\cos(\pi-\theta)\big)\xrightarrow{\theta\to 0}\frac{\sin\nu\pi}{\pi}\ln\theta^2. \tag{2.58}$$

Thus, the result becomes

$$\begin{aligned}A_n&=\frac{1}{N_n}\lim_{\theta\to 0}\frac{1}{P_\nu(\cos(\pi-\theta))}\int_0^h rUF_n(z)\,\mathrm{d}z=-\frac{\mathrm{i}I\,\mathrm{d}l\,F_n(z_s)}{4\omega\varepsilon_0N_n\sin\nu\pi}\\&=-\frac{\mathrm{i}I\,\mathrm{d}l\,F(z_s)}{2\omega\varepsilon_0h\sin\nu\pi}\Lambda_n,\end{aligned} \tag{2.59}$$

where the excitation factor Λ_n is defined by

$$\Lambda_n=\frac{1}{1+\frac{\sin 2kc_nh}{2kc_nh}}. \tag{2.60}$$

Therefore, we have

$$U=-\frac{\mathrm{i}I\,\mathrm{d}l}{2hr\omega\varepsilon_0}\sum_{n=0}^{\infty}\frac{\Lambda_n}{\sin\nu\pi}F_n(z_s)F_n(z_r)P_\nu\big(\cos(\pi-\theta)\big). \tag{2.61}$$

2.2.3 Approximated Formulas of SLF Field

With the substitution of Eq. (2.61) into Eqs. (2.1)–(2.3), the formulas for the three components E_r, E_θ, and H_ϕ can be derived readily:

$$E_r=-\frac{\mathrm{i}I\,\mathrm{d}l}{2\omega\varepsilon_0hr^2}\cdot\sum_{n=0}^{\infty}\frac{\Lambda_n\nu(\nu+1)}{\sin\nu\pi}F_n(z_s)F_n(z_r)P_\nu\big(\cos(\pi-\theta)\big), \tag{2.62}$$

$$E_\theta = -\frac{\mathrm{i}I\,\mathrm{d}l\eta}{2hr}\cdot\sum_{n=0}^{\infty}\Lambda_n F_n(z_s)\frac{\partial F_n(z_r)}{k\partial z}\frac{\partial P_\nu(\cos(\pi-\theta))}{\sin\nu\pi\partial\theta}, \tag{2.63}$$

$$H_\phi = \frac{I\,\mathrm{d}l}{2hr}\cdot\sum_{n=0}^{\infty}\Lambda_n F_n(z_s)F_n(z_r)\frac{\partial P_\nu(\cos(\pi-\theta))}{\sin\nu\pi\partial\theta}. \tag{2.64}$$

In general, the parameter ν has a large positive imaginary part. It follows that

$$\nu \approx kaS_n - \frac{1}{2}; \qquad \sin\nu\pi \approx \frac{1}{2}\mathrm{e}^{-\mathrm{i}(\nu\pi-\frac{\pi}{2})}. \tag{2.65}$$

In SLF range, when the angle θ is not close to 0 and π, the function $P_\nu(\cos(\pi-\theta))$ is approximated as follows:

$$P_\nu\big(\cos(\pi-\theta)\big) \approx \left(\frac{1}{2\pi\nu\sin\theta}\right)^{\frac{1}{2}}\exp\left[-\mathrm{i}\left(\nu+\frac{1}{2}\right)(\pi-\theta)+\mathrm{i}\frac{\pi}{4}\right]. \tag{2.66}$$

Then, the complete formulas for the components E_r, E_θ, and H_ϕ of the electromagnetic field of a VED in the Earth–ionosphere waveguide can be expressed in the following forms:

$$E_r = -\frac{I\,\mathrm{d}l\eta\mathrm{e}^{-\mathrm{i}\frac{\pi}{4}}}{h\sqrt{\lambda a\sin\theta}}\sum_{n=0}^{\infty}\Lambda_n S_n^{\frac{3}{2}}F_n(z_s)F_n(z_r)\mathrm{e}^{\mathrm{i}kaS_n\theta}, \tag{2.67}$$

$$E_\theta = -\frac{I\,\mathrm{d}l\eta\mathrm{e}^{\mathrm{i}\frac{\pi}{4}}}{kh\sqrt{\lambda a\sin\theta}}\sum_{n=0}^{\infty}\Lambda_n S_n^{\frac{1}{2}}F_n(z_s)F_n'(z_r)\mathrm{e}^{\mathrm{i}kaS_n\theta}, \tag{2.68}$$

$$H_\phi = \frac{I\,\mathrm{d}l\mathrm{e}^{-\mathrm{i}\frac{\pi}{4}}}{h\sqrt{\lambda a\sin\theta}}\sum_{n=0}^{\infty}\Lambda_n S_n^{\frac{1}{2}}F_n(z_r)F_n(z_s)\mathrm{e}^{\mathrm{i}kaS_n\theta}. \tag{2.69}$$

Letting

$$E_0 = -\frac{I\,\mathrm{d}l\eta}{\lambda}\frac{\mathrm{e}^{\mathrm{i}ka\theta}}{a\theta}\ (\mathrm{V/m}) = -\frac{\mathrm{e}^{\mathrm{i}ka\theta}}{a\theta}300\sqrt{P_{\mathrm{kW}}}\ (\mathrm{mV/m}), \tag{2.70}$$

the approximated formula for the component E_r is rewritten as follows:

$$E_r = E_0\sqrt{\frac{d/a}{\sin(d/a)}}\frac{\sqrt{d/\lambda}}{h/\lambda}\mathrm{e}^{-\mathrm{i}kd-\mathrm{i}\frac{\pi}{4}}\sum_{n=0}^{\infty}\Lambda_n F_n(z_s)F_n(z_r)S_n^{3/2}\mathrm{e}^{\mathrm{i}kS_n d}, \tag{2.71}$$

where $d = a\theta$ is the propagation distance along the Earth's surface, and P_{kW} is the power in kW. Obviously, the approximated formulas for E_θ and H_ϕ can also be written readily.

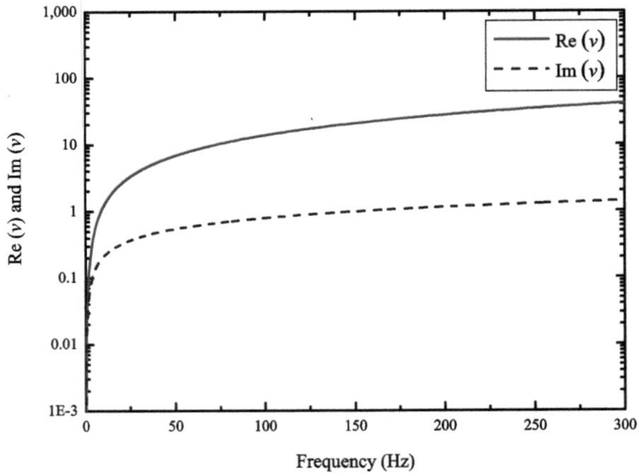

Fig. 2.1 The real and imaginary parts of ν versus the operating frequency

2.2.4 New Algorithm for ELF Field

In the whole ELF range, even in the lower end of SLF range, the wavelength λ is very long, which can be compared with the Earth's circumference. In other words, the parameter ka will be small. From $\nu(\nu+1) = k^2a^2S_n^2$, we find that the eigenvalue ν no longer satisfies the condition of $\nu \gg 1$. The eigenvalue ν of the zero-order TM wave is computed and shown in Fig. 2.1. It is seen that the eigenvalue does not satisfy $\nu \gg 1$ below 50 Hz. The Legendre function of the first kind $P_\nu(\cos(\pi-\theta))$ should not be evaluated by using its asymptotic formula in Eq. (2.29). In what follows, we will attempt to outline a new algorithm for evaluating the Legendre function of the first kind $P_\nu(\cos(\pi-\theta))$ (Peng et al. 2012).

In order to analyze quantitatively the ELF field in the Earth–ionosphere cavity, it is necessary to evaluate accurately the function $P_\nu(\cos(\pi-\theta))$. Generally, the eigenvalue ν is a complex number, and the Legendre function of the first kind $P_\nu(\cos(\pi-\theta))$ can be represented in the following form:

$$P_\nu\big(\cos(\pi-\theta)\big) = -\frac{\sin\nu\pi}{\pi}\sum_{n=0}^{\infty} P_n(\cos\theta)\frac{2n+1}{n(n+1)-\nu(\nu+1)}. \tag{2.72}$$

Another form of the function $P_\nu(\cos\theta)$ is expressed as Gradshteyn and Ryzhik (1980)

$$P_\nu(\cos\theta) = \frac{2}{\pi}\int_0^\theta \frac{\cos[(\nu+0.5)t]}{\sqrt{2(\cos t-\cos\theta)}}\,\mathrm{d}t. \tag{2.73}$$

Letting

$$\Phi(\nu,\theta) = \frac{1}{\sin\nu\pi}P_\nu\big(\cos(\pi-\theta)\big), \tag{2.74}$$

by using the following relation:

$$P'_\nu(z) = \frac{\nu}{z^2 - 1}\big[zP_\nu(z) - P_{\nu-1}(z)\big], \tag{2.75}$$

we write

$$\frac{\partial}{\partial\theta}\Phi(\nu,\theta) = \frac{\nu}{\sin\theta}\big[\Phi(\nu,\theta)\cos\theta - \Phi(\nu-1,\theta)\big]. \tag{2.76}$$

In practical numerical calculations, the function $\Phi(\nu,\theta)$ can be expressed in the integrated form

$$\Phi(\nu,\theta) = \frac{2}{\pi}\int_0^{\pi-\theta} \frac{M(\nu,t)}{\sqrt{2[\cos t - \cos(\pi-\theta)]}}\,\mathrm{d}t, \tag{2.77}$$

where

$$M(\nu,t) = \begin{cases} M_1(\nu,t); & 0 < \mathrm{Im}(\nu) \le 0.75, \\ M_2(\nu,t); & \mathrm{Im}(\nu) > 0.75, \end{cases} \tag{2.78}$$

and

$$M_1(\nu,t) = \frac{\cos(\nu+0.5)t}{\sin(\nu\pi)}, \tag{2.79}$$

$$\begin{aligned} M_2(\nu,t) = -\mathrm{i}\bigg\{&\exp\bigg[\mathrm{i}\bigg(\mathrm{Re}(\nu)t + \mathrm{Re}(\nu)\pi + \frac{t}{2}\bigg)\bigg] \\ &\times \exp\big[-\mathrm{Im}(\nu)(\pi+t)\big] + \exp\big[-\mathrm{Im}(\nu)(\pi-t)\big] \\ &\times \exp\bigg[-\mathrm{i}\bigg(\mathrm{Re}(\nu)t - \mathrm{Re}(\nu)\pi + \frac{t}{2}\bigg)\bigg]\bigg\}. \end{aligned} \tag{2.80}$$

Due to the singularity at $t = \pi - \theta$ existing in the integral Kernel function in Eq. (2.77), the function $\Phi(\nu,\theta)$ can be divided into two parts. We write

$$\Phi(\nu,\theta) = I_1 + I_2, \tag{2.81}$$

where

$$I_1 = \int_0^{\pi-\theta-\delta} \frac{M(\nu,t)}{\sqrt{2[\cos t - \cos(\pi-\theta)]}}\,\mathrm{d}t, \tag{2.82}$$

$$I_2 = \int_{\pi-\theta-\delta}^{\pi-\theta} \frac{M(\nu,t)}{\sqrt{2[\cos t - \cos(\pi-\theta)]}}\,\mathrm{d}t. \tag{2.83}$$

It is noted that the first integral I_1 can be evaluated readily by using the trapezoidal or Simpson's numerical integration method. For the second integral I_2, with the change of the variable $\pi - \theta - s = t$, it follows that

$$I_2 = \int_0^{\delta} \frac{M(\nu, \pi - \theta - s)}{2\sqrt{\sin(\pi - \theta - \frac{s}{2})\sin(\frac{s}{2})}} \mathrm{d}s. \tag{2.84}$$

When $\delta \to 0$, it is seen that the function $\frac{M(\nu,\pi-\theta-s)}{2\sqrt{\sin(\pi-\theta-\frac{s}{2})}}$ is approximately a constant in the interval $[0, \delta]$. Then, we have

$$\begin{aligned} I_2 &\approx \frac{1}{4}\left[\frac{M(\nu, \pi - \theta)}{\sin(\pi - \theta)} + \frac{M(\nu, \pi - \theta - \delta)}{\sin(\pi - \theta - \frac{\delta}{2})}\right] \cdot \int_0^{\delta} \sqrt{\frac{2}{s}}\, \mathrm{d}s \\ &= \sqrt{\frac{\delta}{2}}\left[\frac{M(\nu, \pi - \theta)}{\sin(\pi - \theta)} + \frac{M(\nu, \pi - \theta - \delta)}{\sin(\pi - \theta - \frac{\delta}{2})}\right]. \end{aligned} \tag{2.85}$$

By now, the approximated formula has been derived for the Legendre function of the first order $P_\nu(\cos(\pi - \theta))$ in the case of $\nu < 1$. In whole ELF range, even in the lower end of SLF range, the three non-zero components E_r, E_θ, and H_ϕ can be evaluated accurately by using Eqs. (2.88)–(2.90).

Additionally, by using the Legendre polynomial expansion of $P_\nu(\cos(\pi - \theta))$, the Legendre function $P_\nu(\cos(\pi - \theta))$ can also be calculated accurately. We write

$$\begin{aligned} \Phi_1(\nu, \theta) &= \frac{P_\nu(\cos(\pi - \theta))}{\sin \nu\pi} \\ &= -\frac{1}{\pi}\sum_{n=0}^{\infty} P_n(\cos\theta)\frac{2n+1}{n(n+1) - \nu(\nu+1)}. \end{aligned} \tag{2.86}$$

As shown in Fig. 2.2, the ratios of the calculated results by using the series algorithm of Eq. (2.86) and those by using the numerical integrated algorithm of Eq. (2.81) are carried out at $n = 100$, 200, and 500, respectively. It is seen that the calculated results $\Phi_1(\nu, \theta)$ by using the series algorithm move closer to the corresponding approximated results $\Phi(\nu, \theta)$ by using the approximated algorithm proposed in this section when the parameter n increases. When n is larger than 200, the results by using the series algorithm are in agreement with those by the numerical integrated algorithm. Obviously, the numerical integrated algorithm in evaluating the Legendre function works more quickly and easily.

In earlier works for the SLF field computation, the asymptotic expansion for the function $P_\nu(\cos(\pi - \theta))$ is usually employed by Eq. (2.29). We write

$$\begin{aligned} \Phi_2(\nu, \theta) &= \frac{P_\nu(\cos(\pi - \theta))}{\sin \nu\pi} \\ &\approx \frac{1}{\sin\nu\pi}\sqrt{\frac{2}{\pi\nu\sin\theta}}\cos\left[\left(\nu + \frac{1}{2}\right)(\pi - \theta) - \frac{\pi}{4}\right]. \end{aligned} \tag{2.87}$$

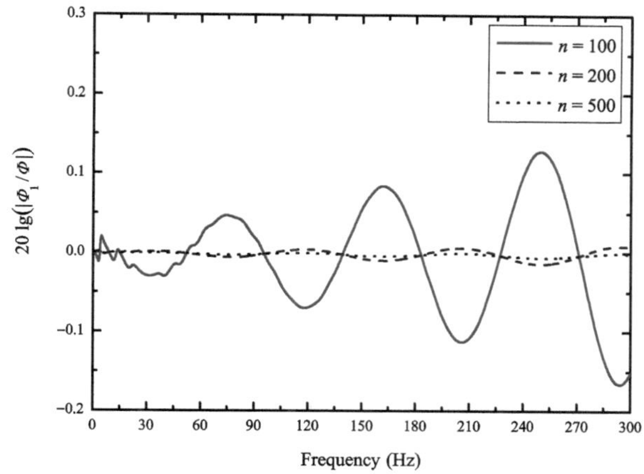

Fig. 2.2 The ratio $\Phi_1(\nu,\theta)/\Phi(\nu,\theta)$ versus the operating frequency

Compared with the other two algorithms, the numerical integrated algorithm has the merits of high accuracy and efficiency, which is suitable for the field computations in ELF range and the lower end of SLF range, which is below 50 Hz or so. Thus, the complete formulas for the field components in Eqs. (2.67)–(2.69) are rewritten in the following forms:

$$E_r = -\frac{\mathrm{i}I\,\mathrm{d}l}{2\omega\varepsilon_0 hr^2}\cdot\sum_{n=0}^{\infty}\left[\Lambda_n(kaS_n)^2\cdot F_n(z_r)F_n(z_s)\Phi(\nu,\theta)\right], \tag{2.88}$$

$$E_\theta = -\frac{\mathrm{i}I\,\mathrm{d}l}{2\omega\varepsilon_0 hr}\cdot\sum_{n=0}^{\infty}\left\{\Lambda_n F_n(z_s)\frac{\mathrm{d}F_n(z_r)}{\mathrm{d}z}\right.$$
$$\left.\times\frac{\nu}{\sin\theta}\left[\Phi(\nu,\theta)\cos\theta-\Phi(\nu-1,\theta)\right]\right\}, \tag{2.89}$$

$$H_\phi = \frac{I\,\mathrm{d}l}{2hr}\cdot\sum_{n=0}^{\infty}\left\{\Lambda_n F_n(z_s)F_n(z_r)\right.$$
$$\left.\times\frac{\nu}{\sin\theta}\left[\Phi(\nu,\theta)\cos\theta-\Phi(\nu-1,\theta)\right]\right\}. \tag{2.90}$$

Evidently, the field components for the ELF range can be evaluated accurately by using Eqs. (2.88)–(2.90).

We assume that the Earth's radius is taken as $a = 6{,}370$ km, the current moment of the dipole is $I\,\mathrm{d}l = 1$ A·m, the ground conductivity is $\sigma_g = 10^{-4}$ S/m, the ionospheric conductivity is $\sigma_i = 10^{-5}$ S/m, and the ionospheric equivalent reflection height is $h = 70$ km. By using Eqs. (2.88)–(2.90), the magnitudes of the three components E_r, E_θ, and H_ϕ are computed at $f = 1$ Hz, 3 Hz, 5 Hz, 10 Hz, 20 Hz, and 30 Hz, and shown in Figs. 2.3, 2.4, 2.5, respectively. From the above computations, the discussions are carried out and conclusions are drawn as follows:

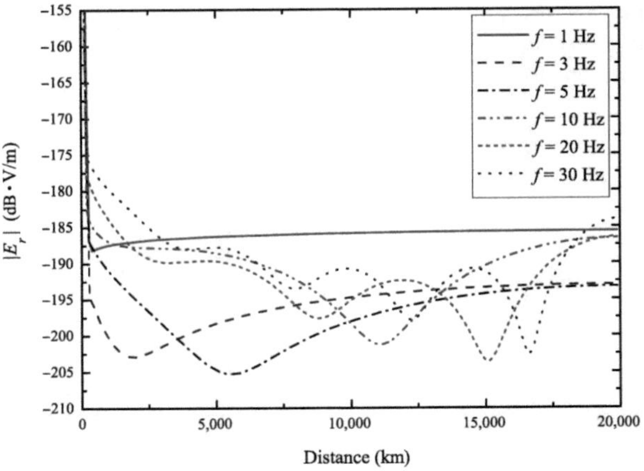

Fig. 2.3 The magnitudes of E_r versus the propagation distances at $f = 1$ Hz, 3 Hz, 5 Hz, 10 Hz, 20 Hz, and 30 Hz

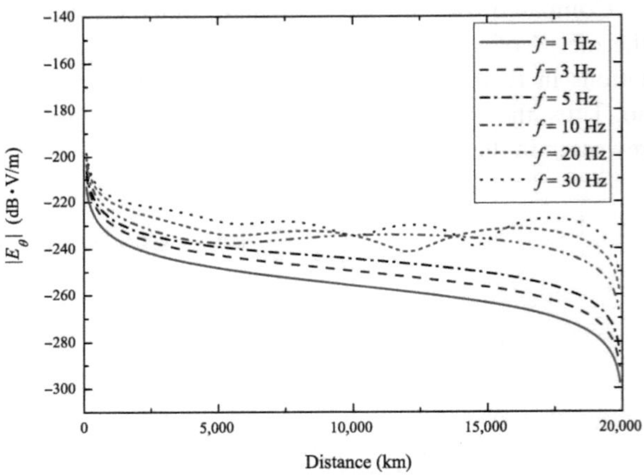

Fig. 2.4 The magnitudes of E_θ versus the propagation distances at $f = 1$ Hz, 3 Hz, 5 Hz, 10 Hz, 20 Hz, and 30 Hz

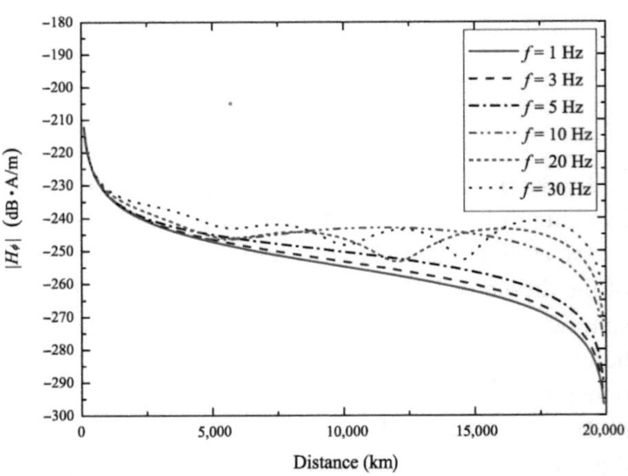

Fig. 2.5 The magnitudes of H_ϕ versus the propagation distances at $f = 1$ Hz, 3 Hz, 5 Hz, 10 Hz, 20 Hz, and 30 Hz

- When the observation point is close to the antipole of the dipole source, the multipath effects should be considered. The interference phenomenon is resulted by the two waves traveling along the small and large circle paths. The higher the operating frequency is, the more severely the field strength by the interference fluctuates. The smaller the distance from the observation point to the antipole, the more smoothing the field strength by the interference fluctuates.
- With the decrease of the operating frequency, the magnitude of the component E_r, which is generated by a VED in the Earth–ionosphere cavity, becomes relatively "flat". When the operating frequency is below 1 Hz, the magnitude tends approximately to a constant at a distance larger than 1,000 km. In the range below 50 Hz, only the quasi-TEM wave can propagate in the Earth–ionosphere cavity. When $\nu \to 0$, we have $P_\nu(\cos(\pi - \theta)) \to 1$, $\phi(\nu, \theta) \to \frac{1}{\nu\pi}$. Namely, the magnitude becomes stable, which varies little with the propagation distance. In what follows, we will attempt to give the physical explanations as follows: In the range below 50 Hz, the wavelength is comparable to the Earth's circumference, and the electromagnetic wave in the Earth–ionosphere cavity exists in the form of "stationary wave". Furthermore, at the frequency range below 1 Hz, the wavelengths can be a dozen or dozens of the Earth's circumference. This means that the phase differences at different propagation paths are small, and the total field is approximately equal to the sum of the electric fields with the same phases on different propagation paths. Therefore, the interference phenomenon is gradually disappearing and the total field becomes stable in the zones which are far away from the source point.
- Due to the spherically symmetry, the propagation characteristics of E_θ and H_ϕ are also similar. Moreover, the magnitudes of the field components E_θ and H_ϕ are equal to 0 at the antipole of the dipole source and decrease drastically in the zone near the antipole.

2.3 SLF/ELF Field of VMD in the Earth–Ionosphere Waveguide/Cavity

In practical applications, it is impossible to employ a VMD to generate TE waves in SLF/ELF ranges. In order to investigate the electromagnetic waves generated by an HED, it is necessary to analyze the propagation characteristics for both TM and TE waves. It is well known that a VMD is a typical excitation source for TE waves. In this section, we will examine the electromagnetic field of VMD in the Earth–ionosphere waveguide or cavity.

If the excitation source in the preceding section is replaced by a VMD with its moment $M = I\,\mathrm{d}a$, and $\mathrm{d}a$ is the area of the loop, the non-zero components E_ϕ, H_r, and H_θ, which the TE waves radiated by a VMD, are represented in the following forms:

$$E_\phi = -\frac{\mathrm{i}\omega\mu}{r}\frac{\partial}{\partial\theta}(Vr), \tag{2.91}$$

$$H_r = \left(\frac{\partial}{\partial r^2} + k^2\right)(Vr), \tag{2.92}$$

$$H_\theta = \frac{1}{r}\frac{\partial^2}{\partial\theta\partial r}(Vr), \tag{2.93}$$

where the potential function V satisfies the scalar Helmholtz equation,

$$\left(\nabla^2 + k^2\right)V = 0. \tag{2.94}$$

The solution for V is represented in the form

$$V = \frac{1}{r}G(r)\Phi(\theta). \tag{2.95}$$

In the same manner as in the preceding section, we obtain readily:

$$\Phi(\theta) = P_\nu\left(\cos(\pi - \theta)\right), \tag{2.96}$$

$$G(r) = A\exp\left[-\mathrm{i}k\int_0^z\left(C^2 + \frac{2z}{a}S^2\right)^{1/2}\mathrm{d}z\right] + B\exp\left[\mathrm{i}k\int_0^z\left(C^2 + \frac{2z}{a}S^2\right)^{1/2}\mathrm{d}z\right], \tag{2.97}$$

where the relations between both the parameters S and C and the eigenvalue ν are represented as follows:

$$S^2 = \frac{\nu(\nu+1)}{k^2a^2}; \qquad C^2 = 1 - S^2. \tag{2.98}$$

From the boundary conditions

$$\left.\frac{E_\phi}{\eta_1 H_\theta}\right|_{r=a} = \Delta_g, \qquad \left.\frac{E_\phi}{\eta_1 H_\theta}\right|_{r=a+h} = -\Delta_i, \tag{2.99}$$

we rewrite

$$\left.\frac{1}{V}\frac{\partial V}{\partial r}\right|_{r=a} = -\frac{\mathrm{i}k}{\Delta_g}, \tag{2.100}$$

$$\left.\frac{1}{V}\frac{\partial V}{\partial r}\right|_{r=a+h} = \frac{\mathrm{i}k}{\Delta_i}. \tag{2.101}$$

Substituting Eq. (2.95) into Eqs. (2.100) and (2.101), the modal equation for TE waves is obtained readily. It is

$$R_g^h R_i^h \exp\left[2\mathrm{i}k \int_0^h \left(C_m^2 + \frac{2z}{a} S_n^2\right)^{1/2} \mathrm{d}z\right] = \mathrm{e}^{2\mathrm{i}(m-1)\pi}, \tag{2.102}$$

where

$$R_g^h = (C_m - \Delta_g^{-1})(C_m + \Delta_g^{-1})^{-1}, \tag{2.103}$$

$$R_i^h = (C_m' - \Delta_i^{-1})(C_m' + \Delta_i^{-1})^{-1}, \tag{2.104}$$

$$C_m' = \left(C_m^2 + \frac{2h}{a} S_m^2\right)^{1/2} \approx C_m. \tag{2.105}$$

If C_m $(m = 1, 2, 3, \ldots)$ are the roots of the modal equation (2.102), the height-gain functions $G_m(z)$ can be represented in the following form:

$$G_m(z) = \frac{1}{(1 + R_g^h)} \left\{\exp\left[-\mathrm{i}k \int_0^z \left(C_m^2 + \frac{2t}{a} S_m^2\right)^{1/2} \mathrm{d}t\right]\right.$$
$$\left. + R_g^h \exp\left[\mathrm{i}k \int_0^z \left(C_m^2 + \frac{2t}{a} S_m^2\right)^{1/2} \mathrm{d}t\right]\right\}. \tag{2.106}$$

In SLF/ELF ranges, we find $1/\Delta_g \gg C_m$ and $1/\Delta_i \gg C_m$. Then, the reflection coefficients R_g^h and R_i^h can be simplified as follows:

$$R_g^h \approx -\exp(-2C_m \Delta_g), \tag{2.107}$$

$$R_i^h \approx -\exp(-2C_m \Delta_i). \tag{2.108}$$

By solving the modal equation (2.102), the roots C_m $(m = 1, 2, 3, \ldots)$ are approximated by

$$C_m \approx \frac{m\pi}{kh} \left(1 + \mathrm{i}\frac{\Delta_g + \Delta_i}{kh}\right)^{-1}, \tag{2.109}$$

$$S_m = (1 - C_m^2)^{1/2}. \tag{2.110}$$

Similar to those for TM waves, the different modes for TE waves are orthogonal each other. Then, we obtain readily:

$$N_{m\mu} = \int_0^h G_m(z) G_\mu(z)\,\mathrm{d}z = 0; \quad \text{when } \mu \neq m, \tag{2.111}$$

and

$$
\begin{aligned}
N_{m\mu} &= \int_0^h G_m(z) G_\mu(z)\,\mathrm{d}z = \lim_{S_\mu \to S_m} \frac{G'_m(h) G_\mu(h) - G_m(h) G'_\mu(h)}{k^2 (S_m^2 - S_\mu^2)} \\
&= -\frac{4 R_g^h k^2 C'_m \frac{\partial H}{\partial S_m} - 2\mathrm{i} k C'_m \frac{\partial R_g^h}{\partial S_m}}{2k^2 S_m (1 + R_g^h)^2} \\
&\quad - \frac{\mathrm{i} k \frac{\partial C'_m}{\partial S_m} [\exp(-2\mathrm{i} k H) - (R_g^h)^2 \exp(2\mathrm{i} k H)]}{2k^2 S_m (1 + R_g^h)^2}; \quad \text{when } \mu = m, \qquad (2.112)
\end{aligned}
$$

where

$$
H = \int_0^h \left(C_m^2 + \frac{2z}{a} S_m^2 \right)^{1/2} \mathrm{d}z = \frac{a}{3 S_m^2} ({C'_m}^3 - C_m^3). \qquad (2.113)
$$

After algebraic manipulation, the factor N_m is written in the following form:

$$
\begin{aligned}
N_m = {} & \frac{4 R_g^h}{(1 + R_g^h)^2} \frac{a C'_m}{3 S_m^4} \left\{ {C'_m}^3 - C_m^3 + \frac{3 S_m^2}{2} \left[C'_m \left(1 - \frac{2h}{a} \right) - C_m \right] \right\} \\
& - \frac{2\mathrm{i} C'_m \Delta_g}{k C_m (1 + C_m \Delta_g)^2 (1 + R_g^h)^2} \\
& + \frac{\mathrm{i}(1 - \frac{2h}{a}) [\exp(-2\mathrm{i} k H) - (R_g^h)^2 \exp(2\mathrm{i} k H)]}{2k C'_m (1 + R_g^h)^2}. \qquad (2.114)
\end{aligned}
$$

In SLF/ELF ranges, R_g^h is close to -1. Then, we have

$$
1 + R_g^h = \frac{2 C_m \Delta_g}{1 + C_m \Delta_g}. \qquad (2.115)
$$

When $C_m \gg \frac{2h}{a}$, we get

$$
{C'_m}^3 - C_m^3 + \frac{3}{2} S_m^2 \left[C'_m \left(1 - \frac{2h}{a} \right) - C_m \right] \approx \frac{3h S_m^4}{2a C_m}. \qquad (2.116)
$$

Then, the factor N_m can be simplified as follows:

$$
\begin{aligned}
N_m = -\frac{h}{2 \Delta_g^2 C_m^2} \Bigg[& 1 - \frac{\sin 2k C_m h}{2k C_m h} - \Delta_g^2 C_m^2 \left(1 + \frac{\sin 2k C_m h}{2k C_m h} \right) \\
& - \mathrm{i} \frac{\Delta_g}{kh} (\cos 2k C_m h - 1) \Bigg]. \qquad (2.117)
\end{aligned}
$$

By now, the potential function V can be written in the form of

$$V = \frac{1}{r} \sum_{m=1}^{\infty} A_m G_m(z) P_\nu\big(\cos(\pi - \theta)\big). \tag{2.118}$$

Correspondingly, the component E_ϕ is expressed as follows:

$$E_\phi = -\frac{\mathrm{i}\omega\mu}{r} \sum_{m}^{\infty} A_m G_m(z) \frac{\partial}{\partial\theta} P_\nu\big(\cos(\pi - \theta)\big). \tag{2.119}$$

By multiplying the factor $r G_m(z)\,\mathrm{d}z$ on both sides of Eq. (2.119), and integrating from 0 to h for z, we have

$$-\mathrm{i}\omega\mu A_m N_m \frac{\partial}{\partial\theta} P_\nu\big(\cos(\pi - \theta)\big) = \int_0^h r E_\phi G_m(z)\,\mathrm{d}z. \tag{2.120}$$

Close to the dipole source, we obtain readily:

$$\lim_{\theta \to 0} 2\pi r_s \theta E_\phi(r_s, \theta) = \mathrm{i}\omega\mu_0 \iint H_z(r_s, \theta)\,\mathrm{d}a = \mathrm{i}\omega\mu_0 \delta(r - r_s) I\,\mathrm{d}a, \tag{2.121}$$

where the moment of the dipole source is $I\,\mathrm{d}a$, and $\mathrm{d}a$ is the loop area. It is noted that Eq. (2.120) is satisfied for any angle θ. Obviously, from Eq. (2.121), we have

$$\lim_{\theta \to 0} \int_0^h r E_\phi G_m(z)\,\mathrm{d}z = \frac{\mathrm{i}\omega\mu_0}{2\pi\theta} I\,\mathrm{d}a G_m(z_s). \tag{2.122}$$

Thus, it follows that

$$A_m = -\frac{I\,\mathrm{d}a G_m(z_s)}{2\pi N_m} \lim_{\theta \to 0} \left\{\theta \frac{\partial}{\partial\theta} P_\nu\big(\cos(\pi - \theta)\big)\right\}^{-1}. \tag{2.123}$$

By using the following asymptotic formula in the case of $\theta \to 0$:

$$P_\nu\big(\cos(\pi - \theta)\big) \approx \frac{2\sin\nu\pi}{\pi} \ln\theta, \tag{2.124}$$

we have

$$\lim_{\theta \to 0} \theta \frac{\partial}{\partial\theta} P_\nu\big(\cos(\pi - \theta)\big) \approx \frac{2\sin\nu\pi}{\pi}. \tag{2.125}$$

Thus, the coefficient A_m can be determined readily. We write

$$A_m = -\frac{I\,\mathrm{d}a G_m(z_s)}{4\sin\nu\pi N_m}. \tag{2.126}$$

The three non-zero components E_ϕ, H_r, and H_θ can be expressed in the following forms:

$$E_\phi = \frac{\mathrm{i}\omega\mu_0 I\,\mathrm{d}a}{4r}\sum_m^\infty \frac{G_m(z_s)}{\sin(\mu\pi)N_m} G_m(z_r)\frac{\partial}{\partial\theta}P_\nu\big(\cos(\pi-\theta)\big), \tag{2.127}$$

$$H_r = -\frac{I\,\mathrm{d}a}{4r^2}\sum \frac{\nu(\nu+1)}{\sin(\mu\pi)N_m} G_m(z_s)G_m(z_r)P_\nu\big(\cos(\pi-\theta)\big), \tag{2.128}$$

$$H_\theta = -\frac{I\,\mathrm{d}a}{4r}\sum \frac{G_m(z_s)}{\sin(\mu\pi)N_m}\frac{\partial}{\partial z} G_m(z_r)\frac{\partial}{\partial\theta}P_\nu\big(\cos(\pi-\theta)\big). \tag{2.129}$$

For TE waves in SLF/ELF ranges, because of $\nu(\nu+1) = k^2a^2S_m^2$, in general the factor S_m is a purity imaginary number. Thus the parameter $\nu+\frac{1}{2}$ is a complex number with a large imaginary part. Then, in the case that the angle θ is not close to 0 and π, we have

$$\frac{P_\nu(\cos(\pi-\theta))}{\sin(\nu\pi)} \approx -\sqrt{\frac{2}{\pi k a S_m \sin\theta}}\cdot \exp\left(\mathrm{i}kaS_m\theta + \frac{\mathrm{i}\pi}{4}\right), \tag{2.130}$$

$$\frac{\partial P_\nu(\cos(\pi-\theta))}{\sin(\nu\pi)\partial\theta} \approx -\mathrm{i}\sqrt{\frac{2kaS_m}{\pi\sin\theta}}\cdot \exp\left(\mathrm{i}kaS_m\theta + \frac{\mathrm{i}\pi}{4}\right). \tag{2.131}$$

The three non-zero components E_ϕ, H_r, and H_θ for TE waves can be written as follows:

$$E_\phi = -\frac{I\,\mathrm{d}ak\eta}{h}\sqrt{\frac{1}{\lambda a\sin\theta}}\sum_{m=1}^\infty \Lambda_m G_m(z_s)G_m(z_r)S_m^{1/2}\mathrm{e}^{\mathrm{i}kaS_m\theta+\mathrm{i}\frac{\pi}{4}}, \tag{2.132}$$

$$H_r = -\frac{I\,\mathrm{d}ak}{h}\sqrt{\frac{1}{\lambda a\sin\theta}}\sum_{m=1}^\infty \Lambda_m G_m(z_s)G_m(z_r)S_m^{3/2}\mathrm{e}^{\mathrm{i}kaS_m\theta+\mathrm{i}\frac{\pi}{4}}, \tag{2.133}$$

$$H_\theta = -\frac{\mathrm{i}I\,\mathrm{d}ak}{h}\sqrt{\frac{1}{\lambda a\sin\theta}}\sum_{m=1}^\infty \Lambda_m G_m(z_s)S_m^{1/2}\frac{\partial G_m(z_r)}{k\partial z}\mathrm{e}^{\mathrm{i}kaS_m\theta+\mathrm{i}\frac{\pi}{4}}, \tag{2.134}$$

where the excitation factor Λ_m for TE waves is defined by

$$\Lambda_m = \Delta_g^2 C_m^2\left[1-\frac{\sin 2kC_m h}{2kC_m h} - \Delta_g^2 C_m^2\left(1+\frac{\sin 2kC_m h}{2kC_m h}\right)\right.$$
$$\left. -\mathrm{i}\frac{\Delta_g}{kh}(\cos 2kC_m h - 1)\right]^{-1}. \tag{2.135}$$

In general, in SLF/ELF ranges, the normalized surface impedance Δ_g is very small, so that the excitation efficiency of VMD is very low.

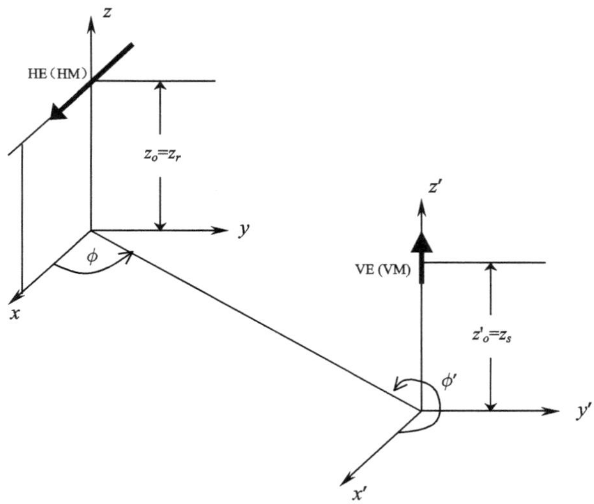

Fig. 2.6 The diagram for the geometric relation between the transmitting and receiving antennas

2.4 SLF/ELF Field of HED in the Earth–Ionosphere Waveguide/Cavity

In practical applications, a SLF/ELF radiation source is usually employed a horizontal linear antenna. So that it is necessary to investigate SLF/ELF electromagnetic field of an HED in the anisotropic Earth–ionosphere waveguide or cavity. In the preceding sections, the electromagnetic field of a VED and that of VMD are addressed, respectively. The complete formulas can be derived for the electromagnetic field of an HED in the presence of the Earth–ionosphere cavity by the reciprocity theorem (Wait 1970; Galejs 1972a). It is stated that the voltage V_2 excited in antenna 2 by current I_1 of antenna 1 is equal to the voltage V_1 excited in antenna 1 by an identical current I_2 in antenna 2.

The geometry for a VED with its length $\mathrm{d}l^{\mathrm{ve}}$ at $z_0' = z_s$ and an HED with its length $\mathrm{d}s^{\mathrm{he}}$ parallel to the x-axis at $z_0 = z_r$ are illustrated in Fig. 2.6, where the subscripts r and s refer to the observation point and dipole source, respectively. It is known that the magnitude of the voltage V^{he} excited in HED is maximum when $E_{\rho'}^{\mathrm{ve}}$ is parallel to the $\hat{x}$ axis ($\phi = 0$ and π). Then, we have

$$V^{\mathrm{he}} = -E_{\rho'}^{\mathrm{ve}}\left(z_0' = z_s,\, z_0 = z_r\right)\cos\phi\,\mathrm{d}s^{\mathrm{he}}, \tag{2.136}$$

where the superscripts *ve* and *he* refer to the VEDs and HEDs, respectively. Similarly, the magnitude of the voltage Vve in the VED is contributed by the field component E_z^{he} excited by the horizontal electric dipole. We write

$$V^{\mathrm{ve}} = E_z^{\mathrm{he}}\left(z_0 = z_s,\, z_0' = z_r\right)\mathrm{d}l^{\mathrm{ve}}. \tag{2.137}$$

With $\mathrm{d}s^{\mathrm{he}} = \mathrm{d}l^{\mathrm{ve}}$, from Eqs. (2.136) and (2.137), it follows that

$$E_z^{\mathrm{he}}\left(z_0 = z_s, z_0' = z_r\right) = -E_{\rho'}^{\mathrm{ve}}\left(z_0' = z_s, z_0 = z_r\right)\cos\phi. \tag{2.138}$$

Next, we consider the geometry for a VMD with its loop area da^{vm} parallel to the z'-axis at $z_0' = z_s$ and an HED parallel to the x-axis at $z_0 = z_r$, which is shown in Fig. 2.6. The magnitude of voltage in the HED V^{he} by the VMD is represented as follows:

$$V^{\mathrm{he}} = E_{\phi'}^{\mathrm{vm}}\left(z_0' = z_s, z_0 = z_r\right)\sin\phi\,\mathrm{d}s^{\mathrm{he}}, \tag{2.139}$$

where the superscript vm refers to the VMD. The magnitude of voltage V^{vm} in the VMD is contributed by the component H_z^{he} excited by the HED with the same current. We write

$$V^{\mathrm{vm}} = \mathrm{i}\omega\mu_0 H_z^{\mathrm{he}}\left(z_0 = z_s, z_0' = z_r\right)\mathrm{d}a^{\mathrm{vm}}. \tag{2.140}$$

With $V^{\mathrm{he}} = V^{\mathrm{vm}}$, from Eqs. (2.139) and (2.140), it follows that

$$H_z^{\mathrm{he}}\left(z_0 = z_s, z_0' = z_r\right) = \frac{1}{\mathrm{i}\omega\mu_0}E_\phi\left(z_0' = z_s, z_0 = z_r\right)\sin\phi\frac{\mathrm{d}s^{\mathrm{he}}}{\mathrm{d}a^{\mathrm{vm}}}. \tag{2.141}$$

For spherical coordinates the z and ρ components are replaced by the r and θ components, respectively. With Eq. (2.138), and considering that the sign should be changed for θ derivatives when interchanging the coordinates of the receiver and source, it follows that

$$E_r^{\mathrm{he}}(r,\theta,\phi) = -\frac{\mathrm{i}I\,\mathrm{d}s^{\mathrm{he}}\eta\cos\phi}{2ha} \times \sum_{n=0}^{\infty}\Lambda_n^e F_n(z_r)\frac{\partial F_n(z_s)}{k\partial z}\frac{\partial P_\nu(\cos(\pi-\theta))}{\sin\nu\pi\partial\theta}. \tag{2.142}$$

Similarly, with the substitution of Eq. (2.132) into Eq. (2.141), it follows that

$$H_r^{\mathrm{he}}(r,\theta,\phi) = \frac{I\,\mathrm{d}s^{\mathrm{he}}\sin\phi}{2ha}\sum_{m=1}^{\infty}\Lambda_m^h G_m(z_r)G_m(z_s)\frac{\partial P_\mu(\cos(\pi-\theta))}{\sin\mu\pi\partial\theta}. \tag{2.143}$$

Here $F_n(z)$ and $G_m(z)$ refer to the height-gain function for TM_n wave and that for TE_m wave, respectively. The excitation factor Λ_n^e for TM_n wave and the excitation factor Λ_m^h for TE_m wave are represented by (2.60) and (2.135), respectively. We rewrite

$$\Lambda_n^e = \frac{1}{1+\frac{\sin 2kC_nh}{2kC_nh}}, \tag{2.144}$$

$$\Lambda_m^h = \Delta_g^2 C_m^2 \left[1 - \frac{\sin 2kC_m h}{2kC_m h} - \Delta_g^2 C_m^2 \left(1 + \frac{\sin 2kC_m h}{2kC_m h}\right) - \mathrm{i}\frac{\Delta_g}{kh}(\cos 2kC_m h - 1)\right]^{-1}. \tag{2.145}$$

From the expanded representations of Maxwell's equations in the spherical coordinate system, the components E_θ, E_ϕ, H_θ, and H_ϕ can be expressed in terms of E_r and H_r. We write

$$\left(k^2 + \frac{\partial^2}{\partial r^2}\right)(r\sin\theta H_\theta) = -\mathrm{i}\omega\varepsilon\frac{\partial}{\partial\phi}E_r + \sin\theta\frac{\partial^2}{\partial\theta\partial r}H_r, \tag{2.146}$$

$$\left(k^2 + \frac{\partial^2}{\partial r^2}\right)(r\sin\theta E_\phi) = \frac{\partial^2}{\partial\phi\partial r}E_r - \mathrm{i}\omega\mu_0\sin\theta\frac{\partial}{\partial\theta}H_r, \tag{2.147}$$

$$\left(k^2 + \frac{\partial^2}{\partial r^2}\right)(r\sin\theta H_\phi) = \mathrm{i}\omega\varepsilon_0\sin\theta\frac{\partial}{\partial\theta}E_r + \frac{\partial^2}{\partial\phi\partial r}H_r, \tag{2.148}$$

$$\left(k^2 + \frac{\partial^2}{\partial r^2}\right)(r\sin\theta E_\theta) = \sin\theta\frac{\partial^2}{\partial\theta\partial r}E_r + \mathrm{i}\omega\mu_0\frac{\partial}{\partial\phi}H_r. \tag{2.149}$$

Taking into account the following relations:

$$\left(\frac{\partial^2}{\partial r^2} + k^2\right)Z_n(z) = \frac{\nu(\nu+1)}{r^2}Z_n(z), \tag{2.150}$$

the complete formulas for the components E_θ^{he}, E_ϕ^{he}, H_θ^{he}, and H_ϕ^{he} can be written in the following forms:

$$H_\theta^{\mathrm{he}} = \frac{I\,\mathrm{d}s^{\mathrm{he}}\sin\phi}{2ha}\left\{\frac{1}{ka\sin\theta}\sum_n \Lambda_n^e S_n^{-2} F_n(z_r)\frac{\partial F_n(z_s)}{k\partial z}\frac{\partial P_\nu(\cos(\pi-\theta))}{\sin\nu\pi\partial\theta} + \frac{1}{ka}\sum_m \frac{\Lambda_m^h G_m(z_s)}{S_m^2}\frac{\partial G_m(z_r)}{k\partial z}\frac{\partial^2 P_\mu(\cos(\pi-\theta))}{\sin\mu\pi\partial^2\theta}\right\}, \tag{2.151}$$

$$H_\phi^{\mathrm{he}} = \frac{I\,\mathrm{d}s^{\mathrm{he}}\cos\phi}{2ha}\left\{\frac{1}{ka}\sum_n \Lambda_n^e S_n^{-2} F_n(z_r)\frac{\partial F_n(z_s)}{k\partial z}\frac{\partial^2 P_\nu(\cos(\pi-\theta))}{\sin\nu\pi\partial^2\theta} + \frac{1}{ka\sin\theta}\sum_m \frac{\Lambda_m^h G_m(z_s)}{S_m^2}\frac{\partial G_m(z_r)}{k\partial z}\frac{\partial P_\mu(\cos(\pi-\theta))}{\sin\mu\pi\partial\theta}\right\}, \tag{2.152}$$

$$E_\theta^{\mathrm{he}} = -\frac{\mathrm{i}I\,\mathrm{d}s^{\mathrm{he}}\eta\cos\phi}{2ha}\left\{\frac{1}{ka}\sum_n \Lambda_n^e S_n^{-2}\frac{\partial F_n(z_r)}{k\partial z}\frac{\partial F_n(z_s)}{k\partial z}\frac{\partial^2 P_\nu(\cos(\pi-\theta))}{\sin\nu\pi\partial\theta^2} - \frac{1}{ka\sin\theta}\sum_m \Lambda_m^h G_m(z_s)G_m(z_r)S_m^{-2}\frac{\partial P_\mu[\cos(\pi-\theta)]}{\sin\mu\pi\partial\theta}\right\}, \tag{2.153}$$

$$E_{\phi}^{\text{he}} = \frac{\mathrm{i}I\,\mathrm{d}s^{\text{he}}\eta}{2ha}\sin\phi\left\{\frac{1}{ka\sin\theta}\sum_{n}\Lambda_n^e S_n^{-2}\frac{\partial F_n(z_r)}{k\partial z}\frac{\partial F_n(z_s)}{k\partial z}\frac{\partial P_\nu(\cos(\pi-\theta))}{\sin\nu\pi\partial\theta}\right.$$
$$\left.-\frac{1}{ka}\sum_{m}\Lambda_m^h G_m(z_s)G_m(z_r)S_m^{-2}\frac{\partial^2 P_\mu(\cos(\pi-\theta))}{\sin\mu\pi\partial^2\theta}\right\}. \qquad (2.154)$$

In SLF/ELF ranges, the horizontal antenna is usually placed on the ground. Then, we take $z_s = z_r = 0$, $F_n(0) = G_m(0) = 1$, $\frac{\partial F_n(z)}{k\partial z}|_{z=0} = -\mathrm{i}\Delta_g$, and $\frac{\partial G_m(z)}{k\partial z}|_{z=0} = -\frac{\mathrm{i}}{\Delta_g}$. We write

$$E_r^{\text{he}}(a,\theta,\phi) = -\frac{I\,\mathrm{d}s^{\text{he}}\eta\Delta_{gs}\cos\phi}{2ha}\sum_{n}\Lambda_n^e\frac{\partial P_\nu(\cos(\pi-\theta))}{\sin\nu\pi\partial\theta}, \qquad (2.155)$$

$$E_{\theta}^{\text{he}}(a,\theta,\phi) = -\frac{\mathrm{i}I\,\mathrm{d}s^{\text{he}}\eta\cos\phi}{2ha}\left\{-\frac{\Delta_{gs}\Delta_{gr}}{ka}\sum_{n}\Lambda_n^e S_n^{-2}\frac{\partial^2 P_\nu(\cos(\pi-\theta))}{\sin\nu\pi\partial\theta^2}\right.$$
$$\left.-\frac{1}{ka\sin\theta}\sum_{m}\Lambda_m^h S_m^{-2}\frac{\partial^2 P_\mu(\cos(\pi-\theta))}{\sin\mu\pi\partial\theta}\right\}, \qquad (2.156)$$

$$E_{\phi}^{\text{he}}(a,\theta,\phi) = \frac{\mathrm{i}I\,\mathrm{d}s^{\text{he}}\eta\sin\phi}{2ha}\left\{-\frac{\Delta_{gr}\Delta_{gs}}{ka\sin\theta}\sum_{n}\Lambda_n^e S_n^{-2}\frac{\partial P_\nu(\cos(\pi-\theta))}{\sin\nu\pi\partial\theta}\right.$$
$$\left.-\frac{1}{ka}\sum_{m}\Lambda_m^h S_m^{-2}\frac{\partial^2 P_\mu(\cos(\pi-\theta))}{\sin\mu\pi\partial\theta^2}\right\}, \qquad (2.157)$$

$$H_r^{\text{he}}(a,\theta,\phi) = \frac{I\,\mathrm{d}s^{\text{he}}\sin\phi}{2ha}\sum_{m}\Lambda_n^h\frac{\partial P_\mu(\cos(\pi-\theta))}{\sin\mu\pi\partial\theta}, \qquad (2.158)$$

$$H_{\theta}^{\text{he}}(a,\theta,\phi) = \frac{I\,\mathrm{d}s^{\text{he}}\sin\phi}{2ha}\left\{-\mathrm{i}\frac{\Delta_{gs}}{ka\sin\theta}\sum_{n}\Lambda_n^e S_n^{-2}\frac{\partial P_\nu(\cos(\pi-\theta))}{\sin\phi\pi\partial\theta}\right.$$
$$\left.-\frac{\mathrm{i}}{ka\Delta_{gr}}\sum_{m}\Lambda_m^h S_m^{-2}\frac{\partial^2 P_\mu(\cos(\pi-\theta))}{\sin\mu\pi\partial\theta^2}\right\}, \qquad (2.159)$$

$$H_{\phi}^{\text{he}}(a,\theta,\phi) = \frac{I\,\mathrm{d}s^{\text{he}}\cos\phi}{2ha}\left\{-\frac{\mathrm{i}\Delta_{gs}}{ka}\sum_{n}\Lambda_n^e S_n^{-2}\frac{\partial^2 P_\mu(\cos(\pi-\theta))}{\sin\upsilon\pi\partial\theta^2}\right.$$
$$\left.-\frac{\mathrm{i}}{ka\sin\theta\Delta_{gr}}\sum_{m}\Lambda_m^h S_m^{-2}\frac{\partial P_\mu(\cos(\pi-\theta))}{\sin\mu\pi\partial\theta}\right\}. \qquad (2.160)$$

From the above six formulas for the field components, it is seen that field components consist of both TM and TE modes. It is well known that only a TM_0 wave can propagate while the rest of the waves are evanescent, which play roles in the far-field regions.

At large distance between the observation point and the dipole source, the components in the far-field regions are simplified as follows:

$$E_r^{\text{he}}(a,\theta,\phi) = -\frac{I\,\mathrm{d}s^{\text{he}}\eta\Delta_{gs}\cos\phi}{2ha}\Lambda_0^e\frac{\partial P_\nu(\cos(\pi-\theta))}{\sin\nu_0\pi\partial\theta}, \tag{2.161}$$

$$E_\theta^{\text{he}}(a,\theta,\phi) = \frac{\mathrm{i}I\,\mathrm{d}s^{\text{he}}\eta\Delta_{gs}}{2ha}\Delta_{gr}\cos\phi\frac{\Lambda_0^e}{ka}S_0^{-2}\frac{\partial^2 P_{\nu_0}(\cos(\pi-\theta))}{\sin\nu_0\pi\partial^2\theta}, \tag{2.162}$$

$$E_\phi^{\text{he}}(a,\theta,\phi) = -\frac{\mathrm{i}I\,\mathrm{d}s^{\text{he}}\eta}{2ha}\sin\phi\Delta_{gs}\Delta_{gr}\frac{\Lambda_0^e}{ka\sin\theta}S_0^{-2}\frac{\partial P_{\nu_0}(\cos(\pi-\theta))}{\sin\nu_0\pi\partial\theta}, \tag{2.163}$$

$$H_\theta^{\text{he}}(a,\theta,\phi) = -\frac{\mathrm{i}I\,\mathrm{d}s^{\text{he}}}{2ha}\Delta_{gs}\frac{\sin\phi}{ka\sin\theta}\Lambda_0^e S_0^{-2}\frac{\partial P_{\nu_0}(\cos(\pi-\theta))}{\sin\nu_0\pi\partial\theta}, \tag{2.164}$$

$$H_\phi^{\text{he}}(a,\theta,\phi) = -\frac{\mathrm{i}I\,\mathrm{d}s^{\text{he}}}{2ha}\Delta_{gs}\frac{\cos\phi}{ka}\Lambda_0^e S_0^{-2}\frac{\partial^2 P_{\nu_0}(\cos(\pi-\theta))}{\sin\nu_0\pi\partial^2\theta}, \tag{2.165}$$

where both Δ_{gs} and Δ_{gr} refer to the normalized surface impedances at the points for the transmitting antenna and the receiving antenna, respectively. It is seen that SLF/ELF field of an HED in the Earth–ionosphere waveguide or cavity is proportional to the surface impedance at the source point. In order to improve the radiation efficiency of the transmitting antenna, SLF/ELF antenna should be chosen to be located at the region where the Earth's surface has large surface impedance or low conductivity.

At large distance between the dipole source and the observation point, and where the observation point is not close to the antipole, for the range of $f > 50$ Hz, we have

$$\begin{aligned}\frac{P_\nu(\cos(\pi-\theta))}{\sin\nu\pi} &\approx -\sqrt{\frac{2}{\pi(\nu+0.5)\sin\theta}}\exp\left[\mathrm{i}(\nu+0.5)\theta+\mathrm{i}\frac{\pi}{4}\right] \\ &\approx -\sqrt{2}(\pi kaS_n\sin\theta)^{-1/2}\exp\left(\mathrm{i}kaS_n\theta+\mathrm{i}\frac{\pi}{4}\right).\end{aligned} \tag{2.166}$$

Then, it follows that

$$\frac{\partial P_\nu(\cos(\pi-\theta))}{\sin\nu\pi\partial\theta} \approx -\mathrm{i}\sqrt{\frac{2kaS_n}{\pi\sin\theta}}\exp\left(\mathrm{i}kaS_n\theta+\mathrm{i}\frac{\pi}{4}\right), \tag{2.167}$$

$$\frac{\partial^2 P_\nu(\cos(\pi-\theta))}{\sin\nu\pi\partial^2\theta} \approx \sqrt{\frac{2}{\pi\sin\theta}}(kaS_n)^{3/2}\exp\left(\mathrm{i}kaS_n\theta+\mathrm{i}\frac{\pi}{4}\right). \tag{2.168}$$

Thus, we obtain readily:

$$\begin{aligned}E_r^{\text{he}}(a,\theta,\phi) = {} & \frac{\mathrm{i}I\,\mathrm{d}s^{\text{he}}\eta\Delta_{gs}\cos\phi}{2ha}\sqrt{\frac{2}{\pi\sin\theta}} \\ & \times (kaS_0)^{\frac{1}{2}}\Lambda_0^e\exp\left(\mathrm{i}kaS_0\theta+\mathrm{i}\frac{\pi}{4}\right),\end{aligned} \tag{2.169}$$

$$E_\theta^{\mathrm{he}}(a,\theta,\phi)=\frac{\mathrm{i}I\,\mathrm{d}s^{\mathrm{he}}\eta\Delta_{gs}}{2ha}\Delta_{gr}\cos\phi\sqrt{\frac{2}{\pi\sin\theta}}$$
$$\times(ka)^{\frac{1}{2}}S_0^{-\frac{1}{2}}\Lambda_0^e\exp\left(\mathrm{i}kaS_0\theta+\mathrm{i}\frac{\pi}{4}\right), \tag{2.170}$$

$$E_\phi^{\mathrm{he}}(a,\theta,\phi)=-\frac{I\,\mathrm{d}s^{\mathrm{he}}\eta}{2ha}\sin\phi\Delta_{gs}\Delta_{gr}\sqrt{\frac{2}{\pi\sin\theta}}$$
$$\times(ka)^{-\frac{1}{2}}S_0^{-\frac{3}{2}}\Lambda_0^e\frac{\exp(\mathrm{i}kaS_0\theta+\mathrm{i}\frac{\pi}{4})}{ka\sin\theta}, \tag{2.171}$$

$$H_\theta^{\mathrm{he}}(a,\theta,\phi)=-\frac{I\,\mathrm{d}s^{\mathrm{he}}}{2ha}\Delta_{gs}\frac{\sin\phi}{ka\sin\theta}\sqrt{\frac{2}{\pi\sin\theta}}$$
$$\times(ka)^{-\frac{1}{2}}S_0^{-\frac{3}{2}}\Lambda_0^e\exp\left(\mathrm{i}kaS_0\theta+\mathrm{i}\frac{\pi}{4}\right), \tag{2.172}$$

$$H_\phi^{\mathrm{he}}(a,\theta,\phi)=-\frac{\mathrm{i}I\,\mathrm{d}s^{\mathrm{he}}}{2ha}\Delta_{gs}\sqrt{\frac{2}{\pi\sin\theta}}$$
$$\times(ka)^{\frac{1}{2}}S_0^{-\frac{1}{2}}\Lambda_0^e\cos\phi\exp\left(\mathrm{i}kaS_0\theta+\mathrm{i}\frac{\pi}{4}\right). \tag{2.173}$$

At small distance between the dipole source and the observation point, namely, $\theta\ll 1$, we have

$$\frac{P_\nu(\cos(\pi-\theta))}{\sin\nu\pi}\approx-\mathrm{i}H_0^{(1)}\left[\left(\nu+\frac{1}{2}\right)\theta\right], \tag{2.174}$$

$$\frac{\partial P_\nu(\cos(\pi-\theta))}{\sin\nu\pi\partial\theta}\approx\mathrm{i}\left(\nu+\frac{1}{2}\right)H_1^{(1)}\left[\left(\nu+\frac{1}{2}\right)\theta\right], \tag{2.175}$$

$$\frac{\partial^2 P_\nu(\cos(\pi-\theta))}{\sin\nu\pi\partial^2\theta}\approx-\frac{\mathrm{i}(\nu+\frac{1}{2})H_1^{(1)}[(\nu+\frac{1}{2})\theta]}{\theta}. \tag{2.176}$$

Then, the formulas for the components in the near-field regions are rewritten in the following forms:

$$E_r^{\mathrm{he}}(a,\theta,\phi)=-\frac{I\,\mathrm{d}s^{\mathrm{he}}\eta\Delta_{gs}}{2ha}\cos\phi\cdot\left[\mathrm{i}ka\sum_n\Lambda_n^eS_nH_1^{(1)}(kaS_n\theta)\right], \tag{2.177}$$

$$E_\theta^{\mathrm{he}}(a,\theta,\phi)=-\frac{\mathrm{i}I\,\mathrm{d}s^{\mathrm{he}}\eta}{2ha}\cos\phi\cdot\left[\frac{\mathrm{i}\Delta_{gs}\Delta_{gr}}{\theta}\sum_n\Lambda_n^eS_n^{-1}H_1^{(1)}(kaS_n\theta)\right.$$
$$\left.-\frac{\mathrm{i}}{\sin\theta}\sum_m\Lambda_m^hS_m^{-1}H_1^{(1)}(kaS_m\theta)\right], \tag{2.178}$$

$$E_\phi^{\mathrm{he}}(a,\theta,\phi) = \frac{\mathrm{i}I\,\mathrm{d}s^{\mathrm{he}}\eta}{2ha}\sin\phi\cdot\left[-\frac{\mathrm{i}\Delta_{gs}\Delta_{gr}}{\sin\theta}\sum_n \Lambda_n^e S_n^{-1}H_1^{(1)}(kaS_n\theta)\right.$$
$$\left.+\frac{\mathrm{i}}{\theta}\sum_m \Lambda_m^h S_m^{-1}H_1^{(1)}(kaS_m\theta)\right], \tag{2.179}$$

$$H_r^{\mathrm{he}}(a,\theta,\phi) = \frac{\mathrm{i}I\,\mathrm{d}s^{\mathrm{he}}}{2ha}\sin\phi\cdot\left[ka\sum_m \Lambda_m^h S_m H_1^{(1)}(kaS_m\theta)\right], \tag{2.180}$$

$$H_\theta^{\mathrm{he}}(a,\theta,\phi) = \frac{I\,\mathrm{d}s^{\mathrm{he}}}{2ha}\sin\phi\left[\frac{\Delta_{gs}}{\sin\theta}\sum_n \Lambda_n^e S_n^{-1}H_1^{(1)}(kaS_n\theta)\right.$$
$$\left.-\frac{1}{\Delta_{gr}}\sum_m \Lambda_m^h S_m^{-1}H_1^{(1)}(kaS_m\theta)\right], \tag{2.181}$$

$$H_\phi^{\mathrm{he}}(a,\theta,\phi) = \frac{I\,\mathrm{d}s^{\mathrm{he}}}{2ha}\cos\phi\left[-\frac{\Delta_{gs}}{\theta}\sum_n S_n^{-1}\Lambda_n^e H_1^{(1)}(kaS_n\theta)\right.$$
$$\left.+\frac{1}{\sin\theta\,\Delta_{gr}}\sum_m \Lambda_m^h S_m^{-1}H_1^{(1)}(haS_m\theta)\right]. \tag{2.182}$$

2.5 Effect of Phase Velocity and Attenuation Rate by Gradual Inhomogeneous Anisotropic Ionosphere in SLF/ELF Ranges

In the above sections, the region of interest is treated as ideal homogeneous isotropic spherical Earth–ionosphere cavity or waveguide. Under practical propagation conditions, the electron density and collision frequency in the ionosphere vary as the height, and the ionosphere is usually regarded as a gradient plasma. Because the wavelengths of SLF/ELF waves are very long, the radio waves can penetrate into the ionosphere deeply, even up to F_2-layer. Therefore, it is necessary to consider the effects on the attenuation and phase velocity by the ionospheric profile in SLF/ELF ranges. In practical computations on the attenuation rate and phase velocity, the ionosphere should be treated as a gradient layered plasma, while the effect by the geomagnetic field should also be considered. Obviously, the reflection characteristics of actual inhomogeneous anisotropic ionosphere can be equivalent to those of a reflection boundary at a certain reference height with an equivalent impedance matrix.

In SLF/ELF ranges, only the TM_0 wave can propagate in the Earth–ionosphere waveguide or cavity. Thus, it is only necessary to carry out the computations on the attenuation rate and the phase velocity for the TM_0 wave. With second-order spherical approximation, the calculation accuracy for the modal equation can be guaranteed. Therefore, the modal equation of SLF/ELF waves in the Earth–ionosphere waveguide or cavity is obtained readily. It is

$$\begin{aligned}&\left[\left(1+R_h e^{2ikH}\right)-C'\Delta_{22}\left(1-R_h e^{2ikH}\right)\right]\\&\times\left[C'\left(1-R_g e^{2ikH}\right)+\Delta_{11}\left(1+R_g e^{2ikH}\right)\right]\\&+C'\Delta_{12}\Delta_{21}\left(1+R_g e^{2ikH}\right)\left(1-R_h e^{2ikH}\right)=0,\end{aligned}\tag{2.183}$$

where Δ_{ij} $(i,j=1,2)$ refer to the elements of the normalized surface impedance matrix of the ionosphere, which are satisfied to the following equation. We write

$$\begin{bmatrix}E_\theta\\E_\phi\end{bmatrix}=\eta\begin{bmatrix}\Delta_{11}\Delta_{12}\\\Delta_{21}\Delta_{22}\end{bmatrix}\begin{bmatrix}H_\theta\\H_\phi\end{bmatrix}.\tag{2.184}$$

Then, the modal equation (2.183) can be rewritten in the form

$$\begin{aligned}&\left(1-R_{g\parallel}R_{\parallel}e^{2ik_0H}\right)\left(1-R_{h\ \perp}R_{\perp}e^{2ik_0H}\right)\\&-{}_{\parallel}R_{\perp}\ {}_{\perp}R_{\parallel}R_hR_g e^{4ik_0H}=0,\end{aligned}\tag{2.185}$$

where

$$H=\int_0^h\left(C^2+\frac{2t}{a}S^2\right)^{1/2}dt,\tag{2.186}$$

$$R_g=\frac{C-\Delta_g}{C-\Delta_g},\tag{2.187}$$

$$R_h=\frac{C-\Delta_g^{-1}}{C+\Delta_g^{-1}},\tag{2.188}$$

$$C'=\left(C^2+\frac{2h}{a}S^2\right)^{1/2}.\tag{2.189}$$

It is noted that the relations between the reflection coefficient matrix and the normalized surface impedance matrix are written as follows:

$${}_{\parallel}R_{\parallel}=\frac{(C'-\Delta_{11})(C'\Delta_{22}-1)+C'\Delta_{12}\Delta_{21}}{(C'+\Delta_{11})(C'\Delta_{22}-1)-C'\Delta_{12}\Delta_{21}},\tag{2.190}$$

$${}_{\parallel}R_{\perp}=\frac{-2C'\Delta_{21}}{(C'+\Delta_{11})(C'\Delta_{22}-1)-C'\Delta_{12}\Delta_{21}},\tag{2.191}$$

$${}_{\perp}R_{\parallel}=\frac{-2C'\Delta_{12}}{(C'+\Delta_{11})(C'\Delta_{22}-1)-C'\Delta_{12}\Delta_{21}},\tag{2.192}$$

$${}_{\perp}R_{\perp}=\frac{(C'+\Delta_{11})(1+C'\Delta_{22})-C'\Delta_{12}\Delta_{21}}{(C'+\Delta_{11})(C'\Delta_{22}-1)-C'\Delta_{12}\Delta_{21}}.\tag{2.193}$$

Next, the procedures for solving the modal equation are addressed as follows:

(1) First we assume that the Earth and the ionosphere are simplified as homogeneous isotropic media with sharp boundaries. From Eqs. (2.38) and (2.39), the roots (C_0, S_0) of the modal equation for TM_0 wave can be obtained readily. Then, we set the roots (C_0, S_0) as initial approximated values.
(2) With the root C_0, the incident angle cosine C' for the TM_0 wave can be obtained readily by Eq. (2.189).
(3) With the incident angle cosine C', and the profiles for both ionospheric electron density and collision frequency varying with the height, the equivalent surface impedance matrix of the ionosphere can be computed easily.
(4) Substituting the parameters of the ionospheric surface impedance matrix into the transcendental equation (2.183), the new incident angle cosine $C_i = C_0 + \triangle C$ can be obtained by using Newton's iteration method with the initial value C_0.
(5) The first-order approximated parameter C_i' for the corresponding ionospheric incident angle cosine can be solved by using Eq. (2.105) with the parameter C_i.
(6) Generally speaking, the ionospheric surface impedance matrix has little change to the incident angle in SLF/ELF ranges, so that the iteration computation repeating the procedures 3 to 5 for two or three times is enough to guarantee the accuracy.

With the roots of the modal equation, both the phase velocity and the attenuation rate of the TM_0 wave can be obtained readily:

$$\frac{c}{v_p} = \operatorname{Re} S, \tag{2.194}$$

$$\alpha = 8.6858k \operatorname{Im} S = 0.02895\omega \operatorname{Im} S \ (\mathrm{dB}/1000\ \mathrm{km}). \tag{2.195}$$

In SLF/ELF ranges, the ground surface impedance is much smaller than the ionospheric surface impedance. Evidently, the change of the ground conductivity has little effect on the phase velocity and attenuation rate of the TM_0 wave. In the practical computation, the ground is usually idealized to the sea surface.

In order to compare quantitatively the effects of the phase velocity and the attenuation rate by different ionosphere structure in SLF/ELF ranges, we choose the three ionospheric models to carry out the numerical calculations. The first ionospheric model is that of the ionosphere idealized as a homogeneous non-ideal plasma, for which the equivalent conductivity is $\sigma = 10^{-6}$ S/m. The second model is the reference structure of the lower ionosphere by the suggestions in the CCIR-895 report, which is characterized by

$$N(z) = 1.43 \times 10^7 \cdot \mathrm{e}^{-0.15H} \cdot \mathrm{e}^{(\beta-0.15)(z-H)}, \tag{2.196}$$

$$\nu(z) = 1.816 \times 10^{11} \cdot \mathrm{e}^{-0.15z}. \tag{2.197}$$

In the above formulas, N is the electron density in $1/\mathrm{cm}^3$, ν is the electron collision frequency in s^{-1}, z is the height of the observation point in km, and H is the

Table 2.1 The recommended values of H and β in mid-latitude regions

	Summer	Winter
Daytime	$\beta = 0.3,\ H = 70$	$\beta = 0.3,\ H = 72$
Nighttime	$\beta = 0.0077f + 0.31,\ H = 87$	$\beta = 0.0077f + 0.31,\ H = 87$

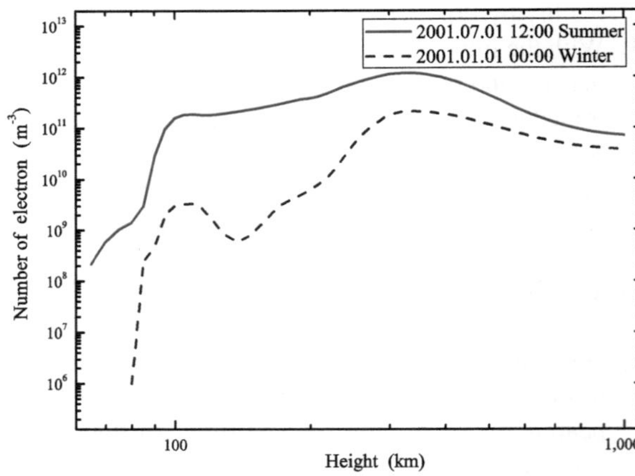

Fig. 2.7 For the international reference ionosphere model, the ionospheric electron density varying as the height in summer daytime and winter nighttime

ionospheric reference height in km. The recommended values of H in km and β in $1/\text{km}$ for mid-latitude regions are shown in Table 2.1. Here f is the operating frequency in kHz.

The third model is the international reference ionosphere model, in which the electron density profiles vary with the height in summer daytime and winter nighttime as shown in Fig. 2.7 (Rawer et al. 1978). The collision frequency varying with the height is taken as Eq. (2.197) for the exponential model.

For the above three ionosphere models, following the computational method and process addressed in this section, the modal equation for the SLF/ELF ranges is solved readily, and the characteristic parameters of each mode can be calculated easily. The attenuation rate on the sea surface versus the operating frequency for the fundamental mode (TM_0 mode or quasi-TEM mode) is computed for daytime and shown in Fig. 2.8. In Fig. 2.9, the corresponding calculated results in nighttime are given. The attenuation rates in nighttime on the sea surface versus the operating frequency for TM_1 and TE_1 modes are computed and shown in Figs. 2.10 and 2.11, respectively.

From the above computations, it is concluded as follows:

- In SLF/ELF ranges, the attenuation rate of the fundamental mode will increase as the operating frequency increasing. However, the attenuation rate is very small overall. Thus SLF/ELF waves can propagate to thousands kilometers or over ten thousand kilometers away from the transmitter, even for any place in the Earth–ionosphere waveguide or cavity.

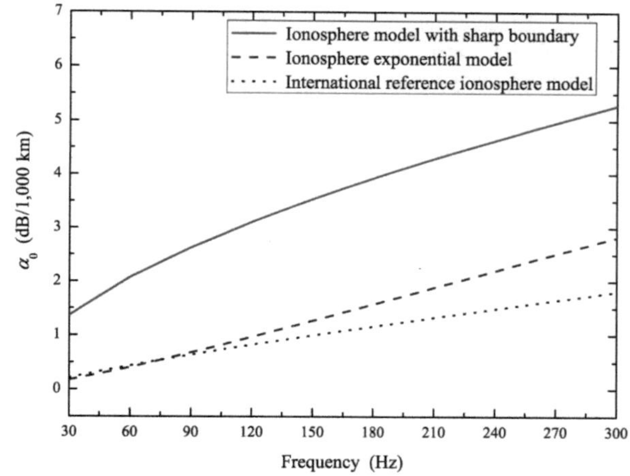

Fig. 2.8 The attenuation rate in daytime versus the operating frequency for the fundamental mode

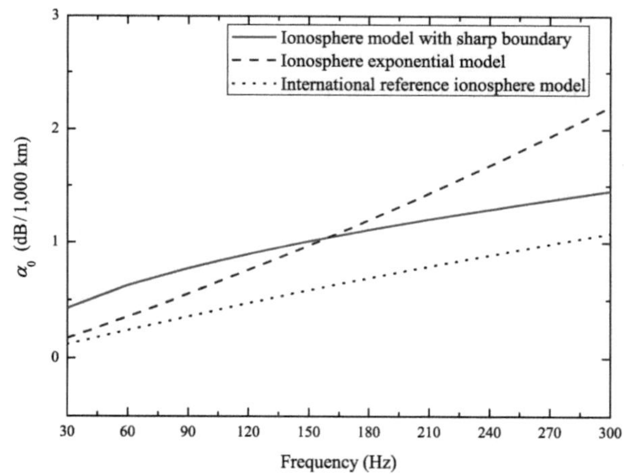

Fig. 2.9 The attenuation rate in nighttime versus the operating frequency for the fundamental mode

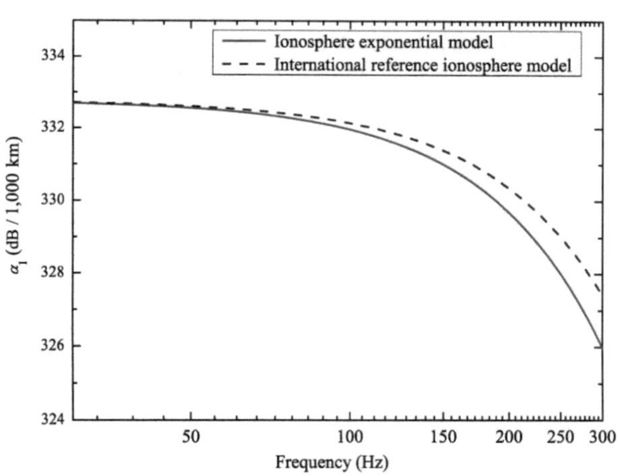

Fig. 2.10 The attenuation rate in nighttime versus the operating frequency for the TM_1 mode

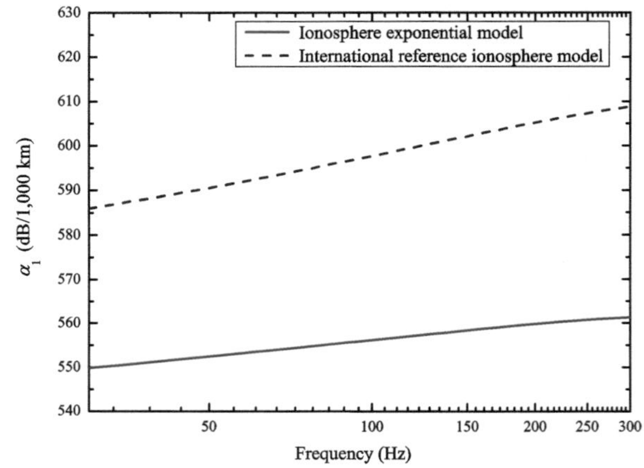

Fig. 2.11 The attenuation rate in nighttime versus the operating frequency for the TE_1 mode

- The attenuation rates for the high-order modes in SLF/ELF ranges are very large, generally several hundred dB/Mm. Thus, there only exists one propagating mode in SLF/ELF ranges.
- For the three different ionospheric models, although the parameters for each order modes are different from each other, the overall trends are consistent. In the case that the observer is located on or near the Earth's surface, it is acceptable to use a relatively simple model in engineering.
- From the computed results, it is seen that the propagation loss of SLF/ELF waves in daytime is larger than that in nighttime. It is resulted by the equivalent reflection height of the ionosphere in nighttime being higher than that in daytime.

The attenuation rates of the fundamental mode versus the ground conductivity are computed for the two different modes and shown in Fig. 2.12, respectively. It is seen that the attenuation rates decrease with the Earth's conductivity increasing, but the changing amplitudes are not large. This is resulted by the fact that the higher the ground conductivity is, the larger the reflections are, and the less the absorption loss of SLF/ELF waves are.

It is noted that the effect by the geomagnetic field is not considered in the computations in Figs. 2.8–2.12. Namely, the ionosphere is regarded as a one-dimensionally planar stratified isotropic plasma in the above computations. In lower frequency ranges, it is necessary to consider the effects by the geomagnetic field, and the ionosphere is regarded as an anisotropic plasma, which is characterized by using a 3×3 matrix. Especially in SLF/ELF ranges, the electromagnetic waves in the ionosphere will show significant anisotropic properties. In order to address the effects of the SLF/ELF wave propagation by the geomagnetic field, the computations for the relative phase velocity versus the propagation direction are carried out at $f = 75$ Hz and shown in Fig. 2.13. Similarly, with the same operating frequency,

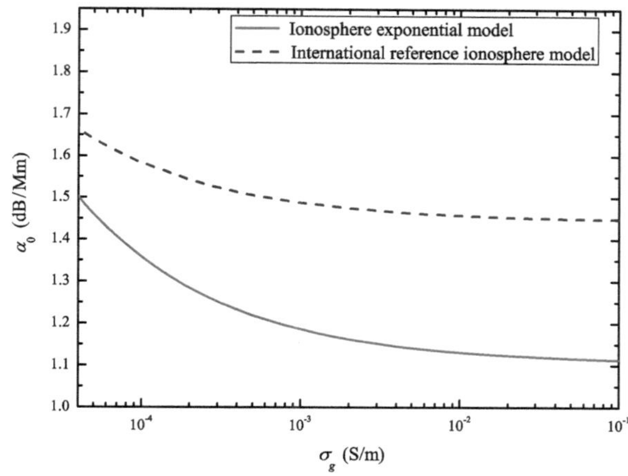

Fig. 2.12 The attenuation rate in nighttime versus the operating frequency for the TEM mode

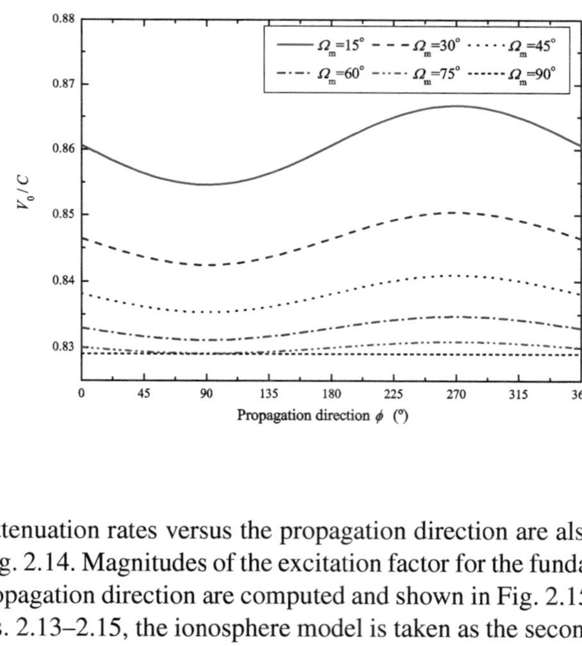

Fig. 2.13 The relative phase velocity versus the propagation direction at $f = 75$ Hz

the computations for the attenuation rates versus the propagation direction are also carried out and shown in Fig. 2.14. Magnitudes of the excitation factor for the fundamental mode versus the propagation direction are computed and shown in Fig. 2.15. In the computations in Figs. 2.13–2.15, the ionosphere model is taken as the second model as addressed in this section, the ionosphere reference height is $H = 70$ km, the ground conductivity is taken as = 3 S/m, and the geomagnetic field is taken as $B_0 = 0.5 \times 10^{-4}$ T.

From the above computations, we conclude as follows:

- For the attenuation rate and phase velocity of the fundamental mode, there exist directional actions. Obviously, the attenuation rate propagating eastward is smaller than that propagating westward, while the relative phase velocity propagating eastward is also smaller than that propagating westward.

Fig. 2.14 The attenuation rates versus the propagation direction at $f = 75$ Hz

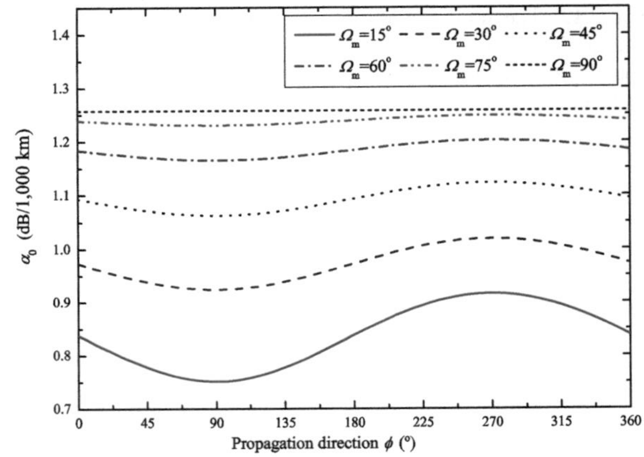

Fig. 2.15 The excitation factor for the fundamental mode versus the propagation direction at $f = 80$ Hz

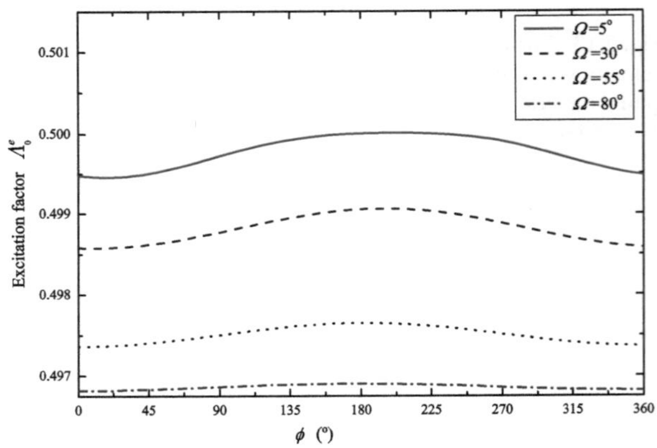

- The attenuation rate and the phase velocity of the fundamental mode are affected by the geomagnetic inclination angle Ω. When the angle Ω is larger, the attenuation rate becomes larger, and the relative phase velocity is reduced correspondingly. Meanwhile, the effects of the attenuation rates and the phase velocity by the propagating direction are weakened.
- The attenuation rate and the phase velocity of the fundamental mode change slowly with the propagation direction and the geomagnetic inclination angle. Specifically, the effect of the attenuation rate is in the range of 0.1–0.4 dB/Mm, while the effect of the relative phase velocity is in the range of 1 %–3 %. Thus, the effects by the geomagnetic field are generally neglected.
- For the excitation factor of the fundamental mode, the effects by the geomagnetic field, which are usually neglected, was not significant with respect to that of the attenuation rate.

2.6 SLF/ELF Fields of Ground-Based Horizontal Transmitting Antenna

2.6.1 SLF Field in Far-Field Region

For practical ground-based SLF/ELF transmitting system, the transmitting antenna is generally an electrode antenna with both grounded ends and low-level frame. The antenna length is usually dozens of kilometers, even over 100 km. At large distance between the observation point and the transmitting antenna over 1,000 km, the transmitting antenna can be taken as an HED. In this case, along the whole of the propagating paths, both the ground and the ionosphere will be no longer homogeneous. Thus, the whole propagating path can be divided into several short uniform paths, of which each uniform path is a homogeneous waveguide. For each section of the Earth–ionosphere waveguide, the root S_0 can be obtained by solving the modal equation. The electromagnetic field in the Earth–ionosphere waveguide can be obtained by using the WKB solution. When the higher-order modes are neglected in the far-field region, the field components can be expressed in the following forms:

$$E_r = \frac{\mathrm{i}I\,\mathrm{d}l\,\eta\Delta_{gs}}{2ha}\sqrt{\frac{2ka}{\pi\sin\theta}}\cos\phi\Lambda_0^e(T)\big[S_0(T)S_0(R)\big]^{\frac{1}{4}} \times \exp\left[\mathrm{i}\int_0^\theta kaS_0\,\mathrm{d}\theta + \mathrm{i}\frac{\pi}{4}\right], \tag{2.198}$$

$$E_\theta = \frac{\mathrm{i}I\,\mathrm{d}l\,\eta}{2ha}\Delta_{gs}\Delta_{gr}\sqrt{\frac{2ka}{\pi\sin\theta}}\cos\phi\Lambda_e^0(T)\big[S_0(T)S_0(R)\big]^{-\frac{1}{4}} \times \exp\left[\mathrm{i}\int_0^\theta kaS_0\,\mathrm{d}\theta + \mathrm{i}\frac{\pi}{4}\right], \tag{2.199}$$

$$E_\phi = \frac{-I\,\mathrm{d}l\,\eta}{2ha}\sqrt{\frac{2}{\pi\sin\theta}}\frac{\Delta_{gs}\Delta_{gr}}{\sqrt{ka}\sin\theta}\sin\phi\Lambda_0^e(T)\big[S_0(T)S_0(R)\big]^{-\frac{3}{4}} \times \exp\left[\mathrm{i}\int_0^\theta kaS_0\,\mathrm{d}\theta + \mathrm{i}\frac{\pi}{4}\right], \tag{2.200}$$

$$H_\theta = \frac{-I\,\mathrm{d}l}{2ha}\sqrt{\frac{2}{\pi\sin\theta}}\frac{\Delta_{gs}\sin\phi}{\sqrt{ka}\sin\theta}\Lambda_0^e(T)\big[S_0(T)S_0(R)\big]^{-\frac{3}{4}} \times \exp\left[\mathrm{i}\int_0^\theta kaS_0\,\mathrm{d}\theta + \mathrm{i}\frac{\pi}{4}\right], \tag{2.201}$$

$$H_\phi = \frac{-\mathrm{i}I\,\mathrm{d}l}{2ha}\sqrt{\frac{2ka}{\pi\sin\theta}}\Delta_{gs}\cos\phi\Lambda_0^e(T)\big[S_0(T)S_0(R)\big]^{-\frac{1}{4}} \times \exp\left[\mathrm{i}\int_0^\theta kaS_0\,\mathrm{d}\theta + \mathrm{i}\frac{\pi}{4}\right], \tag{2.202}$$

where $S_0(T)$ and $S_0(R)$ represent the values of S_0 on the source point and the observation point, respectively. $\Lambda_0^e(T)$ represents the excitation factor on the source point, and both Δ_{gr} and Δ_{gs} represent the normalized surface impedance of the ground and that of the sea surface, respectively.

We assume that the local time of the transmitting point and that of the receiving point are in daytime and nighttime, respectively. Then, it is seen that the part of the ionosphere of the propagating path is in daytime, while the other part is in nighttime. The corresponding equivalent reflection height of the ionosphere in the circadian boundaries of the day-and-night transition period, there will be an obvious mutation, and there exists conversion between different modes. In the circadian boundaries of the unevenly transition period, the mode conversion coefficient is expressed in Eq. (2.203):

$$S_{nm} = \frac{\int_0^{H_{\min}} F_n^d(z) G_m^n(z)\,\mathrm{d}z}{\sqrt{\int_0^{H_1} [F_n^d(z)]^2\,\mathrm{d}z \cdot \int_0^{H_2} [G_n^d(z)]^2\,\mathrm{d}z}}, \tag{2.203}$$

where F_n^d and G_m^n represent the height-gain function for the nth-order mode of daytime and the mth-order mode of nighttime, respectively. H is the ionospheric reference height, $H_{\min} = \min(H_1, H_2)$.

With the mode conversion, in the circadian transitional period, the analytical formulas for SLF field components are obtained readily. We write

$$E_r = \frac{\mathrm{i}I\,\mathrm{d}l\eta\Delta_{gs}}{2\sqrt{h_d \times h_n a}}\sqrt{\frac{2ka}{\pi\sin\theta}}\cos\phi\cdot\left\{\sum_{n=1}^{N}\sum_{m=1}^{M}\Lambda_n^e(T)S_n^{\frac{1}{2}}(T)\right.$$
$$\left.\times S_{nm}S_m^{\frac{1}{2}}(R)\exp\left[\mathrm{i}k\left(S_n^d R_d + S_m^n R_n\right) + \mathrm{i}\frac{\pi}{4}\right]\right\}, \tag{2.204}$$

$$E_\theta = \frac{\mathrm{i}I\,\mathrm{d}l\eta\Delta_{gs}\Delta_{gr}}{2\sqrt{h_d \times h_n a}}\sqrt{\frac{2ka}{\pi\sin\theta}}\cos\phi\cdot\left\{\sum_{n=1}^{N}\sum_{m=1}^{M}\Lambda_n^e(T)S_n^{\frac{1}{2}}(T)\right.$$
$$\left.\times S_{nm}S_m^{\frac{1}{2}}(R)\exp\left[\mathrm{i}k\left(S_n^d R_d + S_m^n R_n\right) + \mathrm{i}\frac{\pi}{4}\right]\right\}, \tag{2.205}$$

$$E_\phi = \frac{-I\,\mathrm{d}l\eta}{2\sqrt{h_d \times h_n a}}\sqrt{\frac{2}{\pi\sin\theta}}\frac{\Delta_{gs}\Delta_{gr}}{\sqrt{ka}\sin\theta}\sin\phi\cdot\left\{\sum_{n=1}^{N}\sum_{m=1}^{M}\Lambda_n^e(T)S_n^{\frac{1}{2}}(T)\right.$$
$$\left.\times S_{nm}S_m^{\frac{1}{2}}(R)\exp\left[\mathrm{i}k\left(S_n^d R_d + S_m^n R_n\right) + \mathrm{i}\frac{\pi}{4}\right]\right\}, \tag{2.206}$$

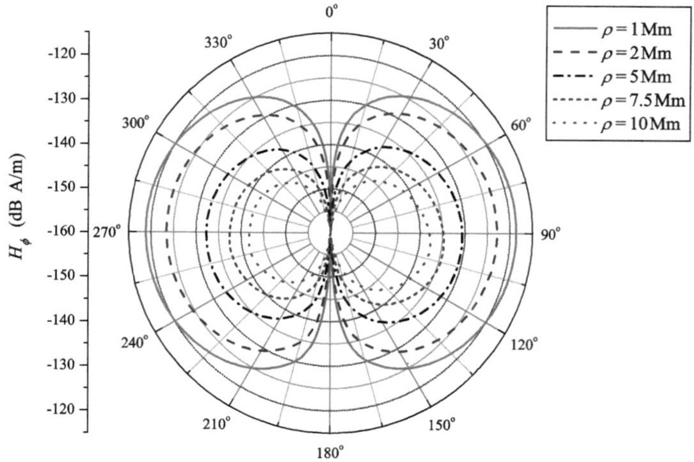

Fig. 2.16 The radiation pattern of the horizontal magnetic field components H_ϕ excited by an HED in East–West direction: $I = 350$ A, d$l = 100$ km, $f = 80$ Hz, $\sigma_g(T) = 0.2 \times 10^{-3}$ S/m, and $\sigma_g(R) = 10^{-3}$

$$H_\theta = \frac{-I\,\mathrm{d}l}{2\sqrt{h_d \times h_n a}}\sqrt{\frac{2}{\pi \sin\theta}}\frac{\Delta_{gs}\sin\phi}{\sqrt{ka}\sin\theta}\cdot\left\{\sum_{n=1}^{N}\sum_{m=1}^{M}\Lambda_n^e(T)S_n^{\frac{1}{2}}(T)\right.$$

$$\left.\times\, S_{nm}S_m^{\frac{1}{2}}(R)\exp\left[\mathrm{i}k\left(S_n^d R_d + S_m^n R_n\right) + \mathrm{i}\frac{\pi}{4}\right]\right\}, \tag{2.207}$$

$$H_\phi = \frac{-\mathrm{i}I\,\mathrm{d}l}{2\sqrt{h_d \times h_n a}}\sqrt{\frac{2ka}{\pi \sin\theta}}\Delta_{gs}\cos\phi\cdot\left\{\sum_{n=1}^{N}\sum_{m=1}^{M}\Lambda_n^e(T)S_n^{\frac{1}{2}}(T)\right.$$

$$\left.\times\, S_{nm}S_m^{\frac{1}{2}}(R)\exp\left[\mathrm{i}k\left(S_n^d R_d + S_m^n R_n\right) + \mathrm{i}\frac{\pi}{4}\right]\right\}, \tag{2.208}$$

where h_d and h_n represent the ionospheric reference height of the daytime and that of the nighttime, respectively. R_d and R_n are the propagation distance of the daytime and that of the nighttime, respectively. N and M represent the order of the wave modes of the daytime and that of the nighttime, respectively.

Following the above method, when the ground conductivities and all parameters of the ionosphere in the regions of the whole propagating paths are given, the distribution of the electromagnetic field on the Earth's surface can be determined and computed readily.

The radiation pattern of the horizontal magnetic field components H_ϕ and H_ρ excited by a horizontal line antenna, which is placed in East–West direction, are shown in Figs. 2.16 and 2.17, respectively. In these computations, the antenna length is d$l = 100$ km, the antenna current is assumed to be $I = 350$ A, the operating frequency is $f = 80$ Hz, and the ground conductivity of the transmitting point and

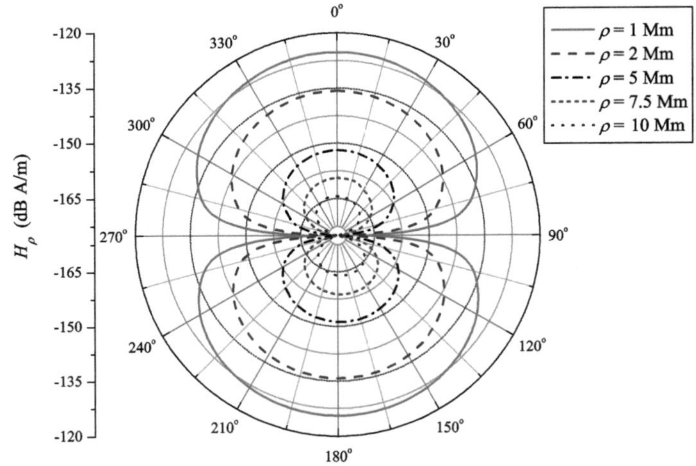

Fig. 2.17 The radiation pattern of the horizontal magnetic field components H_ρ excited by an HED in East–West direction: $I = 350$ A, $\mathrm{d}l = 100$ km, $f = 80$ Hz, $\sigma_g(T) = 0.2 \times 10^{-3}$ S/m, and $\sigma_g(R) = 10^{-3}$

that of the observation point are taken as $\sigma_g(T) = 0.2 \times 10^{-3}$ S/m and $\sigma_g(R) = 10^{-3}$ S/m, respectively.

From Figs. 2.16 and 2.17, we conclude as follows:

- There is obviously directionality for the SLF field components excited by a horizontal line antenna. For the horizontal line antenna in the East–West direction, the main radiation direction of the component H_ϕ is in the East–West direction, while that of the component H_ρ is in the North–South direction.
- At approximately the same propagating distance, the magnitude of the component H_ϕ is much larger than that of the component H_ρ.
- The attenuation rate for the wave propagating eastward is smaller than that for the wave propagating westward, and considering the effects by the uneven ground conductivities, the field strength of the antenna east is larger than that of the west. Obviously, the radiation contour diagram is not completely symmetrical.

2.6.2 SLF Field in the Near-Field Region

In practice, the SLF/ELF transmitting antenna is fairly large in size. At small distances between the observation point and the transmitting antenna, especially the distance being comparable to the actual antenna length, it is unsuitable that the antenna is idealized as an HED. In this case, the transmitting antenna is usually divided into many small segments, of which each segment is regarded as an HED, and the total field can be understood as the superposition of the fields excited by all these HEDs. The radiation pattern of the horizontal magnetic field components

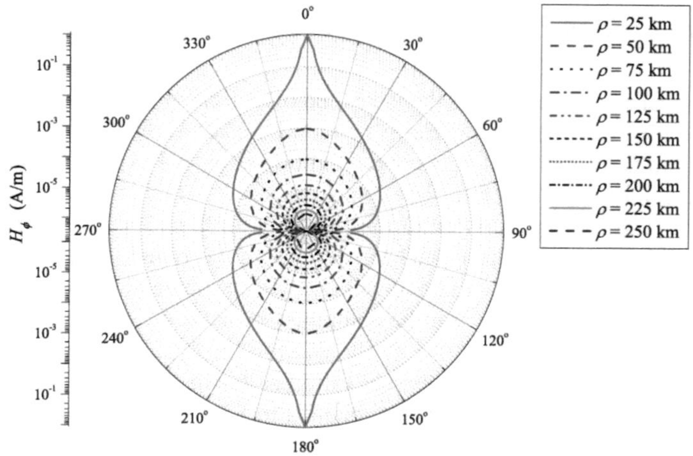

Fig. 2.18 The radiation pattern of the horizontal magnetic field components H_ϕ excited by an HED in South–North direction: $I = 62.5$ A, d$l = 80$ km, $f = 91$ Hz, $\sigma_g(T) = 0.2 \times 10^{-3}$ S/m, and $\sigma_g(R) = 10^{-3}$ S/m

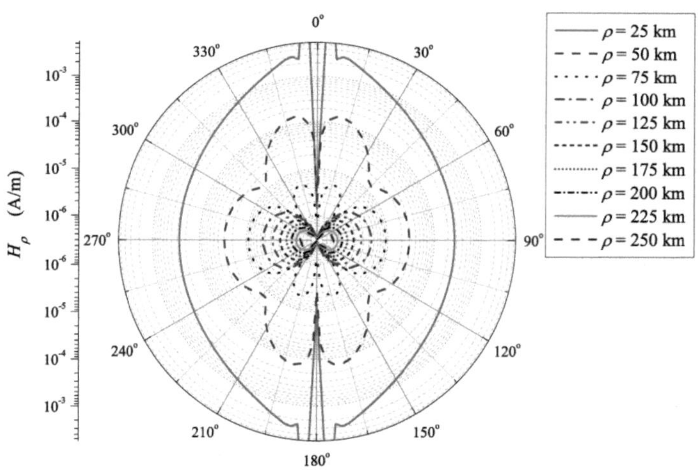

Fig. 2.19 The radiation pattern of the horizontal magnetic field components H_ρ excited by an HED in South–North direction: $I = 62.5$ A, d$l = 80$ km, $f = 91$ Hz, $\sigma_g(T) = 0.2 \times 10^{-3}$ S/m, and $\sigma_g(R) = 10^{-3}$ S/m

H_ϕ and H_ρ excited by a horizontal line antenna with its length 80 km in South–North direction are shown in Figs. 2.18 and 2.19, respectively. In these computations, we take $I = 62.5$ A, d$l = 80$ km, $f = 91$ Hz, $\sigma_g(T) = 0.2 \times 10^{-3}$ S/m, and $\sigma_g(R) = 10^{-3}$ S/m. From Figs. 2.16–2.19, it is seen that the radiation pattern in the near-field region is significantly different from that in the far-field region.

2.6.3 The Field in ELF Range and the Lower End of SLF Range

In the whole ELF range as well as the lower end of SLF range, the wavelength can be compared with, even exceeds, the Earth's circumference. At this time, for the eigenvalue ν of the zero-order TM wave, the condition of $\nu \gg 1$ is no longer satisfied. Thus the Legendre function of the first kind $P_\nu(\cos(\pi-\theta))$ should not be evaluated by using the traditional asymptotic formula in Eq. (2.29). In Sect. 2.2.4, the numerical integrated algorithm is proposed for evaluating the function $P_\nu(\cos(\pi-\theta))$. Then, the field in ELF range and the lower end of SLF range generated by an HED in the Earth–ionosphere cavity can be computed by using the new proposed algorithm (Peng et al. 2013).

In the case of an HED, the analytical formulas for the electromagnetic field in the Earth–ionosphere cavity have been obtained in Sect. 2.4. In these formulas, all components are expressed in terms of $\frac{\partial P_\nu(\cos(\pi-\theta))}{\partial\theta}$ and $\frac{\partial^2 P_\nu(\cos(\pi-\theta))}{\partial\theta^2}$. In order to evaluate accurately these components, it is necessary to give the new algorithm for evaluating the functions $\frac{\partial P_\nu(\cos(\pi-\theta))}{\partial\theta}$ and $\frac{\partial^2 P_\nu(\cos(\pi-\theta))}{\partial\theta^2}$.

As mentioned in Sect. 2.4, we denote

$$\phi(\nu,\theta)=\frac{1}{\sin\nu\pi}P_\nu\big(\cos(\pi-\theta)\big), \tag{2.209}$$

and

$$\begin{aligned}\frac{\partial}{\partial\theta}\phi(\nu,\theta)&=\frac{1}{\sin\nu\pi}\frac{\partial}{\partial\theta}P_\nu\big(\cos(\pi-\theta)\big)\\&=\frac{\nu}{\sin\theta}\big[\phi(\nu,\theta)\cos\theta-\phi(\nu-1,\theta)\big].\end{aligned} \tag{2.210}$$

From the above equations, we obtain readily:

$$\begin{aligned}\frac{\partial^2}{\partial\theta^2}\phi(\nu,\theta)=\frac{\nu}{\sin^2\theta}\big[&\big(\nu\cos^2\theta-1\big)\phi(\nu,\theta)\cos\theta-2(\nu-1)\cos\theta\phi(\nu-1,\theta)\\&+(\nu-1)\phi(\nu-2,\theta)\big].\end{aligned} \tag{2.211}$$

With the substitutions of Eqs. (2.209)–(2.211) into Eqs. (2.155)–(2.160), we readily derive the following. We write

$$E_r=-\frac{I\,\mathrm{d}l\eta\Delta_g}{2ha}\cos\phi\cdot\sum_{n=0}^{\infty}\Lambda_n^e\frac{\partial P_{\nu_n}(\cos(\pi-\theta))}{\sin(\nu_n\pi)\partial\theta}, \tag{2.212}$$

$$\begin{aligned}E_\theta=\frac{\mathrm{i}I\,\mathrm{d}l\eta}{2ha}\cos\phi\cdot\Bigg\{&\frac{\Delta_g^2}{ka}\sum_{n=0}^{\infty}\frac{\Lambda_n^e}{S_n^2}\frac{\partial^2 P_{\nu_n}(\cos(\pi-\theta))}{\sin(\nu_n\pi)\partial\theta^2}\\&+\frac{1}{ka\sin\theta}\sum_{m=1}^{\infty}\frac{\Lambda_m^h}{S_m^2}\frac{\partial P_{\mu_m}(\cos(\pi-\theta))}{\sin(\mu_m\pi)\partial\theta}\Bigg\},\end{aligned} \tag{2.213}$$

$$E_\phi = -\frac{\mathrm{i}I\,\mathrm{d}l\eta}{2ha}\sin\phi\cdot\left\{\frac{\Delta_g^2}{ka\sin\theta}\sum_{n=0}^{\infty}\frac{\Lambda_n^e}{S_n^2}\frac{\partial P_{\nu_n}(\cos(\pi-\theta))}{\sin(\nu_n\pi)\partial\theta}\right.$$
$$\left.+\frac{1}{ka}\sum_{m=1}^{\infty}\frac{\Lambda_m^h}{S_m^2}\frac{\partial^2 P_{\mu_m}(\cos(\pi-\theta))}{\sin(\mu_m\pi)\partial\theta^2}\right\}, \tag{2.214}$$

$$H_r = \frac{I\,\mathrm{d}l}{2ha}\sin\phi\cdot\sum_{m=1}^{\infty}\Lambda_m^h\frac{\partial P_{\mu_m}(\cos(\pi-\theta))}{\sin(\mu_m\pi)\partial\theta}, \tag{2.215}$$

$$H_\theta = -\frac{\mathrm{i}I\,\mathrm{d}l}{2ha}\sin\phi\cdot\left\{\frac{\Delta_g}{ka\sin\theta}\sum_{n=0}^{\infty}\frac{\Lambda_n^e}{S_n^2}\frac{\partial P_{\nu_n}(\cos(\pi-\theta))}{\sin(\nu_n\pi)\partial\theta}\right.$$
$$\left.+\frac{1}{ka\Delta_g}\sum_{m=1}^{\infty}\frac{\Lambda_m^h}{S_m^2}\frac{\partial^2 P_{\mu_m}(\cos(\pi-\theta))}{\sin(\mu_m\pi)\partial\theta^2}\right\}, \tag{2.216}$$

$$H_\phi = -\frac{\mathrm{i}I\,\mathrm{d}l}{2ha}\cos\phi\cdot\left\{\frac{\Delta_g}{ka}\sum_{n=0}^{\infty}\frac{\Lambda_n^e}{S_n^2}\frac{\partial^2 P_{\nu_n}(\cos(\pi-\theta))}{\sin(\nu_n\pi)\partial\theta^2}\right.$$
$$\left.+\frac{1}{ka\Delta_g\sin\theta}\sum_{m=1}^{\infty}\frac{\Lambda_m^h}{S_m^2}\frac{\partial P_{\mu_m}(\cos(\pi-\theta))}{\sin(\mu_m\pi)\partial\theta}\right\}. \tag{2.217}$$

In these equations, the functions $\frac{\partial P_\mu(\cos(\pi-\theta))}{\partial\theta}$ and $\frac{\partial^2 P_\mu(\cos(\pi-\theta))}{\partial\theta^2}$ can be computed by using Eqs. (2.77), (2.210), and (2.211).

We assume that the unit HED is placed at $\theta = 0°$ in the direction of $\phi_T = 0°$ and the operating frequency is taken as $f = 10$ Hz. As shown in Figs. 2.20 and 2.21, with the orientation of the observation point $\phi = 0°$, the magnitudes of the components E_θ and H_ϕ are computed by using the numerical integrated algorithm proposed in this chapter and the traditional method (Bannister 1984), respectively. With the orientation of the observation point $\phi = 90°$, similar computations are carried out for the components E_ϕ and H_θ and also plotted in Figs. 2.20 and 2.21, respectively.

From Figs. 2.20 and 2.21, it is seen that the computed results by using the two methods are significantly different from each other in ELF range and the lower end of SLF range (below 50 Hz). The traditional method by Bannister (1984) is based on the assumption of $\nu \gg 0$, so that it is only suitable for the SLF range. Correspondingly, the results by using the traditional method is uncorrect for the range below 50 Hz. In ELF range and the lower end of SLF range, the distributions of the electromagnetic field on the ground can be evaluated accurately by using the numerical integrated method proposed in this chapter.

We assume that the current moment of the transmitting antenna, which is placed horizontally in the $\hat{x}$ direction, is taken as 1 A·m, the ionospheric reference height is 70 km, and the conductivities of the ground and the ionosphere are assumed as $\sigma_g = 10^{-3}$ S/m and $\sigma_i = 10^{-5}$ S/m, respectively. The contour diagram of the tan-

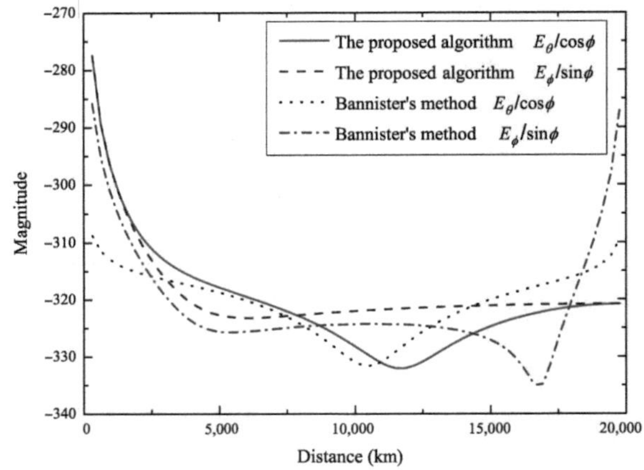

Fig. 2.20 The magnitudes of the components E_θ in dB·V/m at $\phi = 90°$ and E_ϕ in dB·V/m at $\phi = 0°$ versus the propagation distance: $f = 10$ Hz, $\phi_T = 0°$

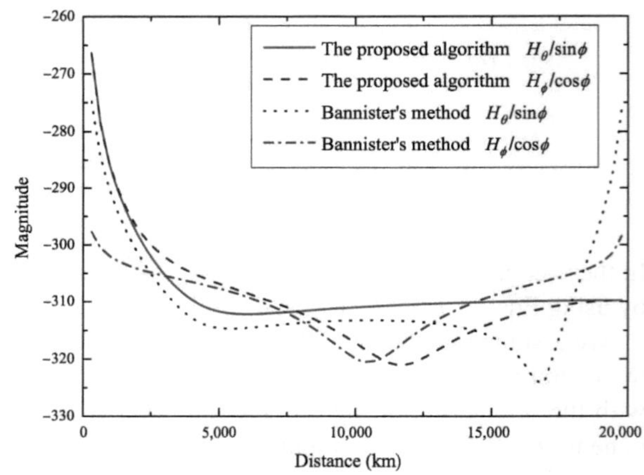

Fig. 2.21 The magnitudes of the components H_ϕ in dB·A/m at $\phi = 0°$ and H_ϕ in dB·A/m at $\phi = 90°$ versus the propagation distance: $f = 10$ Hz, $\phi_T = 0°$

gential electric field and that of the tangential magnetic field at $f = 1$ Hz are shown in Figs. 2.22 and 2.23, respectively. It is noted that all numbers in the above contour diagrams indicate the field strength in dB. For the electric field, 1 V/m is corresponded to 0 dB. For the magnetic field, 1 A/m is corresponded to 0 dB.

At the operating frequency below 2 Hz, the wavelength is much larger than the Earth's circumference, and the space between the Earth's surface and the lower boundary of the ionosphere is actually a concentric spherical shell. Obviously, the eigenvalue ν is smaller than 1. With Eq. (836.7) in the mathematics handbook by Gradshteyn and Ryzhik (1980), we write

$$\left.\frac{\partial P_\nu(\cos(\pi-\theta))}{\partial \nu}\right|_{\nu=0} = 2\ln\sin\left(\frac{\theta}{2}\right). \tag{2.218}$$

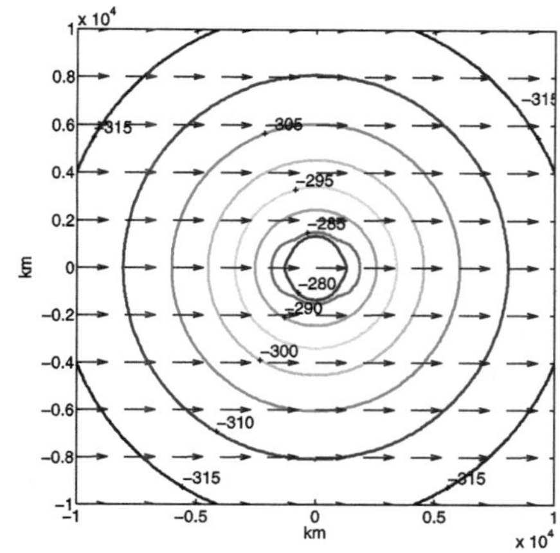

Fig. 2.22 The contour diagram of the tangential electric field at $f = 1$ Hz

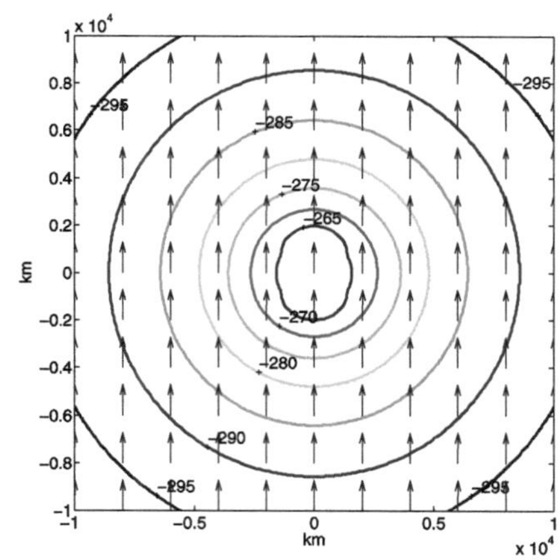

Fig. 2.23 The contour diagram of the tangential magnetic field at $f = 1$ Hz

When $\nu \ll 1$, we have

$$P_\nu\big(\cos(\pi-\theta)\big) = 1 + 2\nu \ln \sin\left(\frac{\theta}{2}\right) + O\big(\nu^2\big). \tag{2.219}$$

If a first-order approximation is taken, we obtain readily:

$$\frac{1}{\sin\theta}\frac{\partial P_\nu(\cos(\pi-\theta))}{\partial\theta} \approx \frac{\partial^2 P_\nu(\cos(\pi-\theta))}{\partial\theta^2}. \tag{2.220}$$

Thus, at the operating frequency below 2 Hz, in the far-field region, the higher modes attenuate rapidly, and the electromagnetic field excited by an HED essentially satisfies the following conditions:

$$\frac{E_\theta}{\cos\phi} = -\frac{E_\phi}{\sin\phi}; \qquad \frac{H_\theta}{\sin\phi} = \frac{H_\phi}{\cos\phi}. \tag{2.221}$$

Then, the horizontal field components are expressed in the following forms:

$$\mathbf{E_t} = E_\theta(\hat{\theta}\cos\phi - \hat{\phi}\sin\phi) = E_\theta\hat{x}, \tag{2.222}$$

$$\mathbf{H_t} = H_\theta(\hat{\theta}\sin\phi + \hat{\phi}\cos\phi) = H_\theta\hat{y}. \tag{2.223}$$

Evidently, the horizontal electric field is essentially in the $\hat{x}$ direction, while the horizontal magnetic field is essentially in the $\hat{y}$ direction. Furthermore, it is seen that both the horizontal electric field and the horizontal magnetic field will not change with the azimuth angle ϕ.

From Figs. 2.22 and 2.23, we conclude as follows:

- At the operating frequency below 2 Hz, except for the multimode region, which the transmitting antenna is surrounded nearby, all curves in the contour diagrams are approximately concentric circles. Namely, at the same distance away from the transmitting antenna, the tangential components of the electromagnetic field are essentially equal.
- At the operating frequency below 2 Hz, the direction of the tangential electric field component is essentially the same as that of the current moment, while the direction of the tangential magnetic field component is essentially perpendicular to that of the current moment.

When the operating frequency increases, the condition of $\nu \ll 1$ is no longer satisfied. For example, at $f = 10$ Hz, the parameter ν is comparable to 1, $\nu \sim 1$. The direction of the tangential electric field component is not completely the same as that of the current moment, while the direction of the tangential magnetic field component is not completely perpendicular to that of the current moment.

References

Bannister PR (1984) ELF propagation update. IEEE J Ocean Eng 9(3):179–188

Bannister PR, Wolkoff EA, Katan JR, Williams FJ (1973) Far-field, extremely-low-frequency propagation measurements. Radio Sci 8(7):623–631

Barrick DE (1999) Exact ULF/ELF dipole field strengths in the Earth–ionosphere cavity over the Schumann resonance region: idealized boundaries. Radio Sci 34(1):209–227

Bouwkamp CJ, Casimir HBG (1954) On multipole expansions in the theory of electromagnetic radiation. Physica 20:539–554

Budden KG (1961) Radio waves in the ionosphere. Cambridge University Press, Cambridge

Carroll KJ, Ferraro AJ (1990) Computer simulation of ELF injection in the Earth–ionosphere waveguide. Radio Sci 25(6):1363–1367

Cummer SA (2000) Modeling electromagnetic propagation in the Earth–ionosphere waveguide. IEEE Trans Antennas Propag 48(9):468–474

Felsen LB, Marcuvitz N (1973) Radiation and scattering of waves, Englewood Cliffs. Prentice Hall, New York

Fraser-Smith AC, Bannister PR (1998) Reception of ELF signals at antipodal distance. Radio Sci 33(1):83–88

Galejs J (1964) Terrestrial extremely-low-frequency propagation. In: Bleil DF (ed) Natural electromagnetic phenomena below 30 kc/s. Plenum, New York, pp 205–258

Galejs J (1968) ELF and VLF fields of a horizontal electric dipole. IEEE Trans Antennas Propag 16(6):689–700
Galejs J (1970) ELF and VLF propagation for models of a perturbed ionosphere. Radio Sci 5(7):1041–1044
Galejs J (1972a) Terrestrial propagation of long electromagnetic waves. Pergamon, New York
Galejs J (1972b) Stable solutions of ionospheric fields in the propagation of ELF and VLF waves. Radio Sci 7(6):549–561
Gradshteyn IS, Ryzhik IM (1980) Table of integrals, series, and products. Academic Press, New York
Kirillov VV, Pronin AE (2007) Normal waves of the anisotropic Earth–ionosphere waveguide in the VLF-ULF range. Int J Geomagn Aeron 7:GI2006. doi:10.1029/2005GI000126G
Li GZ (2012) Researches on mechanism of SLF and ELF wave propagation in an anisotropic Earth-ionosphere cavity. MS thesis, Zhejiang University, Hangzhou, China. (In Chinese)
Nickolaenko AP, Hayakawa M (2002) Resonances in the Earth–ionosphere cavity. Kluwer Academic, Dordrecht
Pan WY (2004) LF VLF ELF wave propagation. UESTC Press, Chengdu. (In Chinese)
Peng HY, Tao W, Pan WY, Guo LX (2012) Numerical integral method for ELF fields excited by vertical electric dipole in asymmetric Earth–ionosphere cavity. Chin J Radio Sci 22(2):333–338. (In Chinese)
Peng HY, Wu SH, Pan WY, Guo LX (2013) ELF fields excited by horizontal electric dipole in asymmetric Earth–ionosphere cavity. Submitted to Chin J Radio Sci. (In Chinese)
Rawer K, Bilitza D, Ramakrishnan S (1978) Goals and status of the international reference ionosphere. Rev Geophys 16:177–181
Rybachek ST, Ponomariev MJ (2007) Efficiency of the Earth–ionosphere waveguide excitation by ELF sources located in an anisotropic ionosphere. Int J Geomagn Aeron 7:GI1005. doi:10.1029/2005GI000123
Suchumann WO (1952a) On the radiation free selfoscillations of a conducting sphere, which is surrounded by an air layer and an ionosphere shell. Z Naturforsch 72:149–154. (In German)
Suchumann WO (1952b) On the damping of electromagnetic selfoscillations of the system Earth–air–ionosphere. Z Naturforsch 72:250–252. (In German)
Tripathi VK, Chang CL, Papadopoulos K (1982) Excitation of the Earth–ionosphere waveguide by an ELF source in the ionosphere. Radio Sci 17(5):1321–1326
Wait JR (1957) The mode theory of VLF ionospheric propagation for finite ground conductivity. Proc IRE 45(6):760–767
Wait JR (1960) Terrestrial propagation of very-low-frequency radio waves, a theoretical investigation. J Res Natl Bur Stand 64D:153–204
Wait JR (1970) Electromagnetic waves in stratified media, 2nd edn. Pergamon, New York
Wait JR (1992) Reflection of VLF radio waves at a junction in the Earth-ionosphere waveguide. IEEE Trans Electromagn Compat 34:4–8
Wait JR, Spies KP (1965) Influence of finite ground conductivity on the propagation of VLF radio waves. Radio Sci 3:787–791
Wang YX, Peng Q, Pan WY, Zhang HQ, Zhang ZW (2005) The fields excited by SLF/ELF horizontal electric dipole in Earth–ionosphere cavity. Chin J Radio Sci 22(5). (In Chinese)
Wang YX, Fan WS, Pan WY, Zhang HQ (2008) Spherical harmonic series solution of fields excited by vertical electric dipole in Earth–ionosphere cavity. Front Electr Electron Eng China 3(1):61–69

Chapter 3
Spherical Harmonic Series Solution for SLF/ELF Field in the Earth–Ionosphere Waveguide/Cavity

The spherical harmonic series solutions are obtained for the SLF/ELF electromagnetic fields of VEDs and HEDs in the non-ideal Earth–ionosphere waveguide or cavity. Especially, the speed-up numerical convergence algorithm is used to improve the convergence speed for the spherical harmonic series solutions.

3.1 Introduction

In the past century, SLF/ELF electromagnetic wave propagation in the Earth-ionosphere waveguide or cavity was intensively investigated (Wait 1962; Johler and Berry 1962; Galejs 1971, 1972; Bannister 1974, 1979; Chang and Wait 1974; Chu 1974; Carroll and Ferraro 1990; Fraser-Smith and Bannister 1998; Cummer 2000). In preceding chapter, this problem has been addressed specifically. It is noted that the great progress on the FDTD modeling of SLF/ELF wave propagation in Earth-ionosphere waveguide has been made in the past decade (Simpson and Taflove 2004, 2007; Simpson 2009; Yang and Pasko 2005, 2006). In what follows, we will attempt to outline a new spherical harmonic series algorithm, which is proposed for SLF/ELE field radiated by a VED in the ideal Earth-ionosphere waveguide or cavity (Barrick 1999). Unfortunately, this algorithm was not extended to the non-ideal case, and the changes of SLF/ELF field with the height and operating frequencies were not addressed specifically. With extensions of the work by Barrick, a spherical harmonic series solutions are obtained for SLF/ELF electromagnetic field radiated by VEDs and HEDs in the non-ideal Earth–ionosphere waveguide or cavity (Wang et al. 2007a, 2007b, 2008). In those works by Wang et al., the speed-up numerical convergence algorithm is proposed for improving the convergence speed of the spherical harmonic series solutions.

In this chapter, we will attempt to summarize the spherical harmonic series solutions for SLF/ELF electromagnetic field in the presence of a non-ideal Earth–ionosphere waveguide or cavity and the speed-up numerical convergence algorithm. The results obtained are in agreement with available results (Pan 2004).

W. Pan, K. Li, *Propagation of SLF/ELF Electromagnetic Waves*,
Advanced Topics in Science and Technology in China,
DOI 10.1007/978-3-642-39050-0_3,

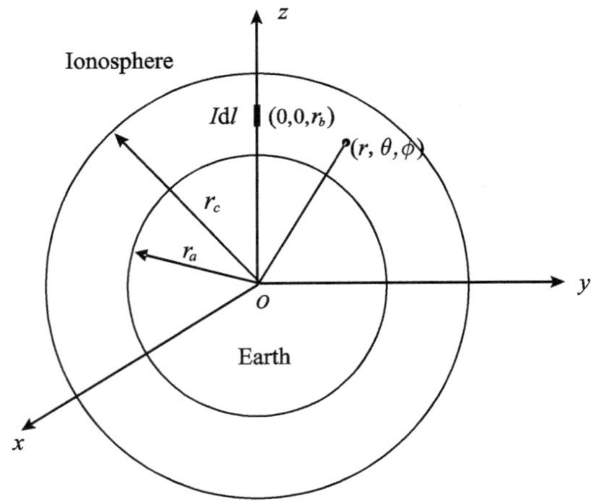

Fig. 3.1 VED at $(0, 0, r_b)$ in the Earth–ionosphere waveguide or cavity

3.2 SLF/ELF Fields of VED in the Earth–Ionosphere Waveguide/Cavity

In this section, we will treat SLF/ELF electromagnetic field of a VED in the Earth–ionosphere waveguide or cavity in detail.

3.2.1 Spherical Harmonic Series Solution for SLF/ELF Field of VED in the Earth–Ionosphere Cavity

The Earth–ionosphere geometry and spherical coordinate system are shown in Fig. 3.1. The two reflection walls, the ground and the ionosphere, are characterized by the normalized surface impedances Δ_g and Δ_i, respectively. Assuming that a VED is represented by its current density $\hat{z}I\,\mathrm{d}l\delta(x)\delta(y)\delta(z-r_b)$, where $r_b = r_a + z_s$, r_a is the Earth's radius, and z_s denotes the height of the dipole above the ground. $r_c = r_a + h$, where h denotes the height of the lower boundary of the ionosphere. Then, the electromagnetic field in the Earth–ionosphere waveguide or cavity can be represented in terms of two potential functions U and V (Barrick 1999):

$$E_r = \frac{1}{-\mathrm{i}\omega\varepsilon}\left(\frac{\partial^2}{\partial r^2} + k^2\right)U, \tag{3.1}$$

$$H_r = \frac{1}{-\mathrm{i}\omega\mu}\left(\frac{\partial^2}{\partial r^2} + k^2\right)V, \tag{3.2}$$

$$E_\theta = \frac{-1}{r\sin\theta}\frac{\partial V}{\partial\phi} + \frac{1}{-\mathrm{i}\omega\varepsilon r}\frac{\partial^2 U}{\partial r\partial\theta}, \tag{3.3}$$

$$H_\theta = \frac{1}{r\sin\theta}\frac{\partial U}{\partial\phi} + \frac{1}{-\mathrm{i}\omega\mu r}\frac{\partial^2 V}{\partial r\partial\theta}, \tag{3.4}$$

$$E_\phi = \frac{1}{r}\frac{\partial V}{\partial \theta} + \frac{1}{-\mathrm{i}\omega\varepsilon r \sin\theta}\frac{\partial^2 U}{\partial r \partial \phi}, \tag{3.5}$$

$$H_\phi = \frac{-1}{r}\frac{\partial U}{\partial \theta} + \frac{1}{-\mathrm{i}\omega\mu r \sin\theta}\frac{\partial^2 V}{\partial r \partial \phi}, \tag{3.6}$$

where the two potential functions U and V satisfy the following equations (Barrick 1999):

$$\left(\nabla^2 + k^2\right)\frac{U}{r} = 0, \tag{3.7}$$

$$\left(\nabla^2 + k^2\right)\frac{V}{r} = 0. \tag{3.8}$$

When the effects by the ground and the ionosphere are not considered, the potential functions for the primary field radiated by VED can be represented as follows:

$$U_0^{\mathrm{VED}} = \frac{\mathrm{i}kI\,\mathrm{d}l}{4\pi v_b^2} T_0, \tag{3.9}$$

$$V_0^{\mathrm{VED}} = 0. \tag{3.10}$$

In the above equation, T_0 is written in the form

$$T_0 = \begin{cases} \sum_{n=0}^{\infty} (2n+1)\hat{H}_n(\nu_b)\hat{J}_n(\nu)P_n(\cos\theta); & \nu < \nu_b, \\ \sum_{n=0}^{\infty} (2n+1)\hat{J}_n(\nu_b)\hat{H}_n(\nu)P_n(\cos\theta); & \nu > \nu_b, \end{cases} \tag{3.11}$$

where $\nu = kr$ and $\nu_b = kr_b$. P_n is the Legendre function of order n. The function $\hat{J}_n(\nu)$ can be expressed in terms of half-order Bessel function of the first kind, while the function $\hat{H}_n(\nu)$ is its corresponding outgoing first-kind Hankel function. They are defined as follows:

$$\hat{J}_n(\nu) = \sqrt{\frac{\pi\nu}{2}} J_{n+\frac{1}{2}}(\nu), \tag{3.12}$$

$$\hat{H}_n(\nu) = \sqrt{\frac{\pi\nu}{2}} H^{(1)}_{n+\frac{1}{2}}(\nu). \tag{3.13}$$

It is noted that the functions $\hat{J}_n(\nu)$ and $\hat{H}_n(\nu)$ should satisfy the following differential equation:

$$\left[\frac{\mathrm{d}^2}{\mathrm{d}\nu^2} + 1 - \frac{n(n+1)}{\nu^2}\right] B_n(\nu) = 0. \tag{3.14}$$

Due to the reflection of the air–Earth boundary and that of the air–ionosphere boundary, the secondary disturbance field in the cavity will be produced. The potential functions of the secondary disturbance field is represented in terms of the spherical harmonic functions. We write

$$U_p^{\mathrm{VED}} = \frac{\mathrm{i}kI\,\mathrm{d}l}{4\pi\nu_b^2}\sum_{n=0}^{\infty}(2n+1)\big[b_n\hat{J}_n(\nu)+c_n\hat{H}_n(\nu)\big]P_n(\cos\theta), \tag{3.15}$$

$$V_p^{\mathrm{VED}} = 0. \tag{3.16}$$

Thus, the total potential functions should be represented as follows:

$$\begin{aligned} U^{\mathrm{VED}} &= U_0^{\mathrm{VED}} + U_p^{\mathrm{VED}} \\ &= \frac{\mathrm{i}kI\,\mathrm{d}l}{4\pi\nu_b^2}\cdot\begin{cases} \sum_{n=0}^{\infty}(2n+1)P_n(\cos\theta) \\ \quad\times\big\{\big[\hat{H}_n(\nu_b)+b_n\big]\hat{J}_n(\nu)+c_n\hat{H}_n(\nu)\big\}; & \nu<\nu_b, \\ \sum_{n=0}^{\infty}(2n+1)P_n(\cos\theta) \\ \quad\times\big\{b_n\hat{J}_n(\nu)+\hat{H}_n(\nu)\big[c_n+\hat{J}_n(\nu_b)\big]\big\}; & \nu>\nu_b, \end{cases} \end{aligned} \tag{3.17}$$

$$V^{\mathrm{VED}} = V_0^{\mathrm{VED}} + V_p^{\mathrm{VED}} = 0. \tag{3.18}$$

On the Earth's surface, $r = r_a$, the field components satisfy the impedance boundary conditions. We have

$$E_\theta = -\Delta_g\eta H_\phi|_{r=r_a}; \qquad E_\phi = \Delta_g\eta H_\theta|_{r=r_a}. \tag{3.19}$$

At the lower boundary of the ionosphere, $r = r_c$, we have

$$E_\theta = \Delta_i\eta H_\phi|_{r=r_c}; \qquad E_\phi = -\Delta_i\eta H_\theta|_{r=r_c}. \tag{3.20}$$

In the above equations, η is the wave impedance in the air.

Substituting Eqs. (3.17) and (3.18) into Eqs. (3.1)–(3.6), the six field components in the Earth–ionosphere waveguide or cavity can be represented in the following forms:

$$E_r = \frac{-\eta k^2 I\,\mathrm{d}l}{4\pi\nu_b^2\nu^2}\begin{cases} \sum_{n=0}^{\infty}n(n+1)(2n+1)P_n(\cos\theta) \\ \quad\times\big\{\big[\hat{H}_n(\nu_b)+b_n\big]\hat{J}_n(\nu)+c_n\hat{H}_n(\nu)\big\}; & \nu<\nu_b, \\ \sum_{n=0}^{\infty}n(n+1)(2n+1)P_n(\cos\theta) \\ \quad\times\big\{b_n\hat{J}_n(\nu)+\hat{H}_n(\nu)\big[c_n+\hat{J}_n(\nu_b)\big]\big\}; & \nu>\nu_b, \end{cases} \tag{3.21}$$

$$H_r = 0, \tag{3.22}$$

$$E_\theta = \frac{\eta k^2 I\,\mathrm{d}l}{4\pi \nu_b^2 \nu} \begin{cases} \sum_{n=0}^{\infty} (2n+1) P_n^1(\cos\theta) \\ \quad \times \left\{ \left[\hat{H}_n(\nu_b) + b_n \right] \hat{J}_n'(\nu) + c_n \hat{H}_n'(\nu) \right\}; \quad \nu < \nu_b, \\ \sum_{n=0}^{\infty} (2n+1) P_n^1(\cos\theta) \\ \quad \times \left\{ b_n \hat{J}_n'(\nu) + \left[\hat{J}_n(\nu_b) + c_n \right] \hat{H}_n'(\nu) \right\}; \quad \nu > \nu_b, \end{cases} \tag{3.23}$$

$$H_\theta = 0, \tag{3.24}$$

$$E_\phi = 0, \tag{3.25}$$

$$H_\phi = \frac{\mathrm{i} k^2 I\,\mathrm{d}l}{4\pi \nu_b^2 \nu} \begin{cases} \sum_{n=0}^{\infty} (2n+1) P_n^1(\cos\theta) \\ \quad \times \left\{ \left[\hat{H}_n(\nu_b) + b_n \right] \hat{J}_n(\nu) + c_n \hat{H}_n(\nu) \right\}; \quad \nu < \nu_b, \\ \sum_{n=0}^{\infty} (2n+1) P_n^1(\cos\theta) \\ \quad \times \left\{ b_n \hat{J}_n(\nu) + \left[\hat{J}_n(\nu_b) + c_n \right] \hat{H}_n(\nu) \right\}; \quad \nu > \nu_b. \end{cases} \tag{3.26}$$

Considering Eqs. (3.19) and (3.20) of the impedance boundary conditions at $r = r_a$ and $r = r_c$, we readily have

$$\begin{aligned} &\left[\hat{H}_n(\nu_b) + b_n \right] \hat{J}_n'(\nu_a) + c_n \hat{H}_n'(\nu_a) \\ &\quad = -\mathrm{i}\Delta_g \left\{ \left[\hat{H}_n(\nu_b) + b_n \right] \hat{J}_n(\nu_a + c_n \hat{H}_n(\nu_a) \right\}, \end{aligned} \tag{3.27}$$

$$\begin{aligned} &b_n \hat{J}_n'(\nu_c) + \left[\hat{J}_n(\nu_b) + c_n \right] \hat{H}_n'(\nu_c) \\ &\quad = \mathrm{i}\Delta_i \left\{ b_n \hat{J}_n(\nu_c) + \left[\hat{J}_n(\nu_b) + c_n \right] \hat{H}_n(\nu_c) \right\}. \end{aligned} \tag{3.28}$$

Then, from the above two equations, the coefficients b_n and c_n may be written as follows:

$$b_n = \frac{\left\{ \begin{array}{l} -\hat{H}_n(\nu_b)[\hat{H}_n'(\nu_c) - \mathrm{i}\Delta_i \hat{H}_n(\nu_c)][\mathrm{i}\Delta_g \hat{J}_n(\nu_a) + \hat{J}_n'(\nu_a)] \\ \quad - [\mathrm{i}\Delta_g \hat{H}_n(\nu_a) + \hat{H}_n'(\nu_a)][\mathrm{i}\Delta_i \hat{H}_n(\nu_c) - \hat{H}_n'(\nu_c)] \hat{J}_n(\nu_b) \end{array} \right\}}{\left\{ \begin{array}{l} [\mathrm{i}\Delta_i \hat{J}_n(\nu_c) - \hat{J}_n'(\nu_c)][\mathrm{i}\Delta_g \hat{H}_n(\nu_a) + \hat{H}_n'(\nu_a)] \\ \quad + [\mathrm{i}\Delta_g \hat{J}_n(\nu_a) + \hat{J}_n'(\nu_a)][\hat{H}_n'(\nu_c) - \mathrm{i}\Delta_i \hat{H}_n(\nu_c)] \end{array} \right\}}, \tag{3.29}$$

$$c_n = \frac{\left\{ \begin{array}{l} \hat{J}_n(\nu_b)[\mathrm{i}\Delta_i \hat{H}_n(\nu_c) - \hat{H}_n'(\nu_c)][\mathrm{i}\Delta_g \hat{J}_n(\nu_a) + \hat{J}_n'(\nu_a)] \\ \quad - [\mathrm{i}\Delta_i \hat{J}_n(\nu_c) - \hat{J}_n'(\nu_c)][\hat{J}_n'(\nu_a) + \mathrm{i}\Delta_g \hat{J}_n(\nu_a)] \hat{H}_n(\nu_b) \end{array} \right\}}{\left\{ \begin{array}{l} [\mathrm{i}\Delta_i \hat{J}_n(\nu_c) - \hat{J}_n'(\nu_c)][\mathrm{i}\Delta_g \hat{H}_n(\nu_a) + \hat{H}_n'(\nu_a)] \\ \quad + [\mathrm{i}\Delta_g \hat{J}_n(\nu_a) + \hat{J}_n'(\nu_a)][\hat{H}_n'(\nu_c) - \mathrm{i}\Delta_i \hat{H}_n(\nu_c)] \end{array} \right\}}. \tag{3.30}$$

So far, the spherical harmonic series solutions have been derived for SLF/ELF electromagnetic field radiated by a VED in the Earth–ionosphere waveguide or cavity. In fact, in order to calculate the numerical summation accurately, it is required to calculate a prohibitively large number terms, even through the series remain absolutely convergent. It is known that the high-frequency asymptotic approximation based on the Watson transform of the series is not appropriate in ELF range and the lower end of SLF range (Barrick 1999). Obviously, in the next step, the main task is to seek a method for improving the convergence speed of the spherical harmonic series.

3.2.2 Speed-up Numerical Convergence Algorithm

In this subsection, a speed-up numerical convergence algorithm, which is an asymptotic extraction technique, is employed to improve the calculation efficiency.

3.2.2.1 Speed-up Numerical Convergence Algorithm for the Component E_r

The electric field component E_r is written in the following form:

$$E_r = \sum_{n=0}^{\infty} (E_{rn} - \tilde{E}_{rn}) P_n(\cos\theta) + \sum_{n=0}^{\infty} \tilde{E}_{rn} P_n(\cos\theta). \tag{3.31}$$

The latter series term in Eq. (3.31) can be expressed in analytical form, which can be calculated readily. Since the term $(E_{rn} - \tilde{E}_{rn})$ reduces rapidly with the increase of n, the first series in Eq. (3.31) will converge rapidly. As a result, the field component can be calculated efficiently with better accuracy. Next, the main task is to find an efficient analytical solution of the latter series in Eq. (3.31).

When $\nu < \nu_0$, the series in Eq. (3.21) consist of two terms: $(2n+1)\hat{J}_n(\nu)\hat{H}_n(\nu_b)$ and $(2n+1)[b_n\hat{J}_n(\nu) + c_n\hat{H}_n(\nu)]$. When $\nu \ll n$, from the series expression of the circular cylinder function near the zero point, we readily have

$$(2n+1)\hat{J}_n(\nu)\hat{H}_n(\nu_b) \Rightarrow -\mathrm{i}\nu_b\left(\frac{\nu}{\nu_b}\right)^{n+1}. \tag{3.32}$$

For mathematical convenience, it is necessary to examine the results at the ideal condition of $\Delta_g = \Delta_i = 0$. At this time, the coefficients b_n and c_n can be simplified as follows:

$$b_n^0 = -\frac{\hat{J}_n'(\nu_a)\hat{H}_n(\nu_b)\hat{H}_n'(\nu_c) - \hat{J}_n(\nu_b)\hat{H}_n'(\nu_a)\hat{H}_n'(\nu_c)}{\hat{J}_n'(\nu_a)\hat{H}_n'(\nu_c) - \hat{J}_n'(\nu_c)\hat{H}_n'(\nu_a)}, \tag{3.33}$$

$$c_n^0 = -\frac{\hat{J}_n'(\nu_a)\hat{J}_n(\nu_b)\hat{H}_n'(\nu_c) - \hat{J}_n'(\nu_a)\hat{J}_n'(\nu_c)\hat{H}_n(\nu_b)}{\hat{J}_n'(\nu_a)\hat{H}_n'(\nu_c) - \hat{J}_n'(\nu_c)\hat{H}_n'(\nu_a)}. \tag{3.34}$$

By using Eqs. (3.33) and (3.34), we readily have

$$
\begin{aligned}
T_n &= (2n+1)\left[b_n^0 \hat{J}_n(\nu) + c_n^0 \hat{H}_n(\nu)\right] \\
&\overset{n\gg\nu}{\Rightarrow} \frac{-\mathrm{i}\nu_a}{n(n+1)} \frac{(n+1)^2\rho^n(\frac{\nu_a}{\nu})^n + n(n+1)\rho^{-(n+1)}\tau^{2n+1}(\frac{\nu_a}{\nu})^n}{1-\tau^{2n+1}} \\
&\quad + \frac{n(n+1)\rho^n\tau^{2n+1}(\frac{\nu}{\nu_a})^{(n+1)} + n^2\rho^{-(n+1)}\tau^{2n+1}(\frac{\nu}{\nu_a})^{(n+1)}}{1-\tau^{2n+1}} \\
&= \frac{-\mathrm{i}\nu_a}{n(n+1)}\left[(n+1)^2\rho^n\left(\frac{\nu_a}{\nu}\right)^n + n(n+1)\rho^{-(n+1)}\tau^{2n+1}\left(\frac{\nu_a}{\nu}\right)^n\right. \\
&\quad + n(n+1)\rho^n\tau^{2n+1}\left(\frac{\nu}{\nu_a}\right)^{(n+1)} \\
&\quad \left. + n^2\rho^{-(n+1)}\tau^{2n+1}\left(\frac{\nu}{\nu_a}\right)^{(n+1)}\right]\sum_{m=0}^{\infty}\tau^{m(2n+1)} \\
&= \frac{-\mathrm{i}\nu_a}{n(n+1)}\left\{\left[(n+1)^2\rho^n\left(\frac{\nu_a}{\nu}\right)^n + n(n+1)\rho^n\left(\frac{\nu}{\nu_a}\right)^{(n+1)}\right.\right. \\
&\quad \left. + n^2\rho^{-(n+1)}\left(\frac{\nu}{\nu_a}\right)^{(n+1)} + n(n+1)\left(\frac{\nu_a}{\nu}\right)^n\rho^{-(n+1)}\right]\sum_{m=1}^{\infty}\tau^{m(2n+1)} \\
&\quad \left. + (n+1)^2\rho^n\left(\frac{\nu_a}{\nu}\right)^n\right\},
\end{aligned} \tag{3.35}
$$

where

$$
\rho = \frac{r_a}{r_b} = \frac{\nu_a}{\nu_b}; \qquad \tau = \frac{\nu_a}{\nu_c}. \tag{3.36}
$$

In the last term in Eq. (3.35), it is assumed that $m_{\max} = 20$, the result becomes

$$
\begin{aligned}
T_n \approx & \frac{-\mathrm{i}\nu_a}{n(n+1)}\left\{\left[(n+1)^2\rho^n\left(\frac{\nu_a}{\nu}\right)^n + n(n+1)\rho^n\left(\frac{\nu}{\nu_a}\right)^{(n+1)}\right.\right. \\
& + n^2\rho^{-(n+1)}\left(\frac{\nu}{\nu_a}\right)^{(n+1)} \\
& \left. + n(n+1)\left(\frac{\nu_a}{\nu}\right)^n\rho^{-(n+1)}\right]\sum_{m=1}^{20}\tau^{m(2n+1)} \\
& \left. + (n+1)^2\rho^n\left(\frac{\nu_a}{\nu}\right)^n\right\}.
\end{aligned} \tag{3.37}
$$

From Eqs. (3.32) and (3.37), we have

$$
\begin{aligned}
&(2n+1)\hat{J}_n(\nu)\hat{H}_n(\nu_b)+T_n\\
&\Rightarrow \frac{-\mathrm{i}\nu_a}{n(n+1)}\cdot\left\{n(n+1)\rho^{-1}\left(\frac{\nu}{\nu_b}\right)^{n+1}+(n+1)^2\rho^n\left(\frac{\nu_a}{\nu}\right)^n\right.\\
&\quad+\left[(n+1)^2\rho^n\left(\frac{\nu_a}{\nu}\right)^n+n(n+1)\rho^n\left(\frac{\nu}{\nu_a}\right)^{(n+1)}\right.\\
&\quad\left.+n^2\rho^{-(n+1)}\left(\frac{\nu}{\nu_a}\right)^{(n+1)}+n(n+1)\left(\frac{\nu_a}{\nu}\right)^n\rho^{-(n+1)}\right]\\
&\quad\left.\times\sum_{m=1}^{20}\tau^{m(2n+1)}\right\}.
\end{aligned}
\tag{3.38}
$$

Letting

$$
E^0_{rn}=\frac{-\eta k^2 I\,\mathrm{d}l}{4\pi\nu_b^2\nu^2}n(n+1)(2n+1)\left\{\left[\hat{H}_n(\nu_b)+b^0_n\right]\hat{J}_n(\nu)+c^0_n\hat{H}_n(\nu)\right\},\tag{3.39}
$$

it is seen that $\tilde{E}_{rn}$ is the asymptotic approximation of E^0_{rn}. We write

$$
E^0_{rn}\big|_{n\to\infty}\Rightarrow\tilde{E}_{rn}.\tag{3.40}
$$

By using Eqs. (3.38)–(3.40), it follows that

$$
\begin{aligned}
\tilde{E}_{rn}&=\frac{\mathrm{i}\eta k^2 I\,\mathrm{d}l}{4\pi\nu_b\nu^2}\rho\\
&\quad\times\left\{n(n+1)\rho^{-1}\left(\frac{\nu}{\nu_b}\right)^{n+1}+(n+1)^2\rho^n\left(\frac{\nu_a}{\nu}\right)^n\right.\\
&\quad+\left[(n+1)^2\rho^n\left(\frac{\nu_a}{\nu}\right)^n+n(n+1)\rho^n\left(\frac{\nu}{\nu_a}\right)^{(n+1)}\right.\\
&\quad\left.+n^2\rho^{-(n+1)}\left(\frac{\nu}{\nu_a}\right)^{(n+1)}+n(n+1)\left(\frac{\nu_a}{\nu}\right)^n\rho^{-(n+1)}\right]\\
&\quad\left.\times\sum_{m=1}^{20}\tau^{m(2n+1)}\right\}.
\end{aligned}
\tag{3.41}
$$

With Eq. (3.41), we readily have

$$\begin{aligned}\sum_{n=0}^{\infty}\tilde{E}_{rn}P_n(\cos\theta) &= \frac{\mathrm{i}\eta k^2 I\,\mathrm{d}l}{4\pi\nu_b\nu^2}\rho \\ &\times\sum_{n=0}^{\infty}\left\{n(n+1)\rho^{-1}\left(\frac{\nu}{\nu_b}\right)^{n+1}+(n+1)^2\rho^n\left(\frac{\nu_a}{\nu}\right)^n\right. \\ &+\left[(n+1)^2\rho^n\left(\frac{\nu_a}{\nu}\right)^n+n(n+1)\rho^n\left(\frac{\nu}{\nu_a}\right)^{(n+1)}\right. \\ &+n(n+1)\left(\frac{\nu_a}{\nu}\right)^n\rho^{-(n+1)} \\ &\left.+n^2\rho^{-(n+1)}\left(\frac{\nu}{\nu_a}\right)^{(n+1)}\right] \\ &\left.\times\sum_{m=1}^{20}\tau^{m(2n+1)}\right\}P_n(\cos\theta). \end{aligned} \tag{3.42}$$

Considering the following relations:

$$\sum_{n=0}^{\infty}\tau^n P_n(x)=\frac{1}{g};\qquad \sum_{n=0}^{\infty}n\tau^n P_n(x)=\frac{\tau(x-\tau)}{g^3}, \tag{3.43}$$

$$\sum_{n=0}^{\infty}n^2\tau^n P_n(x)=\frac{\tau(x+x^2\tau-x\tau^2-2\tau+\tau^3)}{g^5}, \tag{3.44}$$

where

$$x=\cos\theta;\qquad g=\sqrt{1-2x\tau+\tau^2}, \tag{3.45}$$

we readily have

$$\begin{aligned}\sum_{n=0}^{\infty}\tilde{E}_{rn}P_n(\cos\theta) &= \frac{\mathrm{i}\eta k^2 I\,\mathrm{d}l}{4\pi\nu_b\nu^2}\rho\left\{\left[\frac{\tau_1(x+x^2\tau_1-x\tau_1^2-2\tau_1+\tau_1^3)}{g_1^5}\right.\right. \\ &\left.+\frac{2\tau_1(x-\tau_1)}{g_1^3}+\frac{1}{g_1}\right] \\ &+\frac{\nu}{\nu_a}\left[\frac{\tau_2(x+x^2\tau_2-x\tau_2^2-2\tau_2+\tau_2^3)}{g_2^5}\right. \\ &\left.\left.+\frac{\tau_2(x-\tau_2)}{g_2^3}\right]+M\right\}, \end{aligned} \tag{3.46}$$

where

$$M = \sum_{m=1}^{20} \tau^m \left\{ \frac{\tau_3(x + x^2\tau_3 - x\tau_3^2 - 2\tau_3 + \tau_3^3)}{g_3^5} + \frac{2\tau_3(x - \tau_3)}{g_3^3} + \frac{1}{g_3} \right.$$
$$+ \frac{\nu}{\nu_a}\left[\frac{\tau_4(x + x^2\tau_4 - x\tau_4^2 - 2\tau_4 + \tau_4^3)}{g_4^5} + \frac{\tau_4(x - \tau_4)}{g_4^3} \right]$$
$$+ \frac{\nu}{\rho\nu_a} \frac{\tau_5(x + x^2\tau_5 - x\tau_5^2 - 2\tau_5 + \tau_5^3)}{g_5^5}$$
$$\left. + \frac{1}{\rho}\left[\frac{\tau_6(x + x^2\tau_6 - x\tau_6^2 - 2\tau_6 + \tau_6^3)}{g_6^5} + \frac{\tau_6(x - \tau_6)}{g_6^3} \right] \right\}, \tag{3.47}$$

and

$$\tau_1 = \frac{\rho\nu_a}{\nu}; \qquad g_1 = \sqrt{1 - 2x\tau_1 + \tau_1^2}, \tag{3.48}$$

$$\tau_2 = \frac{\nu}{\nu_b}; \qquad g_2 = \sqrt{1 - 2x\tau_2 + \tau_2^2}, \tag{3.49}$$

$$\tau_3 = \frac{\rho\nu_a}{\nu}\tau^{2m}; \qquad g_3 = \sqrt{1 - 2x\tau_3 + \tau_3^2}, \tag{3.50}$$

$$\tau_4 = \frac{\rho\nu}{\nu_a}\tau^{2m}; \qquad g_4 = \sqrt{1 - 2x\tau_4 + \tau_4^2}, \tag{3.51}$$

$$\tau_5 = \frac{\nu}{\rho\nu_a}\tau^{2m}; \qquad g_5 = \sqrt{1 - 2x\tau_5 + \tau_5^2}, \tag{3.52}$$

$$\tau_6 = \frac{\nu_a}{\rho\nu}\tau^{2m}; \qquad g_6 = \sqrt{1 - 2x\tau_6 + \tau_6^2}. \tag{3.53}$$

With a similar procedure, when $\nu > \nu_b$, the asymptotic approximation of $\tilde{E}_{rn}$ can be derived readily:

$$\tilde{E}_{rn} = \frac{\mathrm{i}\eta k^2 I\,\mathrm{d}l}{4\pi\nu_b\nu^2}\rho$$
$$\times \left\{ n(n+1)\rho^{-1}\left(\frac{\nu_b}{\nu}\right)^n + (n+1)^2\rho^n\left(\frac{\nu_a}{\nu}\right)^n + \left[(n+1)^2\rho^n\left(\frac{\nu_a}{\nu}\right)^n \right.\right.$$
$$+ n(n+1)\rho^n\left(\frac{\nu}{\nu_a}\right)^{(n+1)} + n^2\rho^{-(n+1)}\left(\frac{\nu}{\nu_a}\right)^{(n+1)}$$
$$\left.\left. + n(n+1)\left(\frac{\nu_a}{\nu}\right)^n \rho^{-(n+1)} \right] \sum_{m=1}^{20} \tau^{m(2n+1)} \right\}. \tag{3.54}$$

Then, it follows that

$$\sum_{n=0}^{\infty} \tilde{E}_{rn} P_n(\cos\theta) = \frac{\mathrm{i}\eta k^2 I\,\mathrm{d}l}{4\pi \nu_b \nu^2}\rho \cdot \left\{ \left[\frac{\tau_1(x + x^2\tau_1 - x\tau_1^2 - 2\tau_1 + \tau_1^3)}{g_1^5} \right.\right.$$
$$\left. + 2\frac{\tau_1(x-\tau_1)}{g_1^3} + \frac{1}{g_1} \right]$$
$$+ \frac{1}{\rho}\left[\frac{\tau_2(x + x^2\tau_2 - x\tau_2^2 - 2\tau_2 + \tau_2^3)}{g_2^5} \right.$$
$$\left.\left. + \frac{\tau_2(x-\tau_2)}{g_2^3} \right] + K \right\}, \tag{3.55}$$

where

$$K = M; \qquad x = \cos\theta, \tag{3.56}$$

$$\tau_1 = \frac{\rho \nu_a}{\nu}; \qquad g_1 = \sqrt{1 - 2x\tau_1 + \tau_1^2}, \tag{3.57}$$

$$\tau_2 = \frac{\nu_b}{\nu}; \qquad g_2 = \sqrt{1 - 2x\tau_2 + \tau_2^2}. \tag{3.58}$$

By using the above method, the approximated formulas for $\sum \tilde{E}_{rn} P_n(\cos\theta)$ are derived readily. It is noted that $\tilde{E}_{rn}$ is obtained from the formulas for E_{rn} in the case of $n \gg \nu$ under the ideal condition of $\Delta_g = \Delta_i = 0$. Obviously, the formulas for $\sum \tilde{E}_{rn} P_n(\cos\theta)$ can be calculated easily. In general, compared with calculating directly the series $\sum E_{rn} P_n(\cos\theta)$, the convergence speed for calculating the series $\sum (E_{rn} - \tilde{E}_{rn}) P_n(\cos\theta)$ is improved greatly. The proposed algorithm has been demonstrated by practical numerical calculation.

3.2.2.2 Speed-up Numerical Convergence Algorithm for the Component E_θ

The electric field component E_θ is written in the following form:

$$E_\theta = \sum_{n=0}^{\infty} (E_{\theta n} - \tilde{E}_{\theta n}) P_n^1(\cos\theta) + \sum_{n=0}^{\infty} \tilde{E}_{\theta n} P_n^1(\cos\theta). \tag{3.59}$$

The latter series term in Eq. (3.59) can be solved analytically. The term $(E_{\theta n} - \tilde{E}_{\theta n})$ is reduced rapidly with the increase of n, and the first series in Eq. (3.59) will converge rapidly. As a result, the field component E_θ can be calculated efficiently with better accuracy.

When $\nu < \nu_b$, the series in Eq. (3.23) consist of two terms: $(2n+1)\hat{J}_n'(\nu)\hat{H}_n(\nu_b)$ and $(2n+1)[b_n \hat{J}_n'(\nu) + c_n \hat{H}_n'(\nu)]$. When $\nu \ll n$, from the series expression of the circular cylinder function near the zero point, we derive readily:

$$(2n+1)\hat{J}_n'(\nu)\hat{H}_n(\nu_b) \Rightarrow -\mathrm{i}(n+1)\left(\frac{\nu}{\nu_b}\right)^n. \tag{3.60}$$

In order to obtain the asymptotic expression of $(2n+1)[b_n\hat{J}'_n(\nu)+c_n\hat{H}'_n(\nu)]$, we take the ideal condition of $\Delta_g=\Delta_i=0$. At this time, the coefficients b_n and c_n are reduced to Eqs. (3.33) and (3.34), respectively.

When $n\gg\nu$, we get

$$\begin{aligned}T_n &= (2n+1)\left[b_n^0\hat{J}'_n(\nu)+c_n^0\hat{H}'_n(\nu)\right]\\ &= \frac{\mathrm{i}(n+1)\nu_a}{\nu}\left(\frac{\nu_a}{\nu}\right)^n\rho^n+\left[-\mathrm{i}(n+1)\left(\frac{\nu}{\nu_c}\right)^n\rho^n+\frac{\mathrm{i}(n+1)\nu_a}{\nu}\left(\frac{\nu_a}{\nu}\right)^n\rho^n\right.\\ &\quad\left.-\,\mathrm{i}n\left(\frac{\nu}{\nu_a}\right)^n\rho^{-(n+1)}+\frac{\mathrm{i}n\nu_a}{\nu}\left(\frac{\nu_a}{\nu}\right)^n\rho^{-(n+1)}\right]\sum_{m=1}^{\infty}\tau^{m(2n+1)}.\end{aligned}\tag{3.61}$$

If the last term in Eq. (3.61) is cut at $m_{\max}=20$, the result becomes

$$\begin{aligned}T_n &\approx \frac{\mathrm{i}(n+1)\nu_a}{\nu}\left(\frac{\nu_a}{\nu}\right)^n\rho^n+\left[-\mathrm{i}(n+1)\left(\frac{\nu}{\nu_c}\right)^n\rho^n+\frac{\mathrm{i}(n+1)\nu_a}{\nu}\left(\frac{\nu_a}{\nu}\right)^n\rho^n\right.\\ &\quad\left.-\,\mathrm{i}n\left(\frac{\nu}{\nu_a}\right)^n\rho^{-(n+1)}+\frac{\mathrm{i}n\nu_a}{\nu}\left(\frac{\nu_a}{\nu}\right)^n\rho^{-(n+1)}\right]\sum_{m=1}^{20}\tau^{m(2n+1)}.\end{aligned}\tag{3.62}$$

From Eqs. (3.60) and (3.62), we readily have

$$(2n+1)\hat{J}'_n(\nu)\hat{H}_n(\nu_b)+T_n\Rightarrow-\mathrm{i}(n+1)\left(\frac{\nu}{\nu_b}\right)^n+T_n.\tag{3.63}$$

Letting

$$E_{\theta n}=\frac{\eta k^2I\,\mathrm{d}l}{4\pi\nu_b^2\nu}(2n+1)\left\{\left[\hat{H}_n(\nu_b)+b_n^0\right]\hat{J}'_n(\nu)+c_n^0h'_n(\nu)\right\},\tag{3.64}$$

it is seen that $\tilde{E}_{\theta n}$ is the asymptotic approximation of $E_{\theta n}$. Namely,

$$E_{\theta n}|_{n\to\infty}\Rightarrow\tilde{E}_{\theta n}.\tag{3.65}$$

From Eqs. (3.63)–(3.65), we write

$$\tilde{E}_{\theta n}=\frac{\eta k^2I\,\mathrm{d}l}{4\pi\nu_b^2\nu}\left[-\mathrm{i}(n+1)\left(\frac{\nu}{\nu_b}\right)^n+T_n\right].\tag{3.66}$$

Thus, we have

$$\sum_{n=0}^{\infty}\tilde{E}_{\theta n}P_n^1(\cos\theta)=\frac{\eta k^2I\,\mathrm{d}l}{4\pi\nu_b^2\nu}\sum_{n=0}^{\infty}\left[-\mathrm{i}(n+1)\left(\frac{\nu}{\nu_b}\right)^n+T_n\right]P_n^1(\cos\theta).\tag{3.67}$$

Considering the following relations:

$$\sum_{n=0}^{\infty}n\tau^nP_n^1(x)=\frac{y\tau}{g^3}+\frac{3y\tau^2(x-\tau)}{g^5},\tag{3.68}$$

$$\sum_{n=0}^{\infty}(n+1)\tau^n P_n^1(x) = \frac{2y\tau}{g^3} + \frac{3y\tau^2(x-\tau)}{g^5}, \tag{3.69}$$

where

$$x = \cos\theta; \qquad y = \sqrt{1-x^2}; \qquad g = \sqrt{1-2x\tau+\tau^2}, \tag{3.70}$$

it follows that

$$\sum_{n=0}^{\infty}\tilde{E}_{\theta n}P_n^1(\cos\theta) = \frac{\eta k^2 I\,\mathrm{d}l}{4\pi\nu_b^2\nu}\left\{-\mathrm{i}\left[\frac{2y\tau_1}{g_1^3} + \frac{3y\tau_1^2(x-\tau_1)}{g_1^5}\right] + N\right\}, \tag{3.71}$$

where

$$\begin{aligned} N = {} & \frac{\mathrm{i}\nu_a}{\nu}\left[\frac{y\tau_2}{g_2^3} + \frac{3y\tau_2^2(x-\tau_2)}{g_2^5}\right] \\ & + \sum_{m=1}^{20}\tau^m\left\{-\mathrm{i}\left[\frac{2y\tau_3}{g_3^3} + \frac{3y\tau_3^2(x-\tau_3)}{g_3^5}\right]\right. \\ & + \frac{\mathrm{i}\nu_a}{\nu}\left[\frac{2y\tau_4}{g_4^3} + \frac{3y\tau_4(x-\tau_4)}{g_4^5}\right] - \frac{\mathrm{i}}{\rho}\left[\frac{y\tau_5}{g_5^3} + \frac{3y\tau_5^2(x-\tau_5)}{g_5^5}\right] \\ & \left. + \frac{\mathrm{i}\nu_a}{\rho\nu}\left[\frac{y\tau_6}{g_6^3} + \frac{3y\tau_6^2(x-\tau_6)}{g_6^5}\right]\right\}, \end{aligned} \tag{3.72}$$

and

$$\tau_1 = \frac{\nu}{\nu_b}; \qquad g_1 = \sqrt{1-2x\tau_1+\tau_1^2}, \tag{3.73}$$

$$\tau_2 = \frac{\rho\nu_a}{\nu}; \qquad g_2 = \sqrt{1-2x\tau_2+\tau_2^2}, \tag{3.74}$$

$$\tau_3 = \frac{\rho\nu}{\nu_a}\tau^{2m}; \qquad g_3 = \sqrt{1-2x\tau_3+\tau_3^2}, \tag{3.75}$$

$$\tau_4 = \frac{\rho\nu_a}{\nu}\tau^{2m}; \qquad g_4 = \sqrt{1-2x\tau_4+\tau_4^2}, \tag{3.76}$$

$$\tau_5 = \frac{\nu}{\rho\nu_a}\tau^{2m}; \qquad g_5 = \sqrt{1-2x\tau_5+\tau_5^2}, \tag{3.77}$$

$$\tau_6 = \frac{\nu_a}{\rho\nu}\tau^{2m}; \qquad g_6 = \sqrt{1-2x\tau_6+\tau_6^2}. \tag{3.78}$$

When $\nu > \nu_b$, the series in Eq. (3.23) consist of two terms: $(2n+1)\hat{J}_n(\nu_b)\hat{H}_n'(\nu)$ and $(2n+1)[b_n^0\hat{J}_n'(\nu) + c_n^0\hat{H}_n'(\nu)]$. When $\nu \ll n$, from the series expression of the circular cylinder function near the zero point, we readily have

$$(2n+1)\hat{J}_n(\nu_b)\hat{H}_n'(\nu) \Rightarrow \mathrm{i}n\left(\frac{\nu_b}{\nu}\right)^{n+1}. \tag{3.79}$$

From Eqs. (3.62) and (3.79), we have

$$(2n+1)\hat{J}_n(\nu_b)\hat{H}_n'(\nu) + T_n \Rightarrow \mathrm{i}n\left(\frac{\nu_b}{\nu}\right)^{n+1} + T_n. \tag{3.80}$$

Considering

$$E_{\theta n} = \frac{\eta k^2 I\,\mathrm{d}l}{4\pi\nu_b^2\nu}(2n+1)\left\{b_n^0\hat{J}_n'(\nu) + \left[\hat{J}_n'(\nu_b) + c_n^0\right]\hat{H}_n'(\nu)\right\}, \tag{3.81}$$

$$E_{\theta n}|_{n\to\infty} \Rightarrow \tilde{E}_{\theta n}, \tag{3.82}$$

and by using (3.80)–(3.82), we readily have

$$\tilde{E}_{\theta n} = \frac{\eta k^2 I\,\mathrm{d}l}{4\pi\nu_b^2\nu}\left[\mathrm{i}n\left(\frac{\nu_b}{\nu}\right)^{n+1} + T_n\right], \tag{3.83}$$

$$\sum_{n=0}^{\infty}\tilde{E}_{\theta n}P_n^1(\cos\theta) = \frac{\eta k^2 I\,\mathrm{d}l}{4\pi\nu_b^2\nu}\sum_{n=0}^{\infty}\left[\mathrm{i}n\left(\frac{\nu_b}{\nu}\right)^{n+1} + T_n\right]P_n^1(\cos\theta). \tag{3.84}$$

Considering the following relations:

$$\sum_{n=0}^{\infty}n\tau^n P_n^1(x) = \frac{y\tau}{g^3} + \frac{3y\tau^2(x-\tau)}{g^5}, \tag{3.85}$$

$$\sum_{n=0}^{\infty}(n+1)\tau^n P_n^1(x) = \frac{2y\tau}{g^3} + \frac{3y\tau^2(x-\tau)}{g^5}, \tag{3.86}$$

where

$$x = \cos\theta; \qquad y = \sqrt{1-x^2}; \qquad g = \sqrt{1-2x\tau+\tau^2}, \tag{3.87}$$

it follows that

$$\sum_{n=0}^{\infty}\tilde{E}_{\theta n}P_n^1(\cos\theta) = \frac{\eta k^2 I\,\mathrm{d}l}{4\pi\nu_b^2\nu}\left\{\frac{\mathrm{i}\nu_b}{\nu}\left[\frac{y\tau_1}{g_1^3} + \frac{3y\tau_1^2(x-\tau_1)}{g_1^5}\right] + N'\right\}, \tag{3.88}$$

where

$$N' = N; \qquad \tau_1 = \frac{\nu_b}{\nu}; \qquad g_1 = \sqrt{1-2x\tau_1+\tau_1^2}. \tag{3.89}$$

So far, $\tilde{E}_{\theta n}$ has been obtained analytically in both cases of $\nu > \nu_b$ and $\nu < \nu_b$, while the series $\sum\tilde{E}_{\theta n}P_n^1(\cos\theta)$ can be expressed in analytical form. As a result, it is seen that the calculation efficiency for the field component can be improved greatly by using the proposed algorithm.

3.2.3 Evaluations for $\hat{J}_n(\nu)$, $\hat{H}_n(\nu)$, $P_n(\nu)$, and $P_n^1(\nu)$

In order to compute the first series terms in Eqs. (3.31) and (3.59), it is necessary to evaluate the functions $\hat{J}_n(\nu)$, $\hat{H}_n(\nu)$, $P_n(\nu)$, and $P_n^1(\nu)$ firstly. It is noted that, for sufficiently large n, the function $\hat{J}_n(\nu)$ becomes exceedingly small, and the function $\hat{H}_n(\nu)$ becomes exceedingly large, but their product $\hat{J}_n(\nu) \times \hat{H}_n(\nu)$ remains well within machine bounds. Next, we will consider the lower-order and higher-order cases, respectively.

For lower-order n, beginning directly with the lowest-order terms, which are known functions:

$$\hat{N}_0(\nu) = -\cos\nu; \qquad \hat{N}_1(\nu) = -\frac{\cos\nu}{\nu} - \sin\nu, \tag{3.90}$$

$$\hat{J}_0(\nu) = \sin\nu; \qquad \hat{J}_1(\nu) = \frac{\sin\nu}{\nu} - \cos\nu, \tag{3.91}$$

the nth terms can be obtained by the following recurrence relations:

$$\hat{N}_{n+1}(\nu) = \frac{2n+1}{\nu}\hat{N}_n(\nu) - \hat{N}_{n-1}(\nu), \tag{3.92}$$

$$\hat{J}_{n+1}(\nu) = \frac{2n+1}{\nu}\hat{J}_n(\nu) - \hat{J}_{n-1}(\nu), \tag{3.93}$$

$$\hat{H}_n(\nu) = \hat{J}_n(\nu) + \mathrm{i}\hat{N}_n(\nu), \tag{3.94}$$

$$\hat{J}_n'(\nu) = \hat{J}_{n-1}(\nu) - \frac{n}{\nu}\hat{J}_n(\nu), \tag{3.95}$$

$$\hat{H}_n'(\nu) = \hat{H}_{n-1}(\nu) - \frac{n}{\nu}\hat{H}_n(\nu). \tag{3.96}$$

For higher-order n, the products of Bessel functions (or their derivatives) can be obtained by using the following formulas (Gradshteyn and Ryzhik 1980):

$$\begin{aligned}
\hat{J}_n(\nu_1)\hat{H}_n(\nu_2) \approx -\mathrm{i}&\left\{\frac{1}{2}\sqrt{\frac{f_n(\nu_1)}{S_n(\nu_1)}} + \frac{1}{16}\left[\frac{f_n(\nu_1)}{S_n(\nu_1)}\right]^{\frac{3}{2}}\nu_1^{-1}\right. \\
&\left. - \frac{5}{48}\left(n+\frac{1}{2}\right)^2\left[\frac{f_n(\nu_1)}{S_n(\nu_1)}\right]^{\frac{7}{2}}\nu_1^{-3}\right\} \\
&\times\left\{\sqrt{\frac{f_n(\nu_2)}{S_n(\nu_2)}} - \frac{1}{8}\left[\frac{f_n(\nu_2)}{S_n(\nu_2)}\right]^{\frac{3}{2}}\nu_2^{-1}\right. \\
&\left. + \frac{5}{24}\left(n+\frac{1}{2}\right)^2\left[\frac{f_n(\nu_2)}{S_n(\nu_2)}\right]^{\frac{7}{2}}\nu_2^{-3}\right\}\cdot \mathrm{e}^{a_n(\nu_1)-a_n(\nu_2)},
\end{aligned} \tag{3.97}$$

$$
\begin{aligned}
\hat{J}_n'(\nu_1)\hat{H}_n(\nu_2) \approx -\mathrm{i}\Bigg\{ & \Bigg[\frac{1}{4}\sqrt{\frac{S_n(\nu_1)}{f_n(\nu_1)}} + \frac{3}{32}\sqrt{\frac{f_n(\nu_1)}{S_n(\nu_1)}}\nu_1^{-1} \\
& - \frac{35}{96}\left(n+\frac{1}{2}\right)^2 \nu_1^{-3}\left(\frac{f_n(\nu_1)}{S_n(\nu_1)}\right)^{\frac{5}{2}}\Bigg] \\
& \times \left[\frac{1}{(n+\frac{1}{2})S_n(\nu_1)} + \frac{f_n^2(\nu_1)}{(n+\frac{1}{2})^2 S_n^3(\nu_1)}\right] \\
& - \frac{1}{16}\left(\frac{f_n(\nu_1)}{S_n(\nu_1)}\right)^{\frac{3}{2}}\nu_1^{-2} + \frac{5}{16}\left(n+\frac{1}{2}\right)^2\nu_1^{-4}\left(\frac{f_n(\nu_1)}{S_n(\nu_1)}\right)^{\frac{7}{2}} \\
& + \Bigg[\frac{1}{2}\sqrt{\frac{f_n(\nu_1)}{S_n(\nu_1)}} + \frac{1}{16}\left(\frac{f_n(\nu_1)}{S_n(\nu_1)}\right)^{\frac{3}{2}}\nu_1^{-1} \\
& - \frac{5}{48}\left(n+\frac{1}{2}\right)^2\nu_1^{-3}\left(\frac{f_n(\nu_1)}{S_n(\nu_1)}\right)^{\frac{7}{2}}\Bigg]\frac{f_n(\nu_1)S_n(\nu_1)}{1-S_n^2(\nu_1)}\Bigg\} \\
& \times \Bigg[\sqrt{\frac{f_n(\nu_2)}{S_n(\nu_2)}} - \frac{1}{8}\left(\frac{f_n(\nu_2)}{S_n(\nu_2)}\right)^{\frac{3}{2}}\nu_2^{-1} \\
& + \frac{5}{24}\left(n+\frac{1}{2}\right)^2\left(\frac{f_n(\nu_2)}{S_n(\nu_2)}\right)^{\frac{7}{2}}\nu_2^{-3}\Bigg]\cdot \mathrm{e}^{a_n(\nu_1)-a_n(\nu_2)},
\end{aligned}
\tag{3.98}
$$

$$
\begin{aligned}
\hat{J}_n(\nu_1)\hat{H}_n'(\nu_2) \approx -\mathrm{i}\Bigg[& \frac{1}{2}\sqrt{\frac{f_n(\nu_1)}{S_n(\nu_1)}} + \frac{1}{16}\left(\frac{f_n(\nu_1)}{S_n(\nu_1)}\right)^{\frac{3}{2}}\nu_1^{-1} \\
& - \frac{5}{48}\left(n+\frac{1}{2}\right)^2\left(\frac{f_n(\nu_1)}{S_n(\nu_1)}\right)^{\frac{7}{2}}\nu_1^{-3}\Bigg] \\
& \times \Bigg\{\Bigg[\frac{1}{2}\sqrt{\frac{S_n(\nu_2)}{f_n(\nu_2)}} - \frac{3}{16}\sqrt{\frac{f_n(\nu_2)}{S_n(\nu_2)}}\nu_2^{-1} \\
& + \frac{35}{48}\left(n+\frac{1}{2}\right)^2\nu_2^{-3}\left(\frac{f_n(\nu_2)}{S_n(\nu_2)}\right)^{\frac{5}{2}}\Bigg] \\
& \times \left[\frac{1}{(n+\frac{1}{2})S_n(\nu_2)} + \frac{f_n^2(\nu_2)}{(n+\frac{1}{2})^2 S_n^3(\nu_2)}\right] \\
& + \frac{1}{8}\left(\frac{f_n(\nu_2)}{S_n(\nu_2)}\right)^{\frac{3}{2}}\nu_2^{-2} - \frac{5}{8}\left(n+\frac{1}{2}\right)^2\nu_2^{-4}\left(\frac{f_n(\nu_2)}{S_n(\nu_2)}\right)^{\frac{7}{2}} \\
& + \Bigg[\sqrt{\frac{f_n(\nu_2)}{S_n(\nu_2)}} - \frac{1}{8}\left(\frac{f_n(\nu_2)}{S_n(\nu_2)}\right)^{\frac{3}{2}}\nu_2^{-1}
\end{aligned}
$$

$$+\frac{5}{24}\left(n+\frac{1}{2}\right)^2\nu_2^{-3}\left(\frac{f_n(\nu_2)}{S_n(\nu_2)}\right)^{\frac{7}{2}}\Bigg]$$

$$\times\frac{f_n(\nu_2)S_n(\nu_2)}{1-S_n^2(\nu_2)}\Bigg\}\cdot \mathrm{e}^{a_n(\nu_1)-a_n(\nu_2)}, \tag{3.99}$$

$$\hat{J}_n'(\nu_1)\hat{H}_n'(\nu_2)\approx -\mathrm{i}\Bigg\{\Bigg[\frac{1}{4}\sqrt{\frac{S_n(\nu_1)}{f_n(\nu_1)}}+\frac{3}{32}\sqrt{\frac{f_n(\nu_1)}{S_n(\nu_1)}}\nu_1^{-1}$$

$$-\frac{35}{96}\left(n+\frac{1}{2}\right)^2\nu_1^{-3}\left(\frac{f_n(\nu_1)}{S_n(\nu_1)}\right)^{\frac{5}{2}}\Bigg]$$

$$\times\left[\frac{1}{(n+\frac{1}{2})S_n(\nu_1)}+\frac{f_n^2(\nu_1)}{(n+\frac{1}{2})^2S_n^3(\nu_1)}\right]$$

$$-\frac{1}{16}\left(\frac{f_n(\nu_1)}{S_n(\nu_1)}\right)^{\frac{3}{2}}\nu_1^{-2}+\frac{5}{16}\left(n+\frac{1}{2}\right)^2\nu_1^{-4}\left(\frac{f_n(\nu_1)}{S_n(\nu_1)}\right)^{\frac{7}{2}}$$

$$+\Bigg[\frac{1}{2}\sqrt{\frac{f_n(\nu_1)}{S_n(\nu_1)}}+\frac{1}{16}\left(\frac{f_n(\nu_1)}{S_n(\nu_1)}\right)^{\frac{3}{2}}\nu_1^{-1}$$

$$-\frac{5}{48}\left(n+\frac{1}{2}\right)^2\nu_1^{-3}\left(\frac{f_n(\nu_1)}{S_n(\nu_1)}\right)^{\frac{7}{2}}\Bigg]\frac{f_n(\nu_1)S_n(\nu_1)}{1-S_n^2(\nu_1)}\Bigg\}$$

$$\times\Bigg\{\Bigg[\frac{1}{2}\sqrt{\frac{S_n(\nu_2)}{f_n(\nu_2)}}-\frac{3}{16}\sqrt{\frac{f_n(\nu_2)}{S_n(\nu_2)}}\nu_2^{-1}$$

$$+\frac{35}{48}\left(n+\frac{1}{2}\right)^2\nu_2^{-3}\left(\frac{f_n(\nu_2)}{S_n(\nu_2)}\right)^{\frac{5}{2}}\Bigg]$$

$$\times\left[\frac{1}{(n+\frac{1}{2})S_n(\nu_2)}+\frac{f_n^2(\nu_2)}{(n+\frac{1}{2})^2S_n^3(\nu_2)}\right]$$

$$+\frac{1}{8}\left(\frac{f_n(\nu_2)}{S_n(\nu_2)}\right)^{\frac{3}{2}}\nu_2^{-2}-\frac{5}{8}\left(n+\frac{1}{2}\right)^2\nu_2^{-4}\left(\frac{f_n(\nu_2)}{S_n(\nu_2)}\right)^{\frac{7}{2}}$$

$$+\Bigg[\sqrt{\frac{f_n(\nu_2)}{S_n(\nu_2)}}-\frac{1}{8}\left(\frac{f_n(\nu_2)}{S_n(\nu_2)}\right)^{\frac{3}{2}}\nu_2^{-1}$$

$$+\frac{5}{24}\left(n+\frac{1}{2}\right)^2\nu_2^{-3}\left(\frac{f_n(\nu_2)}{S_n(\nu_2)}\right)^{\frac{7}{2}}\Bigg]$$

$$\times\frac{f_n(\nu_2)S_n(\nu_2)}{1-S_n^2(\nu_2)}\Bigg\}\cdot \mathrm{e}^{a_n(\nu_1)-a_n(\nu_2)}, \tag{3.100}$$

where

$$f_n(\nu) = \frac{\nu}{n+\frac{1}{2}}; \qquad S_n(\nu) = \sqrt{1 - f_n^2(\nu)}, \tag{3.101}$$

$$a_n(\nu) = \left(n + \frac{1}{2}\right)\left\{S_n(\nu) - \operatorname{arctan} h\left[S_n(\nu)\right]\right\}. \tag{3.102}$$

The Legendre functions $P_n(x)$ and associated Legendre functions $P_n^1(x)$ can be obtained by the following recurrence relations:

$$P_{n+1}(x) = \frac{2n+1}{n+1} x P_n(x) - \frac{n}{n+1} P_{n-1}(x), \tag{3.103}$$

$$P_{n+1}^1(x) = \frac{2n+1}{n} x P_n^1(x) - \frac{n+1}{n} P_{n-1}^1(x). \tag{3.104}$$

For $n = 0, 1, 2$, the lowest-order terms for the Legendre function $P_n(x)$ and associated Legendre function $P_n^1(x)$ are well known, and they can be written as follows:

$$P_0(x) = 1; \qquad P_1(x) = x; \qquad P_2(x) = \frac{1}{2}\left(3x^2 - 1\right), \tag{3.105}$$

$$P_0^1(x) = 0; \qquad P_1^1(x) = y; \qquad P_2^1(x) = 3xy, \tag{3.106}$$

where $x = \cos\theta$, $y = \sin\theta$.

3.2.4 Computations and Discussions

In this section, the computations are carried out for SLF/ELF field components due to the unit VED in the presence of a non-ideal Earth–ionosphere waveguide or cavity. The radius of the Earth is taken to be $a = 6{,}370$ km, so that the Earth's circumference is about 40,000 km or so. That is to say, the propagation distance between the dipole source and its antipole is about 20,000 km or so. Assuming that both the dipole source and the observation point are located on the Earth's surface, magnitudes of the field components $|E_r|$ and $|E_\theta|$ versus the propagation distance between the dipole source and its antipole are computed at $f = 100$ Hz by using the speed-up numerical convergence algorithm and plotted in Figs. 3.2 and 3.3, respectively. In order to verify the correctness of this algorithm, with the same parameters, the computations are also carried out by using the second-order spherical approximation algorithm, this traditional algorithm in SLF range is addressed in the preceding chapter, and plotted in the same figures. From these results, it is seen that the numerical results by the speed-up numerical convergence algorithm are in agreement with those by the second-order spherical approximation algorithm except for the nearby of the antipole. Obviously, there exists mutual interference between the two waves along the short-arc path and large-arc path in the

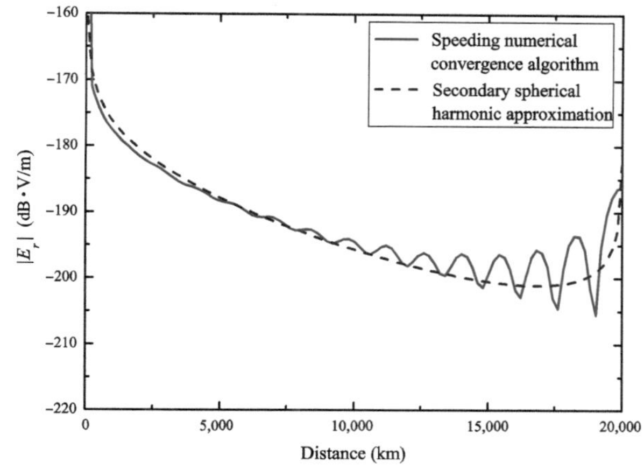

Fig. 3.2 The electric field E_r in dB·V/m due to a VED versus the propagation distance: $f = 100$ Hz, $I\,dl = 1$ A·m, $a = 6{,}370$ km, $h = 80$ km, $\sigma_g = 2 \times 10^{-4}$ S/m, $\sigma_i = 10^{-5}$ S/m

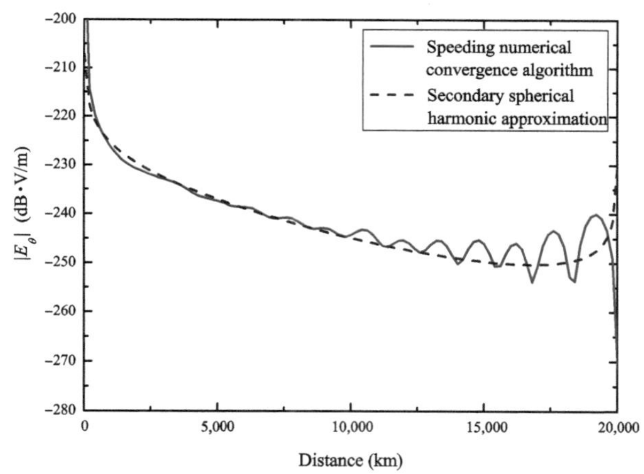

Fig. 3.3 The electric field E_θ in dB·V/m due to a VED versus the propagation distance: $f = 100$ Hz, $I\,dl = 1$ A·m, $a = 6{,}370$ km, $h = 80$ km, $\sigma_g = 2 \times 10^{-4}$ S/m, $\sigma_i = 10^{-5}$ S/m

zone near the antipole of the dipole source. In the asymptotic approximation of the radial propagation factor $P_\nu(\cos(\pi - \theta))$ in the waveguide or cavity for the traditional algorithm, only the short-arc path is considered, which the long circular path is neglected. So that the mutual interference does not appear in the numerical results by using the second-order spherical approximation algorithm. With the same parameters in Figs. 3.2 and 3.3, the magnitudes of the components $|E_r|$ and $|E_\theta|$ versus the height of the observation point are computed and plotted in Figs. 3.4 and 3.5, respectively.

From these results in Figs. 3.3–3.5, it is concluded as follows:

- The numerical results in SLF range by using the above two algorithm are in agreement with each other. And the correctness of the traditional algorithm using the second spherical approximation has been verified by the available experimental results in USA and Russia.
- When the observation point is located in the zone near the antipole of the dipole source, it is unsuitable to neglect the electromagnetic wave along the large-arc

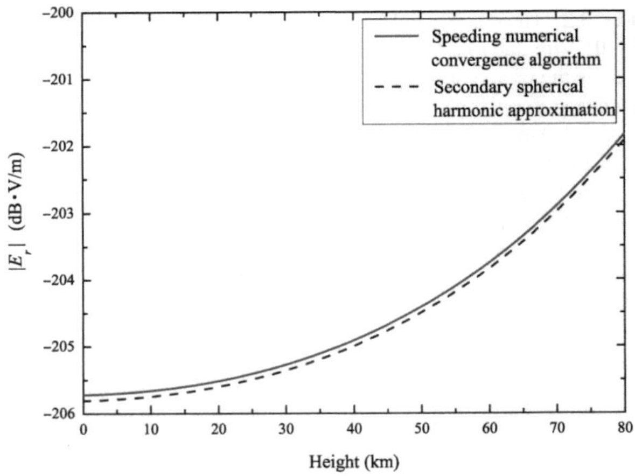

Fig. 3.4 The electric field E_r in dB·V/m due to a VED versus the height of the observation point: $f = 100$ Hz, $I\,dl = 1$ A·m, $\theta = \frac{\pi}{3}$, $a = 6{,}370$ km, $h = 80$ km, $\sigma_g = 2 \times 10^{-4}$ S/m, $\sigma_i = 10^{-5}$ S/m

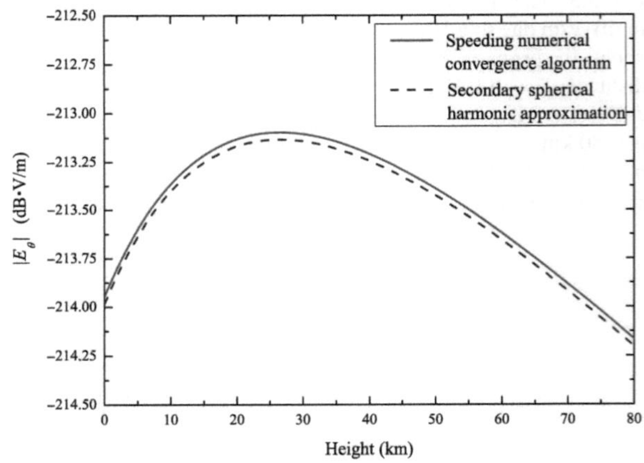

Fig. 3.5 The electric field E_θ in dB·V/m due to a VED versus the height of the observation point: $f = 100$ Hz, $I\,dl = 1$ A·m, $\theta = \frac{\pi}{3}$, $a = 6{,}370$ km, $h = 80$ km, $\sigma_g = 2 \times 10^{-4}$ S/m, $\sigma_i = 10^{-5}$ S/m

path. Therefore, the speed-up numerical convergence algorithm has the advantage to deal with the problem in case of the observation point being in the zone near the antipole of the dipole source.

- For the numerical convergence algorithm, the operating frequency is not limited in the derivation process, so that this algorithm is permitted to use in the lower frequency range, especially in ELF range.

The magnitudes of the components E_r and E_θ are computed at $f = 100$ Hz, 60 Hz, 30 Hz, and 10 Hz as shown in Figs. 3.6 and 3.7, respectively. It is seen that, with decreasing frequency, the wavelength is gradually close to or even exceeds the Earth's circumference, and the electromagnetic wave in the waveguide is a "standing wave", of which the peaks and troughs appear clearly. In Fig. 3.8, magnitudes of the field component E_r versus the operating frequency are shown at $\theta = \frac{\pi}{2}, \frac{\pi}{3}, \frac{\pi}{4}$, and $\frac{\pi}{5}$, respectively. It is found that a maximum of the field intensity appears at the nearby values of $f = 8.8$ Hz, 16.5 Hz, and 22.5 Hz.

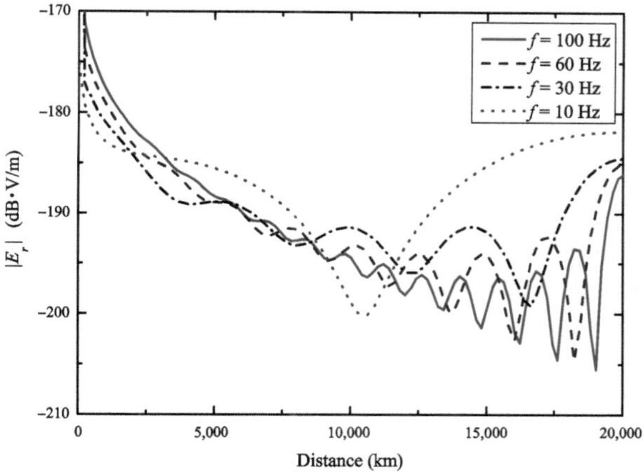

Fig. 3.6 The electric field E_r in dB·V/m due to a VED versus the propagation distance at $f = 100$ Hz, 60 Hz, 30 Hz, and 10 Hz with $I\,dl = 1$ A·m, $a = 6{,}370$ km, $h = 80$ km, $\sigma_g = 2 \times 10^{-4}$ S/m, $\sigma_i = 10^{-5}$ S/m

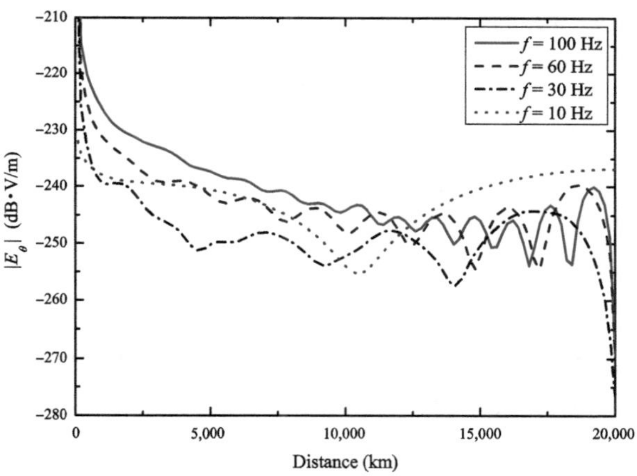

Fig. 3.7 The electric field E_θ in dB·V/m due to a VED versus the propagation distance at $f = 100$ Hz, 60 Hz, 30 Hz, and 10 Hz with $I\,dl = 1$ A·m, $a = 6{,}370$ km, $h = 80$ km, $\sigma_g = 2 \times 10^{-4}$ S/m, $\sigma_i = 10^{-5}$ S/m

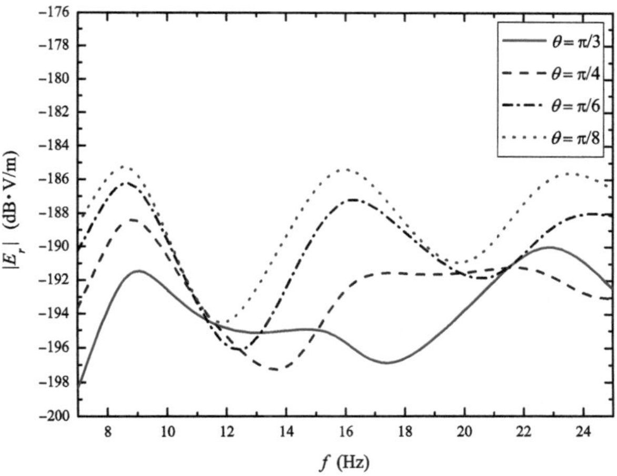

Fig. 3.8 The electric field E_r in dB·V/m due to a VED versus the operating frequency with $I\,dl = 1$ A·m, $a = 6{,}370$ km, $h = 80$ km, $\sigma_g = 2 \times 10^{-4}$ S/m, $\sigma_i = 10^{-5}$ S/m

Table 3.1 The first four resonant frequencies

Number n	1	2	3	4
Ideal lossless cavity	10.6 Hz	18.3 Hz	25.9 Hz	33.5 Hz
Non-ideal cavity	7.6 Hz	13.6 Hz	19.4 Hz	25.4 Hz

In the 1950s, Schumann predicted that there exists a resonant phenomenon in the space between the ground and the lower boundary of the ionosphere. In his pioneering work (Schumann 1957), from the waveguide wave propagation theory, the resonant frequency of the Schumann resonance on the Earth–ionosphere cavity was obtained readily. It is estimated that

$$f_n \approx \frac{7.5 \times \sqrt{n(n+1)}}{\operatorname{Re} S}, \tag{3.107}$$

where n refers to the nth resonant frequency, and S refers to the ratio between the wave number of the main propagation mode and that of free space. According to this theory, the four resonant frequencies are given in Table 3.1.

From atmospheric noise spectrum in ELF range, the measured resonant frequencies are 7.5 Hz, 14.5 Hz, 20 Hz, and 27.5 Hz, respectively (Larsen and Egeland 1968). These measured results are also broadly consistent with the calculated results by using the speed-up numerical convergence algorithm.

3.3 SLF/ELF Fields of HED in the Earth–Ionosphere Waveguide/Cavity

In this section, we will treat the SLF/ELF electromagnetic field of an HED in the Earth–ionosphere waveguide or cavity in detail.

3.3.1 Spherical Harmonic Series Solution for SLF/ELF Fields of HED in the Earth–Ionosphere Waveguide/Cavity

In SLF/ELF ranges, the wavelength is very long, and the height of a vertical antenna is very short compared with its wavelength. So that in practical applications, it is impossible to employ a vertical antenna to generate TM waves in SLF/ELF ranges. In general, a SLF/ELF radiation source is usually designed as a horizontal linear antenna. So it is necessary to investigate SLF/ELF electromagnetic field of an HED in the presence of the Earth–ionosphere waveguide or cavity. In Sect. 3.2, the electromagnetic field of a VED in the Earth–ionosphere cavity is treated by using the speed-up numerical series algorithm. In this subsection, we will attempt to obtain a spherical harmonic series solution for SLF/ELF fields of an HED in the Earth–ionosphere waveguide or cavity.

The two reflection walls, the ground and the ionosphere, are characterized by the normalized surface impedances Δ_g and Δ_i, respectively. We assume that an HED is represented by its current moment $\hat{x} I\, dl\delta(x)\delta(y)\delta(z-r_b)$, where $r_b = r_a + z_s$, r_a is the radius of the Earth, and z_s denotes the height of the dipole above the Earth's surface. $r_c = r_a + h$, where h denotes the height of the lower boundary of the ionosphere. Then, the electromagnetic field of an HED in the Earth–ionosphere waveguide or cavity can be represented in terms of the two potential functions U and V (Barrick 1999):

$$E_r = \frac{1}{-\mathrm{i}\omega\varepsilon}\left(\frac{\partial^2}{\partial r^2} + k^2\right)U, \tag{3.108}$$

$$H_r = \frac{1}{-\mathrm{i}\omega\mu}\left(\frac{\partial^2}{\partial r^2} + k^2\right)V, \tag{3.109}$$

$$E_\theta = \frac{-1}{r\sin\theta}\frac{\partial V}{\partial\phi} + \frac{1}{-\mathrm{i}\omega\varepsilon r}\frac{\partial^2 U}{\partial r\partial\theta}, \tag{3.110}$$

$$H_\theta = \frac{1}{r\sin\theta}\frac{\partial U}{\partial\phi} + \frac{1}{-\mathrm{i}\omega\mu r}\frac{\partial^2 V}{\partial r\partial\theta}, \tag{3.111}$$

$$E_\phi = \frac{1}{r}\frac{\partial V}{\partial\theta} + \frac{1}{-\mathrm{i}\omega\varepsilon r\sin\theta}\frac{\partial^2 U}{\partial r\partial\phi}, \tag{3.112}$$

$$H_\phi = \frac{-1}{r}\frac{\partial U}{\partial\theta} + \frac{1}{-\mathrm{i}\omega\mu r\sin\theta}\frac{\partial^2 V}{\partial r\partial\phi}, \tag{3.113}$$

where the two potential functions U and V satisfy the following equations:

$$(\nabla^2 + k^2)\frac{U}{r} = 0, \tag{3.114}$$

$$(\nabla^2 + k^2)\frac{V}{r} = 0. \tag{3.115}$$

When the effects by the ground and the ionosphere are not considered, the potential functions for the primary field radiated by an HED are written in the forms:

$$U_0^{\mathrm{HED}} = \frac{\mathrm{i}kI\,\mathrm{d}l}{4\pi\nu_b}\cos\phi\cdot T_0^c, \tag{3.116}$$

$$V_0^{\mathrm{HED}} = \frac{\mathrm{i}kI\,\mathrm{d}l\eta}{4\pi\nu_b}\sin\phi\cdot T_0^s, \tag{3.117}$$

where

$$T_0^c = \begin{cases} \sum_{n=1}^{\infty} \frac{(2n+1)}{n(n+1)} H_n'(\nu_b)\hat{J}_n(\nu)P_n^1(\cos\theta); & \nu < \nu_b, \\ \sum_{n=1}^{\infty} \frac{(2n+1)}{n(n+1)} j_n'(\nu_b)\hat{H}_n(\nu)P_n^1(\cos\theta); & \nu > \nu_b, \end{cases} \tag{3.118}$$

$$T_0^s = \begin{cases} \sum_{n=1}^{\infty} \frac{(2n+1)}{n(n+1)} \hat{H}_n(\nu_b)\hat{J}_n(\nu)P_n^1(\cos\theta); & \nu < \nu_b, \\ \sum_{n=1}^{\infty} \frac{(2n+1)}{n(n+1)} \hat{J}_n(\nu_b)\hat{H}_n(\nu)P_n^1(\cos\theta); & \nu > \nu_b, \end{cases} \tag{3.119}$$

where $\nu = kr$ and $\nu_b = kr_b$. $P_n^1(x)$ is the associated Legendre function. The function $\hat{J}_n(\nu)$ can be expressed in terms of cylindrical half-order Bessel function of the first kind, while the function $\hat{H}_n(\nu)$ is its corresponding outgoing first-kind Hankel function. The above two functions are defined by Eqs. (3.12) and (3.13), respectively. It is noted that the functions $\hat{J}_n(\nu)$ and $\hat{H}_n(\nu)$ satisfy the following differential equation:

$$\left[\frac{\mathrm{d}^2}{\mathrm{d}\nu^2} + 1 - \frac{n(n+1)}{\nu^2}\right] B_n(\nu) = 0. \tag{3.120}$$

Considering the reflection of the air–ground boundary and that of the air-ionosphere boundary, the secondary disturbance field in the cavity will be produced. The potential functions of the secondary disturbance field can be represented in terms of the spherical harmonic functions. We have

$$U_p^{\mathrm{HED}} = \frac{\mathrm{i}kI\,\mathrm{d}l}{4\pi\nu_b}\cos\phi \times \sum_{n=1}^{\infty} \frac{(2n+1)}{n(n+1)} \left[b_n\hat{J}_n(\nu) + c_n\hat{H}_n(\nu)\right] P_n^1(\cos\theta), \tag{3.121}$$

$$V_p^{\mathrm{HED}} = \frac{\mathrm{i}\eta kI\,\mathrm{d}l}{4\pi\nu_b}\sin\phi \times \sum_{n=1}^{\infty} \frac{(2n+1)}{n(n+1)} \left[d_n\hat{J}_n(\nu) + e_n\hat{H}_n(\nu)\right] P_n^1(\cos\theta). \tag{3.122}$$

Thus, the total potential functions can be written in the following forms:

$$\begin{aligned}
U^{\mathrm{HED}} &= U_0^{\mathrm{HED}} + U_p^{\mathrm{HED}} \\
&= \frac{\mathrm{i}kI\,\mathrm{d}l}{4\pi\nu_b}\cos\phi \\
&\times \begin{cases}
\displaystyle\sum_{n=1}^{\infty}\frac{(2n+1)}{n(n+1)}P_n^1(\cos\theta) \\
\qquad\times\left\{\left[\hat{H}_n'(\nu_b)+b_n\right]\hat{J}_n(\nu)+c_n\hat{H}_n(\nu)\right\}; & \nu<\nu_b, \\
\displaystyle\sum_{n=1}^{\infty}\frac{(2n+1)}{n(n+1)}P_n^1(\cos\theta) \\
\qquad\times\left\{b_n\hat{J}_n(\nu)+\hat{H}_n(\nu)\left[c_n+\hat{J}_n'(\nu_b)\right]\right\}; & \nu>\nu_b,
\end{cases}
\end{aligned} \tag{3.123}$$

$$\begin{aligned}
V^{\mathrm{HED}} &= V_0^{\mathrm{HED}} + V_p^{\mathrm{HED}} \\
&= \frac{\mathrm{i}\eta kI\,\mathrm{d}l}{4\pi\nu_b}\sin\phi \\
&\times \begin{cases}
\displaystyle\sum_{n=1}^{\infty}\frac{(2n+1)}{n(n+1)}P_n^1(\cos\theta) \\
\qquad\times\left\{\left[\hat{H}_n(\nu_b)+d_n\right]\hat{J}_n(\nu)+e_n\hat{H}_n(\nu)\right\}; & \nu<\nu_b, \\
\displaystyle\sum_{n=1}^{\infty}\frac{(2n+1)}{n(n+1)}P_n^1(\cos\theta) \\
\qquad\times\left\{d_n\hat{J}_n(\nu)+\hat{H}_n(\nu)\left[e_n+\hat{J}_n(\nu_b)\right]\right\}; & \nu>\nu_b.
\end{cases}
\end{aligned} \tag{3.124}$$

On the Earth's surface, $r = r_a$, the field components satisfy the impedance boundary conditions. We have

$$E_\theta = -\Delta_g \eta H_\phi|_{r=r_a}; \qquad E_\phi = \Delta_g \eta H_\theta|_{r=r_a}. \tag{3.125}$$

At the lower boundary of the ionosphere, $r = r_c$, we have

$$E_\theta = \Delta_i \eta H_\phi|_{r=r_c}; \qquad E_\phi = -\Delta_i \eta H_\theta,|_{r=r_c}. \tag{3.126}$$

With the substitutions of Eqs. (3.125) and (3.126) into Eqs. (3.108)–(3.113), the field components in the Earth–ionosphere waveguide or cavity can be expressed as follows:

$$E_r = \frac{-\eta k^2 I\,\mathrm{d}l}{4\pi\nu_b\nu^2}\cos\phi\begin{cases}\displaystyle\sum_{n=1}^{\infty}(2n+1)P_n^1(\cos\theta)\\ \quad\times\left\{\left[\hat{H}_n'(\nu_b)+b_n\right]\hat{J}_n(\nu)+c_n\hat{H}_n(\nu)\right\}; \quad \nu<\nu_b,\\ \displaystyle\sum_{n=1}^{\infty}(2n+1)P_n^1(\cos\theta)\\ \quad\times\left\{b_n\hat{J}_n(\nu)+\hat{H}_n(\nu)\left[c_n+\hat{J}_n'(\nu_b)\right]\right\}; \quad \nu>\nu_b,\end{cases} \tag{3.127}$$

$$H_r = \frac{-k^2 I\,\mathrm{d}l}{4\pi\nu_b\nu^2}\sin\phi\begin{cases}\displaystyle\sum_{n=1}^{\infty}(2n+1)P_n^1(\cos\theta)\\ \quad\times\left\{\left[\hat{H}_n(\nu_b)+d_n\right]\hat{J}_n(\nu)+e_n\hat{H}_n(\nu)\right\}; \quad \nu<\nu_b,\\ \displaystyle\sum_{n=1}^{\infty}(2n+1)P_n^1(\cos\theta)\\ \quad\times\left\{d_n\hat{J}_n(\nu)+\hat{H}_n(\nu)\left[e_n+\hat{J}_n(\nu_b)\right]\right\}; \quad \nu>\nu_b,\end{cases} \tag{3.128}$$

$$\begin{aligned}E_\theta = &-\frac{\eta k^2 I\,\mathrm{d}l}{4\pi\nu\nu_b}\cos\phi\sum_{n=1}^{\infty}\frac{(2n+1)}{n(n+1)}\left[n(n+1)P_n(\cos\theta)-\cot\theta P_n^1(\cos\theta)\right]\\ &\times\begin{cases}\left\{\left[\hat{H}_n'(\nu_b)+b_n\right]\hat{J}_n(\nu)+c_n\hat{H}_n'(\nu)\right\}\\ \left\{b_n\hat{J}_n'(\nu)+\hat{H}_n'(\nu)\left[c_n+\hat{J}_n'(\nu_b)\right]\right\}\end{cases}\\ &-\frac{\mathrm{i}\eta k^2 I\,\mathrm{d}l}{4\pi\sin\theta\nu\nu_b}\cos\phi\sum_{n=1}^{\infty}\frac{(2n+1)}{n(n+1)}P_n^1(\cos\theta)\\ &\times\begin{cases}\left\{\left[\hat{H}_n(\nu_b)+d_n\right]\hat{J}_n(\nu)+e_n\hat{H}_n(\nu)\right\}; & \nu<\nu_b,\\ \left\{d_n\hat{J}_n(\nu)+\hat{H}_n(\nu)\left[e_n+\hat{J}_n(\nu_b)\right]\right\}; & \nu>\nu_b,\end{cases}\end{aligned} \tag{3.129}$$

$$\begin{aligned}H_\theta = &-\frac{k^2 I\,\mathrm{d}l}{4\pi\nu\nu_b}\sin\phi\sum_{n=1}^{\infty}\frac{(2n+1)}{n(n+1)}\left[n(n+1)P_n(\cos\theta)-\cot\theta P_n^1(\cos\theta)\right]\\ &\times\begin{cases}\left\{\left[H_n(\nu_b)+d_n\right]\hat{J}_n'(\nu)+e_n\hat{H}_n'(\nu)\right\}\\ \left\{d_n\hat{J}_n'(\nu)+\hat{H}_n'(\nu)\left[e_n+\hat{J}_n(\nu_b)\right]\right\}\end{cases}\\ &-\frac{\mathrm{i}k^2 I\,\mathrm{d}l}{4\pi\sin\theta\nu\nu_b}\sin\phi\sum_{n=1}^{\infty}\frac{(2n+1)}{n(n+1)}P_n^1(\cos\theta)\\ &\times\begin{cases}\left\{\left[\hat{H}_n'(\nu_b)+b_n\right]\hat{J}_n(\nu)+c_n\hat{H}_n(\nu)\right\}; & \nu<\nu_b,\\ \left\{b_n\hat{J}_n(\nu)+\hat{H}_n(\nu)\left[c_n+\hat{J}_n'(\nu_b)\right]\right\}; & \nu>\nu_b,\end{cases}\end{aligned} \tag{3.130}$$

$$E_\phi = \frac{\mathrm{i}\eta k^2 I\,\mathrm{d}l}{4\pi\nu\nu_b}\sin\phi\sum_{n=1}^{\infty}\frac{(2n+1)}{n(n+1)}\left[n(n+1)P_n(\cos\theta)-\cot\theta P_n^1(\cos\theta)\right]$$

$$\times\begin{cases}\left\{\left[\hat{H}_n(\nu_b)+d_n\right]\hat{J}_n(\nu)+e_n\hat{H}_n(\nu)\right\}\\ \left\{d_n\hat{J}_n(\nu)+\hat{H}_n(\nu)\left[e_n+\hat{J}_n(\nu_b)\right]\right\}\end{cases}$$

$$+\frac{\eta k^2 I\,\mathrm{d}l}{4\pi\sin\theta\nu\nu_b}\sin\phi\sum_{n=1}^{\infty}\frac{(2n+1)}{n(n+1)}P_n^1(\cos\theta)$$

$$\times\begin{cases}\left\{\left[\hat{H}_n'(\nu_b)+b_n\right]\hat{J}_n'(\nu)+c_n\hat{H}_n'(\nu)\right\}; & \nu<\nu_b,\\ \left\{b_n\hat{J}_n'(\nu)+\hat{H}_n'(\nu)\left[c_n+\hat{J}_n'(\nu_b)\right]\right\}; & \nu>\nu_b,\end{cases} \tag{3.131}$$

$$H_\phi = \frac{-\mathrm{i}k^2 I\,\mathrm{d}l}{4\pi\nu\nu_b}\cos\phi\sum_{n=1}^{\infty}\frac{(2n+1)}{n(n+1)}\left[n(n+1)P_n(\cos\theta)-\cot\theta P_n^1(\cos\theta)\right]$$

$$\times\begin{cases}\left\{\left[\hat{H}_n'(\nu_b)+b_n\right]\hat{J}_n(\nu)+c_n\hat{H}_n(\nu)\right\}\\ \left\{b_n\hat{J}_n(\nu)+\hat{H}_n(\nu)\left[c_n+\hat{J}_n'(\nu_b)\right]\right\}\end{cases}$$

$$-\frac{k^2 I\,\mathrm{d}l}{4\pi\sin\theta\nu\nu_b}\cos\phi\sum_{n=1}^{\infty}\frac{(2n+1)}{n(n+1)}P_n^1(\cos\theta)$$

$$\times\begin{cases}\left\{\left[\hat{H}_n(\nu_b)+d_n\right]\hat{J}_n'(\nu)+e_n\hat{H}_n'(\nu)\right\}; & \nu<\nu_b,\\ \left\{d_n\hat{J}_n'(\nu)+\hat{H}_n'(\nu)\left[e_n+\hat{J}_n(\nu_b)\right]\right\}; & \nu>\nu_b.\end{cases} \tag{3.132}$$

With Eqs. (3.125) and (3.126) of the impedance boundary conditions, it follows that

$$d_n\left[\hat{J}_n(\nu_a)-\mathrm{i}\Delta_g\hat{J}_n'(\nu_a)\right]+e_n\left[\hat{H}_n(\nu_a)-\mathrm{i}\Delta_g\hat{H}_n'(\nu_a)\right]$$
$$=-\hat{H}_n(\nu_b)\left[\hat{J}_n(\nu_a)-\mathrm{i}\Delta_g\hat{J}_n'(\nu_a)\right], \tag{3.133}$$

$$d_n\left[\hat{J}_n(\nu_c)+\mathrm{i}\Delta_i\hat{J}_n'(\nu_c)\right]+e_n\left[\hat{H}_n(\nu_c)+\mathrm{i}\Delta_i\hat{H}_n'(\nu_c)\right]$$
$$=-\hat{J}_n(\nu_b)\left[\hat{H}_n(\nu_c)+\mathrm{i}\Delta_i\hat{H}_n'(\nu_c)\right], \tag{3.134}$$

$$b_n\left[\hat{J}_n'(\nu_a)+\mathrm{i}\Delta_g\hat{J}_n(\nu_a)\right]+c_n\left[\hat{H}_n'(\nu_a)+\mathrm{i}\Delta_g\hat{H}_n(\nu_a)\right]$$
$$=-\hat{H}_n'(\nu_b)\left[\hat{J}_n'(\nu_a)+\mathrm{i}\Delta_g\hat{J}_n(\nu_a)\right], \tag{3.135}$$

$$b_n\left[\hat{J}_n'(\nu_c)-\mathrm{i}\Delta_i\hat{J}_n(\nu_c)\right]+c_n\left[\hat{H}_n'(\nu_c)-\mathrm{i}\Delta_i\hat{H}_n(\nu_c)\right]$$
$$=-\hat{J}_n'(\nu_b)\left[\hat{H}_n'(\nu_c)-\mathrm{i}\Delta_i\hat{H}_n(\nu_c)\right]. \tag{3.136}$$

The coefficients b_n, c_n, d_n, and e_n, can be obtained readily by solving the above four equations. We write

$$b_n = \frac{\left\{\begin{array}{l}[\hat{H}_n'(\nu_c) - \mathrm{i}\Delta_i \hat{H}_n(\nu_c)](\hat{J}_n'(\nu_b)[\hat{H}_n'(\nu_a) + \mathrm{i}\Delta_g \hat{H}_n(\nu_a)] \\ \quad - \hat{H}_n'(\nu_b)[\hat{J}_n'(\nu_a) + \mathrm{i}\Delta_g \hat{J}_n(\nu_a)])\end{array}\right\}}{\left\{\begin{array}{l}[\hat{J}_n'(\nu_a) + \mathrm{i}\Delta_g \hat{J}_n(\nu_a)][\hat{H}_n'(\nu_c) - \mathrm{i}\Delta_i \hat{H}_n(\nu_c)] \\ \quad - [\hat{H}_n(\nu_a) + \mathrm{i}\Delta_g \hat{H}_n'(\nu_a)][\hat{J}_n(\nu_c) - \mathrm{i}\Delta_i \hat{J}_n'(\nu_c)]\end{array}\right\}}, \tag{3.137}$$

$$c_n = \frac{\left\{\begin{array}{l}[\hat{J}_n'(\nu_a) + \mathrm{i}\Delta_g \hat{J}_n(\nu_a)](\hat{H}_n'(\nu_b)[\hat{J}_n'(\nu_c) - \mathrm{i}\Delta_i \hat{J}_n(\nu_c)] \\ \quad - \hat{J}_n'(\nu_b)[\hat{H}_n'(\nu_c) - \mathrm{i}\Delta_i \hat{H}_n(\nu_c)])\end{array}\right\}}{\left\{\begin{array}{l}[\hat{J}_n'(\nu_a) + \mathrm{i}\Delta_g \hat{J}_n(\nu_a)][\hat{H}_n'(\nu_c) - \mathrm{i}\Delta_i \hat{H}_n(\nu_c)] \\ \quad - [\hat{H}_n(\nu_a) - \mathrm{i}\Delta_g \hat{H}_n'(\nu_a)][\hat{J}_n(\nu_c) + \mathrm{i}\Delta_i \hat{J}_n'(\nu_c)]\end{array}\right\}}, \tag{3.138}$$

$$d_n = \frac{\left\{\begin{array}{l}\hat{J}_n(\nu_b)[\hat{H}_n(\nu_c) + \mathrm{i}\Delta_i \hat{H}_n'(\nu_c)][\hat{H}_n(\nu_a) - \mathrm{i}\Delta_g \hat{H}_n'(\nu_a)] \\ \quad - \hat{H}_n(\nu_b)[\hat{J}_n(\nu_a) - \mathrm{i}\Delta_g \hat{J}_n'(\nu_a)][\hat{H}_n(\nu_c) + \mathrm{i}\Delta_i \hat{H}_n'(\nu_c)]\end{array}\right\}}{\left\{\begin{array}{l}[\hat{J}_n(\nu_a) - \mathrm{i}\Delta_g \hat{J}_n'(\nu_a)][\hat{H}_n(\nu_c) + \mathrm{i}\Delta_i \hat{H}_n'(\nu_c)] \\ \quad - [\hat{H}_n(\nu_a) - \mathrm{i}\Delta_g \hat{H}_n'(\nu_a)][\hat{J}_n(\nu_c) + \mathrm{i}\Delta_i \hat{J}_n'(\nu_c)]\end{array}\right\}}, \tag{3.139}$$

$$e_n = \frac{\left\{\begin{array}{l}\hat{H}_n(\nu_b)[\hat{J}_n(\nu_a) - \mathrm{i}\Delta_g \hat{J}_n'(\nu_a)][\hat{J}_n(\nu_c) + \mathrm{i}\Delta_i \hat{J}_n'(\nu_c)] \\ \quad - \hat{J}_n(\nu_b)[\hat{H}_n(\nu_c) + \mathrm{i}\Delta_i \hat{H}_n'(\nu_c)][\hat{J}_n(\nu_a) - \mathrm{i}\Delta_g \hat{J}_n'(\nu_a)]\end{array}\right\}}{\left\{\begin{array}{l}[\hat{J}_n(\nu_a) - \mathrm{i}\Delta_g \hat{J}_n'(\nu_a)][\hat{H}_n(\nu_c) + \mathrm{i}\Delta_i \hat{H}_n'(\nu_c)] \\ \quad - [\hat{H}_n(\nu_a) - \mathrm{i}\Delta_g \hat{H}_n'(\nu_a)][\hat{J}_n(\nu_c) + \mathrm{i}\Delta_i \hat{J}_n'(\nu_c)]\end{array}\right\}}. \tag{3.140}$$

In this subsection, the spherical harmonic series solutions have been derived for SLF/ELF electromagnetic field radiated by an HED in the Earth–ionosphere waveguide or cavity.

3.3.2 Speed-up Numerical Convergence Algorithm

In this subsection, we will attempt to treat the speed-up numerical convergence algorithm for SLF/ELF fields due to an HED in the Earth–ionosphere waveguide or cavity.

3.3.2.1 Speed-up Numerical Convergence Algorithm for the Component E_r^{he}

The electric field component E_r^{he} can be written in the following form:

$$E_r^{\mathrm{he}} = \sum_{n=1}^{\infty} (E_{rn}^{\mathrm{he}} - \tilde{E}_{rn}^{\mathrm{he}}) P_n^1(\cos\theta) + \sum_{n=1}^{\infty} \tilde{E}_{rn}^{\mathrm{he}} P_n^1(\cos\theta). \tag{3.141}$$

The latter series term in Eq. (3.141) can be summed up analytically. Since the term $(E_{rn}^{\mathrm{he}} - \tilde{E}_{rn}^{\mathrm{he}})$ reduces rapidly with the increase of n, the first series in Eq. (3.141) will converge rapidly. As a result, the field component is calculated efficiently with better accuracy. With similar procedures as addressed in Sect. 3.2.2.1, an efficient analytical solution of the latter term in Eq. (3.141) can be obtained easily.

When $\nu < \nu_b$, the asymptotic approximation for the series $\sum \tilde{E}^{\rm he}_{rn} P^1_n(\cos\theta)$ is expressed by

$$\sum_{n=1}^{\infty} \tilde{E}^{\rm he}_{rn} P^1_n(\cos\theta) = \frac{-\mathrm{i}\eta k^2 I\,\mathrm{d}l}{4\pi\nu_b\nu^2}\cos\phi \sum_{n=1}^{\infty} P^1_n(\cos\theta)$$
$$\times\left\{ n\left(\frac{\nu}{\nu_b}\right)^{n+1} + \left[(n+1)\left(\rho^{n+1}\left(\frac{\nu_a}{\nu}\right)^n + \tau^{2n+1}\left(\frac{\nu_b}{\nu}\right)^n\right)\right.\right.$$
$$\left.\left. - n\tau^n\left(\rho^{-n} - \rho^{n+1}\right)\left(\frac{\nu}{\nu_c}\right)^{n+1}\right]\sum_{m=0}^{20}\tau^{m(2n+1)}\right\}. \tag{3.142}$$

Correspondingly, the analytical formula for the series $\tilde{E}^{\rm he}_{rn}$ is expressed in the following form:

$$\sum_{n=1}^{\infty} \tilde{E}_{rn} P^1_n(\cos\theta) = \frac{-\mathrm{i}\eta k^2 I\,\mathrm{d}l}{4\pi\nu_b\nu^2}\cos\phi$$
$$\times\left\{\frac{\nu}{\nu_b}\left[\frac{y\tau_1}{g_1^3} + \frac{3y\tau_1^2(x-\tau_1)}{g_1^5}\right] + L\right\}, \tag{3.143}$$

where

$$L = \sum_{m=0}^{20}\tau^m\left\{\rho\left[\frac{2y\tau_2}{g_2^3} + \frac{3y\tau_2^2(x-\tau_2)}{g_2^5}\right] + \tau\left[\frac{2y\tau_3}{g_3^3} + \frac{3y\tau_3^2(x-\tau_3)}{g_3^5}\right]\right.$$
$$-\frac{\nu}{\nu_c}\left[\frac{y\tau_4}{g_4^3} + \frac{3y\tau_4^2(x-\tau_4)}{g_4^5}\right]$$
$$\left. + \rho\frac{\nu}{\nu_c}\left[\frac{y\tau_5}{g_5^3} + \frac{3y\tau_5^2(x-\tau_5)}{g_5^5}\right]\right\}, \tag{3.144}$$

and

$$\tau_1 = \frac{\nu}{\nu_b}; \qquad g_1 = \sqrt{1 - 2x\tau_1 + \tau_1^2}, \tag{3.145}$$

$$\tau_2 = \rho\frac{\nu_a}{\nu}\tau^{2m}; \qquad g_2 = \sqrt{1 - 2x\tau_2 + \tau_2^2}, \tag{3.146}$$

$$\tau_3 = \frac{\nu_b}{\nu}\tau^{2m+2}; \qquad g_3 = \sqrt{1 - 2x\tau_3 + \tau_3^2}, \tag{3.147}$$

$$\tau_4 = \frac{\nu}{\rho\nu_c}\tau^{2m+1}; \qquad g_4 = \sqrt{1 - 2x\tau_4 + \tau_4^2}, \tag{3.148}$$

$$\tau_5 = \rho \frac{\nu}{\nu_c} \tau^{2m+1}; \qquad g_5 = \sqrt{1 - 2x\tau_5 + \tau_5^2}. \tag{3.149}$$

When $\nu > \nu_b$, the asymptotic approximation for the series $\sum \tilde{E}_{rn}^{\text{he}} P_n^1(\cos\theta)$ is also obtained readily. We write

$$\begin{aligned}\sum_{n=1}^{\infty} \tilde{E}_{rn}^{\text{he}} P_n^1(\cos\theta) = & \frac{-\mathrm{i}\eta k^2 I\,\mathrm{d}l}{4\pi\nu_b\nu^2} \cos\phi \sum_{n=1}^{\infty} P_n^1(\cos\theta) \\ & \times \left\{ -(n+1)\left(\frac{\nu_b}{\nu}\right)^{n+1} \right. \\ & + \left[(n+1)\left(\rho^{n+1}\left(\frac{\nu_a}{\nu}\right)^n + \tau^{2n+1}\left(\frac{\nu_b}{\nu}\right)^n\right) \right. \\ & \left. - n\tau^n\left(\rho^{-n} - \rho^{n+1}\right)\left(\frac{\nu}{\nu_c}\right)^{n+1} \right] \\ & \left. \times \sum_{m=0}^{20} \tau^{m(2n+1)} \right\}. \end{aligned} \tag{3.150}$$

The corresponding analytical formula for the series $\sum \tilde{E}_{rn}^{\text{he}} P_n^1(\cos\theta)$ is written in the form

$$\sum_{n=1}^{\infty} \tilde{E}_{rn} P_n^1(\cos\theta) = \frac{-\mathrm{i}\eta k^2 I\,\mathrm{d}l}{4\pi\nu_b\nu^2} \cos\phi \left[-\frac{2y\tau_1}{g_1^3} - \frac{3y\tau_1^2(x-\tau_1)}{g_1^5} + L' \right], \tag{3.151}$$

where

$$L' = L; \qquad \tau_1 = \frac{\nu_b}{\nu}; \qquad g_1 = \sqrt{1 - 2x\tau_1 + \tau_1^2}. \tag{3.152}$$

3.3.2.2 Speed-up Numerical Convergence Algorithm for the Component E_ϕ^{he}

The electric field component E_ϕ^{he} is written in the form

$$E_\phi^{\text{he}} = \sum_{n=1}^{\infty} \left(E_{\phi n}^{\text{he}} - \tilde{E}_{\phi n}^{\text{he}}\right) + \sum_{n=1}^{\infty} \tilde{E}_{\phi n}^{\text{he}}. \tag{3.153}$$

Obviously, the latter series term in Eq. (3.153) can be expressed in an analytical form, while the first series term in Eqs. (3.153) will converge rapidly. With similar researching line as above, an analytical solution of the latter term in Eq. (3.153) can be obtained readily.

When $\nu < \nu_b$, the asymptotic approximation for the series $\sum \tilde{E}^{\text{he}}_{\phi n}$ can be written in the form

$$\begin{aligned}\sum_{n=1}^{\infty} \tilde{E}_{\phi n} &= \frac{\mathrm{i}\eta k^2 I\,\mathrm{d}l}{4\pi \sin\theta \nu \nu_b^2} \sin\phi \\ &\times \sum_{n=1}^{\infty} (2n+1) \left\{ \left(\frac{\nu}{\nu_b}\right)^n - \rho^n \left(\frac{\nu_a}{\nu}\right)^{n+1} + \sum_{m=1}^{20} \tau^{m(2n+1)} \right. \\ &\times \left. \left[\left(\frac{\nu_b}{\nu}\right)^{n+1} - \rho^n \left(\frac{\nu_a}{\nu}\right)^{n+1} \right] \right\} P_n^1(\cos\theta) - \frac{\eta k^2 I\,\mathrm{d}l}{4\pi \nu} \sin\phi \\ &\times \sum_{n=1}^{\infty} (2n+1) \left\{ -\left(\frac{\nu}{\nu_b}\right)^{n+1} + \rho^{n+1} \left(\frac{\nu_a}{\nu}\right)^n + \sum_{m=1}^{20} \tau^{m(2n+1)} \right. \\ &\times \left. \left[\rho^{n+1} \left(\frac{\nu_a}{\nu}\right)^n - \left(\frac{\nu_b}{\nu}\right)^n \right] \right\} P_n(\cos\theta) + \frac{\eta k^2 I\,\mathrm{d}l}{4\pi \nu} \sin\phi \cot\theta \\ &\times \sum_{n=1}^{\infty} \frac{(2n+1)}{n(n+1)} \left\{ \rho^{n+1} \left(\frac{\nu_a}{\nu}\right)^n - \left(\frac{\nu}{\nu_b}\right)^{n+1} + \sum_{m=1}^{20} \tau^{m(2n+1)} \right. \\ &\times \left. \left[\rho^{n+1} \left(\frac{\nu_a}{\nu}\right)^n - \left(\frac{\nu_b}{\nu}\right)^n \right] \right\} P_n^1(\cos\theta). \end{aligned} \tag{3.154}$$

The corresponding analytical formula can also be obtained readily. We write

$$\begin{aligned}\sum_{n=1}^{\infty} \tilde{E}_{\phi n} = \frac{\mathrm{i}\eta k^2 I\,\mathrm{d}l}{4\pi \sin\theta \nu \nu_b^2} \sin\phi \cdot &\left\{ \left[\frac{3y\tau_1}{g_1^3} + \frac{6y\tau_1^2(x-\tau_1)}{g_1^5} \right] \right. \\ &\left. - \frac{\nu_a}{\nu} \left[\frac{3y\tau_2}{g_2^3} + \frac{6y\tau_2^2(x-\tau_2)}{g_2^5} \right] + S \right\} + P, \end{aligned} \tag{3.155}$$

where

$$S = \sum_{m=1}^{20} \tau^m \left\{ \frac{\nu_b}{\nu} \left[\frac{3y\tau_3}{g_3^3} + \frac{6y\tau_3^2(x-\tau_3)}{g_3^5} \right] - \frac{\nu_a}{\nu} \left[\frac{3y\tau_4}{g_4^3} + \frac{6y\tau_4^2(x-\tau_4)}{g_4^5} \right] \right\}, \tag{3.156}$$

$$\begin{aligned} P = &-\frac{\eta k^2 I\,dl}{4\pi \nu} \sin\phi \left\{ \rho \left(\frac{1-\tau_2^2}{g_2^3} - 1 \right) - \frac{\nu}{\nu_b} \left(\frac{1-\tau_1^2}{g_1^3} - 1 \right) \right. \\ &\left. + \sum_{m=1}^{20} \tau^m \left[\rho \left(\frac{1-\tau_4^2}{g_4^3} - 1 \right) - \left(\frac{1-\tau_3^2}{g_3^3} - 1 \right) \right] \right\} \end{aligned}$$

$$+\frac{\eta k^2 I\,\mathrm{d}l}{4\pi\nu}\sin\phi\cot\theta\left\{\rho\left(\frac{x\tau_2+1}{y\tau_2}+\frac{\tau_2^2-1}{y\tau_2 g_2}\right)-\frac{\nu}{\nu_b}\left(\frac{x\tau_1+1}{y\tau_1}+\frac{\tau_1^2-1}{y\tau_1 g_1}\right)\right.$$

$$\left.+\sum_{m=1}^{20}\tau^m\left[\rho\left(\frac{x\tau_4+1}{y\tau_4}+\frac{\tau_4^2-1}{y\tau_4 g_4}\right)-\left(\frac{x\tau_3+1}{y\tau_3}+\frac{\tau_3^2-1}{y\tau_3 g_3}\right)\right]\right\}, \tag{3.157}$$

and

$$\tau_1=\frac{\nu}{\nu_b};\qquad \tau_2=\rho\frac{\nu_a}{\nu};\qquad \tau_3=\frac{\nu_b}{\nu}\tau^{2m};\qquad \tau_4=\rho\frac{\nu_a}{\nu}\tau^{2m}, \tag{3.158}$$

$$g_1=\sqrt{1-2x\tau_1+\tau_1^2};\qquad g_2=\sqrt{1-2x\tau_2+\tau_2^2}, \tag{3.159}$$

$$g_3=\sqrt{1-2x\tau_3+\tau_3^2};\qquad g_4=\sqrt{1-2x\tau_4+\tau_4^2}. \tag{3.160}$$

When $\nu > \nu_b$, the asymptotic approximation for the series $\sum \tilde{E}_{\phi n}^{\mathrm{he}}$ is also obtained readily. We write

$$\begin{aligned}
\sum_{n=1}^{\infty}\tilde{E}_{\phi n} = {} & \frac{\mathrm{i}\eta k^2 I\,\mathrm{d}l}{4\pi\sin\theta\,\nu\nu_b^2}\sin\phi\sum_{n=1}^{\infty}(2n+1)\\
& \times\left\{\left(\frac{\nu_b}{\nu}\right)^n-\rho^n\left(\frac{\nu_a}{\nu}\right)^{n+1}\right.\\
& \left.+\sum_{m=1}^{20}\tau^{m(2n+1)}\left[\left(\frac{\nu_b}{\nu}\right)^{n+1}-\rho^n\left(\frac{\nu_a}{\nu}\right)^{n+1}\right]\right\}P_n^1(\cos\theta)\\
& -\frac{\eta k^2 I\,\mathrm{d}l}{4\pi\nu}\sin\phi\sum_{n=1}^{\infty}(2n+1)\\
& \times\left\{-\left(\frac{\nu}{\nu_b}\right)^{n+1}+\rho^{n+1}\left(\frac{\nu_a}{\nu}\right)^n\right.\\
& \left.+\sum_{m=1}^{20}\tau^{m(2n+1)}\left[\rho^{n+1}\left(\frac{\nu_a}{\nu}\right)^n-\left(\frac{\nu_b}{\nu}\right)^n\right]\right\}P_n(\cos\theta)\\
& +\frac{\eta k^2 I\,\mathrm{d}l}{4\pi\nu}\sin\phi\cot\theta\sum_{n=1}^{\infty}\frac{(2n+1)}{n(n+1)}\\
& \times\left\{\rho^{n+1}\left(\frac{\nu_a}{\nu}\right)^n-\left(\frac{\nu_b}{\nu}\right)^{n+1}\right.\\
& \left.+\sum_{m=1}^{20}\tau^{m(2n+1)}\left[\rho^{n+1}\left(\frac{\nu_a}{\nu}\right)^n-\left(\frac{\nu_b}{\nu}\right)^n\right]\right\}P_n^1(\cos\theta).
\end{aligned} \tag{3.161}$$

The corresponding analytical formula is derived readily. We write

$$\sum_{n=1}^{\infty} \tilde{E}_{\phi n} = \frac{\mathrm{i}\eta k^2 I\,\mathrm{d}l}{4\pi \sin\theta \nu \nu_b^2} \sin\phi \cdot \left\{ \frac{\nu_b}{\nu}\left[\frac{3y\tau_1}{g_1^3} + \frac{6y\tau_1^2(x-\tau_1)}{g_1^5}\right] - \frac{\nu_a}{\nu}\left[\frac{3y\tau_2}{g_2^3} + \frac{6y\tau_2^2(x-\tau_2)}{g_2^5}\right] + S' \right\} + P', \tag{3.162}$$

where

$$S' = S; \qquad P' = P; \qquad \tau_1 = \frac{\nu_b}{\nu}; \qquad \tau_2 = \rho\frac{\nu_a}{\nu}, \tag{3.163}$$

$$g_1 = \sqrt{1 - 2x\tau_1 + \tau_1^2}; \qquad g_2 = \sqrt{1 - 2x\tau_2 + \tau_2^2}. \tag{3.164}$$

3.3.2.3 Speed-up Numerical Convergence Algorithm for the Component E_θ^{he}

The electric field component E_θ^{he} is expressed by

$$E_\theta^{\mathrm{he}} = \sum_{n=1}^{\infty} \left(E_{\theta n}^{\mathrm{he}} - \tilde{E}_{\theta n}^{\mathrm{he}}\right) + \sum_{n=1}^{\infty} \tilde{E}_{\theta n}^{\mathrm{he}}. \tag{3.165}$$

It is seen that the latter series term in Eq. (3.165) can be solved analytically. The first series term in Eq. (3.165) will converge rapidly, and the field component can be calculated efficiently with better accuracy. With similar steps to the above, the latter term in Eq. (3.165) can be solved analytically.

When $\nu < \nu_b$, the asymptotic approximation for the series $\sum \tilde{E}_{\theta n}^{\mathrm{he}}$ can be derived readily. We write

$$\begin{aligned}\sum_{n=1}^{\infty} \tilde{E}_{\theta n} = &-\frac{\mathrm{i}\eta k^2 I\,\mathrm{d}l}{4\pi \sin\theta \nu \nu_b^2} \cos\phi \sum_{n=1}^{\infty} n(n+1)(2n+1) \\ &\times \left\{ \left(\frac{\nu}{\nu_b}\right)^n - \rho^n \left(\frac{\nu_a}{\nu}\right)^{n+1} \right. \\ &\left. + \sum_{m=1}^{20} \tau^{m(2n+1)} \left[\left(\frac{\nu_b}{\nu}\right)^{n+1} - \rho^n \left(\frac{\nu_a}{\nu}\right)^{n+1} \right] \right\} P_n(\cos\theta) \\ &+ \frac{\mathrm{i}\eta k^2 I\,\mathrm{d}l}{4\pi \nu \nu_b^2} \cos\phi \cot\theta \sum_{n=1}^{\infty} (2n+1) \\ &\times \left\{ \left(\frac{\nu}{\nu_b}\right)^n - \rho^n \left(\frac{\nu_a}{\nu}\right)^{n+1} \right.\end{aligned}$$

$$+\sum_{m=1}^{20}\tau^{m(2n+1)}\left[\left(\frac{\nu_b}{\nu}\right)^{n+1}-\rho^n\left(\frac{\nu_a}{\nu}\right)^{n+1}\right]\Bigg\}P_n^1(\cos\theta)$$

$$+\frac{\eta k^2 I\,\mathrm{d}l}{4\pi\sin\theta\nu}\cos\phi\sum_{n=1}^{\infty}\frac{(2n+1)}{n(n+1)}$$

$$\times\left\{\rho^{n+1}\left(\frac{\nu_a}{\nu}\right)^n-\left(\frac{\nu}{\nu_b}\right)^{n+1}\right.$$

$$\left.+\sum_{m=1}^{20}\tau^{m(2n+1)}\left[\rho^{n+1}\left(\frac{\nu_a}{\nu}\right)^n-\left(\frac{\nu_b}{\nu}\right)^n\right]\right\}P_n^1(\cos\theta). \quad (3.166)$$

Correspondingly, the analytical formula can also be obtained easily. It is

$$\sum_{n=1}^{\infty}\tilde{E}_{\theta n}=-\frac{\mathrm{i}\eta k^2 I\,\mathrm{d}l}{4\pi\nu\nu_b^2}\cos\phi$$

$$\times\left\{\left[2W_1+\frac{3\tau_1(x+x^2\tau_1-x\tau_1^2-2\tau_1+\tau_1^3)}{g_1^5}+\frac{\tau_1(x-\tau_1)}{g_1^3}\right]\right.$$

$$-\frac{\nu_a}{\nu}\left[2W_2+\frac{3\tau_2(x+x^2\tau_2-x\tau_2^2-2\tau_2+\tau_2^3)}{g_2^5}+\frac{\tau_2(x-\tau_2)}{g_2^3}\right]$$

$$+\sum_{m=1}^{20}\tau^m\left[\frac{\nu_b}{\nu}\left(2W_3+\frac{3\tau_3(x+x^2\tau_3-x\tau_3^2-2\tau_3+\tau_3^3)}{g_3^5}\right.\right.$$

$$\left.+\frac{\tau_3(x-\tau_3)}{g_3^3}\right)-\frac{\nu_a}{\nu}\left(2W_4+\frac{3\tau_4(x+x^2\tau_4-x\tau_4^2-2\tau_4+\tau_4^3)}{g_4^5}\right.$$

$$\left.\left.\left.+\frac{\tau_4(x-\tau_4)}{g_4^3}\right)\right]\right\}+Q. \quad (3.167)$$

In the above formulas, W_1, W_2, W_3, W_4, and Q are expressed in the following forms:

$$W_1=\frac{\tau_1(x+2x^2\tau_1-3x\tau_1^2-4\tau_1+4\tau_1^3)}{g_1^5}$$

$$+\frac{5\tau_1^2(x-\tau_1)(x+x^2\tau_1-x\tau_1^2-2\tau_1+\tau_1^3)}{g_1^7}, \quad (3.168)$$

$$W_2=\frac{\tau_2(x+2x^2\tau_2-3x\tau_2^2-4\tau_2+4\tau_2^3)}{g_2^5}$$

$$+\frac{5\tau_2^2(x-\tau_2)(x+x^2\tau_2-x\tau_2^2-2\tau_2+\tau_2^3)}{g_2^7}, \quad (3.169)$$

$$W_3 = \frac{\tau_3(x + 2x^2\tau_3 - 3x\tau_3^2 - 4\tau_3 + 4\tau_3^3)}{g_3^5} + \frac{5\tau_3^2(x - \tau_3)(x + x^2\tau_3 - x\tau_3^2 - 2\tau_3 + \tau_3^3)}{g_3^7}, \tag{3.170}$$

$$W_4 = \frac{\tau_4(x + 2x^2\tau_4 - 3x\tau_4^2 - 4\tau_4 + 4\tau_4^3)}{g_4^5} + \frac{5\tau_4^2(x - \tau_4)(x + x^2\tau_4 - x\tau_4^2 - 2\tau_4 + \tau_4^3)}{g_4^7}, \tag{3.171}$$

$$\begin{aligned}
Q = {} & \frac{\mathrm{i}\eta k^2 I\,\mathrm{d}l}{4\pi\nu\nu_b^2}\cos\phi\cot\theta \\
& \times \left\{ \left[\frac{3y\tau_1}{g_1^3} + \frac{6y\tau_1^2(x-\tau_1)}{g_1^5}\right] - \frac{\nu_a}{\nu}\left[\frac{3y\tau_2}{g_2^3} + \frac{6y\tau_2^2(x-\tau_2)}{g_2^5}\right] \right. \\
& + \sum_{m=1}^{20}\tau^m\left[\frac{\nu_b}{\nu}\left(\frac{3y\tau_3}{g_3^3} + \frac{6y\tau_3^2(x-\tau_3)}{g_3^5}\right)\right. \\
& \left.\left. - \frac{\nu_a}{\nu}\left(\frac{3y\tau_4}{g_4^3} + \frac{6y\tau_4^2(x-\tau_4)}{g_4^5}\right)\right]\right\} \\
& + \frac{\eta k^2 I\,\mathrm{d}l}{4\pi\sin\theta\,\nu}\cos\phi \\
& \times \left\{ \rho\left[\frac{x\tau_2+1}{y\tau_2} + \frac{\tau_2^2-1}{y\tau_2 g_2}\right] - \frac{\nu}{\nu_b}\left[\frac{x\tau_1+1}{y\tau_1} + \frac{\tau_1^2-1}{y\tau_1 g_1}\right] \right. \\
& + \sum_{m=1}^{20}\tau^m\left[\rho\left(\frac{x\tau_4+1}{y\tau_4} + \frac{\tau_4^2-1}{y\tau_4 g_4}\right)\right. \\
& \left.\left. - \left(\frac{x\tau_3+1}{y\tau_3} + \frac{\tau_3^2-1}{y\tau_3 g_3}\right)\right]\right\},
\end{aligned} \tag{3.172}$$

where

$$\tau_1 = \frac{\nu}{\nu_b}; \qquad g_1 = \sqrt{1 - 2x\tau_1 + \tau_1^2}, \tag{3.173}$$

$$\tau_2 = \rho\frac{\nu_a}{\nu}; \qquad g_2 = \sqrt{1 - 2x\tau_2 + \tau_2^2}, \tag{3.174}$$

$$\tau_3 = \frac{\nu_b}{\nu}\tau^{2m}; \qquad g_3 = \sqrt{1 - 2x\tau_3 + \tau_3^2}, \tag{3.175}$$

$$\tau_4 = \rho \frac{\nu_a}{\nu} \tau^{2m}; \qquad g_4 = \sqrt{1 - 2x\tau_4 + \tau_4^2}. \tag{3.176}$$

When $\nu > \nu_b$, the asymptotic approximation for the series $\sum \tilde{E}_{\theta n}^{\text{he}}$ is written in the form

$$\begin{aligned}
\sum_{n=1}^{\infty} \tilde{E}_{\theta n} = &-\frac{\mathrm{i}\eta k^2 I\,\mathrm{d}l}{4\pi \sin\theta \nu \nu_b^2} \cos\phi \cdot \sum_{n=1}^{\infty} n(n+1)(2n+1) \\
&\times \left\{ \left(\frac{\nu_b}{\nu}\right)^{n+1} - \rho^n \left(\frac{\nu_a}{\nu}\right)^{n+1} \right. \\
&\left. + \sum_{m=1}^{20} \tau^{m(2n+1)} \left[\left(\frac{\nu_b}{\nu}\right)^{n+1} - \rho^n \left(\frac{\nu_a}{\nu}\right)^{n+1} \right] \right\} P_n(\cos\theta) \\
&+ \frac{\mathrm{i}\eta k^2 I\,\mathrm{d}l}{4\pi \nu \nu_b^2} \cos\phi \cot\theta \cdot \sum_{n=1}^{\infty} (2n+1) \\
&\times \left\{ \left(\frac{\nu_b}{\nu}\right)^{n+1} - \rho^n \left(\frac{\nu_a}{\nu}\right)^{n+1} \right. \\
&\left. + \sum_{m=1}^{20} \tau^{m(2n+1)} \left[\left(\frac{\nu_b}{\nu}\right)^{n+1} - \rho^n \left(\frac{\nu_a}{\nu}\right)^{n+1} \right] \right\} P_n^1(\cos\theta) \\
&+ \frac{\eta k^2 I\,\mathrm{d}l}{4\pi \sin\theta \nu} \cos\phi \cdot \sum_{n=1}^{\infty} \frac{(2n+1)}{n(n+1)} \\
&\times \left\{ \rho^{n+1} \left(\frac{\nu_a}{\nu}\right)^{n} - \left(\frac{\nu_b}{\nu}\right)^{n} \right. \\
&\left. + \sum_{m=1}^{20} \tau^{m(2n+1)} \left[\rho^{n+1} \left(\frac{\nu_a}{\nu}\right)^{n} - \left(\frac{\nu_b}{\nu}\right)^{n} \right] \right\} P_n^1(\cos\theta).
\end{aligned} \tag{3.177}$$

Thus, the corresponding analytical formula is obtained readily. The result becomes

$$\begin{aligned}
\sum_{n=1}^{\infty} \tilde{E}_{\theta n} = &-\frac{\mathrm{i}\eta k^2 I\,\mathrm{d}l}{4\pi \nu \nu_b^2} \cos\phi \\
&\times \left\{ \frac{\nu_b}{\nu} \left[2W_1 + \frac{3\tau_1(x + x^2\tau_1 - x\tau_1^2 - 2\tau_1 + \tau_1^3)}{g_1^5} + \frac{\tau_1(x - \tau_1)}{g_1^3} \right] \right. \\
&- \frac{\nu_a}{\nu} \left[2W_2 + \frac{3\tau_2(x + x^2\tau_2 - x\tau_2^2 - 2\tau_2 + \tau_2^3)}{g_2^5} + \frac{\tau_2(x - \tau_2)}{g_2^3} \right]
\end{aligned}$$

$$+\sum_{m=1}^{20}\tau^m\left[\frac{\nu_b}{\nu}\left(2W_3+\frac{3\tau_3(x+x^2\tau_3-x\tau_3^2-2\tau_3+\tau_3^3)}{g_3^5}\right.\right.$$

$$\left.+\frac{\tau_3(x-\tau_3)}{g_3^3}\right)-\frac{\nu_a}{\nu}\left(2W_4+\frac{3\tau_4(x+x^2\tau_4-x\tau_4^2-2\tau_4+\tau_4^3)}{g_4^5}\right.$$

$$\left.\left.\left.+\frac{\tau_4(x-\tau_4)}{g_4^3}\right)\right]\right\}+Q', \tag{3.178}$$

where

$$Q'=Q; \qquad \tau_1=\frac{\nu_b}{\nu}; \qquad g_1=\sqrt{1-2x\tau_1+\tau_1^2}, \tag{3.179}$$

$$W_1=\frac{\tau_1(x+2x^2\tau_1-3x\tau_1^2-4\tau_1+4\tau_1^3)}{g_1^5}$$
$$+\frac{5\tau_1^2(x-\tau_1)(x+x^2\tau_1-x\tau_1^2-2\tau_1+\tau_1^3)}{g_1^7}. \tag{3.180}$$

3.3.3 Evaluations for $\hat{J}_n(\nu)$, $\hat{H}_n(\nu)$, $P_n(\nu)$, and $P_n^1(\nu)$

The computations for the functions $\hat{J}_n(\nu)$, $\hat{H}_n(\nu)$, $P_n(\nu)$, and $P_n^1(\nu)$ can be treated with same method as used in Sect. 3.2.3 in this book.

3.3.4 Numerical Results and Discussions

The computations are carried out for the SLF/ELF field components with the unit HED in the non-ideal Earth–ionosphere waveguide or cavity. With $I\,\mathrm{d}l = 1$ A·m, $a = 6{,}370$ km, $h = 70$ km, $\sigma_g = 10^{-4}$ S/m, $\sigma_i = 10^{-5}$ S/m, and $\phi = \frac{\pi}{4}$, the magnitudes of the field components $|E_r|$, $|E_\theta|$, and $|E_\phi|$ versus the propagation distance are computed at $f = 100$ Hz, 60 Hz, and 30 Hz by using the speed-up numerical convergence algorithm and they are plotted in Figs. 3.9, 3.10, 3.11, respectively. From these results, it is seen that in the case of the propagation distance being over 10,000 km there exists mutual interference between the two waves along the short and long circular paths.

Assuming that the dipole source is located at the height of $z_s = 10$ m, the observation point is located on the Earth's surface, $z_r = 0$ m, and the angular distance is taken as $\theta = \frac{\pi}{5}$, the magnitudes of the field components $|E_r|$, $|E_\theta|$, and $|E_\phi|$ versus the operating frequency are computed and shown in Fig. 3.12, respectively. From Fig. 3.12, for the components $|E_r|$, $|E_\theta|$, and $|E_\phi|$, the resonant frequency is approximated as $f = 8.8$ Hz, 16.8 Hz, and 22.8 Hz, respectively. Meanwhile, with the

Fig. 3.9 The electric field E_r in dB·V/m due to unit HED versus the propagation distance: $I\,\mathrm{d}l = 1$ A·m, $a = 6{,}370$ km, $h = 80$ km, $\sigma_g = 2 \times 10^{-4}$ S/m, $\sigma_i = 10^{-5}$ S/m, and $\phi = \frac{\pi}{4}$

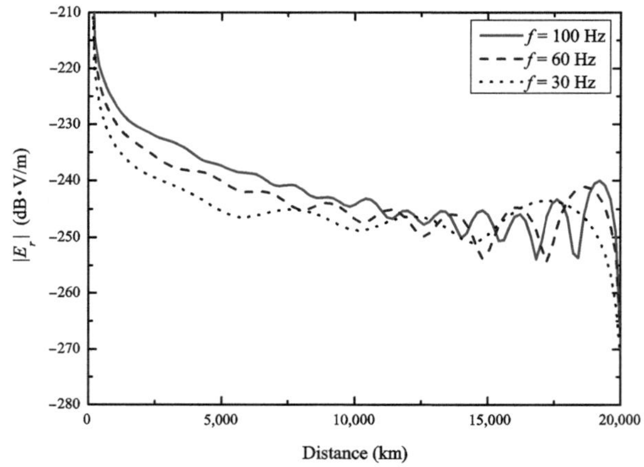

Fig. 3.10 The electric field E_θ in dB·V/m due to the unit HED versus the propagation distance: $I\,\mathrm{d}l = 1$ A·m, $a = 6{,}370$ km, $h = 80$ km, $\sigma_g = 2 \times 10^{-4}$ S/m, $\sigma_i = 10^{-5}$ S/m, and $\phi = \frac{\pi}{4}$

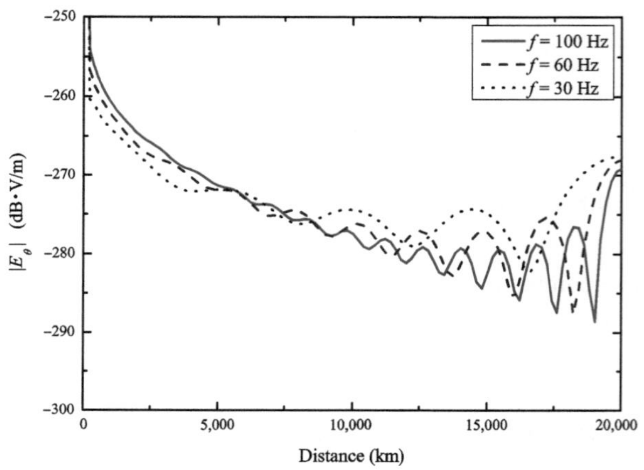

Fig. 3.11 The electric field E_ϕ in dB·V/m due to the unit HED versus the propagation distance: $I\,\mathrm{d}l = 1$ A·m, $a = 6{,}370$ km, $h = 80$ km, $\sigma_g = 2 \times 10^{-4}$ S/m, $\sigma_i = 10^{-5}$ S/m, and $\phi = \frac{\pi}{4}$

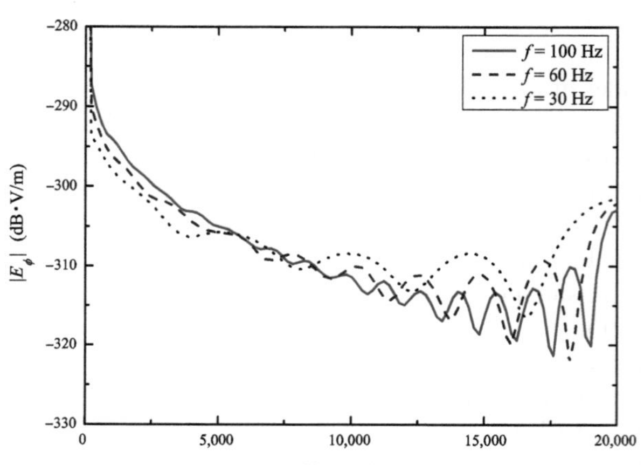

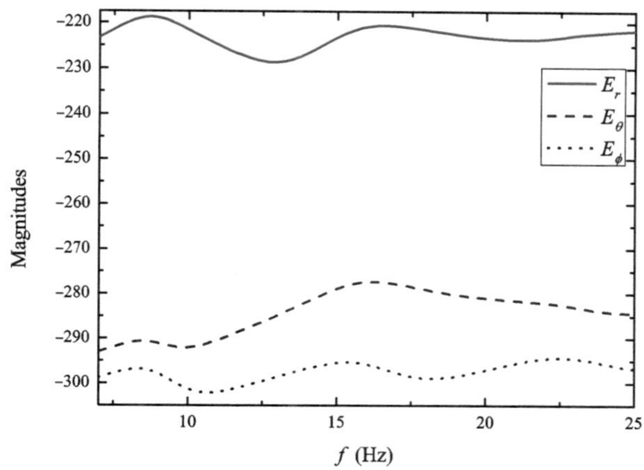

Fig. 3.12 The components E_r, E_θ, and E_ϕ in dB V/m due to the unit HED versus the operating frequency: $I\,\mathrm{d}l = 1$ A m, $a = 6{,}370$ km, $h = 80$ km, $\sigma_g = 2 \times 10^{-4}$ S/m, $\sigma_i = 10^{-5}$ S/m, $\theta = \frac{\pi}{5}$, $z_s = z_r = 0$ m, and $\phi = \frac{\pi}{4}$

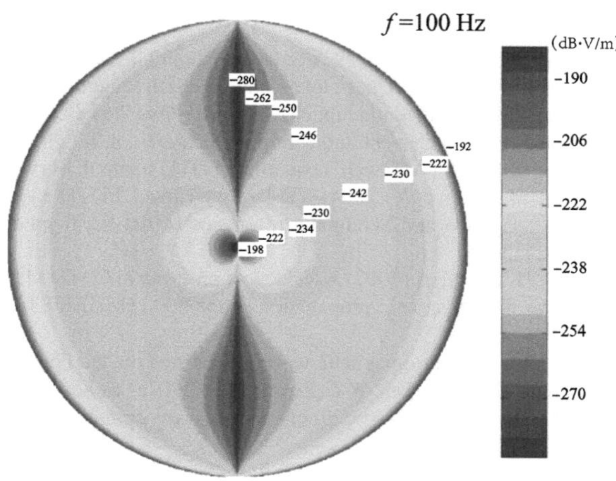

Fig. 3.13 Two-dimensional diagram of the component E_r in dB·V/m due to the unit HED versus the propagation distance: $I\,\mathrm{d}l = 1$ A·m, $a = 6{,}370$ km, $h = 80$ km, $\sigma_g = 2 \times 10^{-4}$ S/m, $\sigma_i = 10^{-5}$ S/m, $z_s = 0$ m, $z_r = 0$ m, $\phi = 0$

absorption loss, the resonance curve is relatively "flat". With the same parameters as in Fig. 3.12, a two-dimensional diagram for the magnitude of the component E_r versus the propagation distance is shown in Fig. 3.13.

References

Bannister PR (1974) Far-field extremely low frequency (ELF) propagation measurements. IEEE Trans Commun 22(4):468–474

Bannister PR (1979) Some notes on ELF Earth–ionosphere waveguide daytime propagation parameters. IEEE Trans Antennas Propag 27(5):696–698

Barrick DE (1999) Exact ULF/ELF dipole field strengths in the Earth–ionosphere cavity over the Schumann resonance region: idealized boundaries. Radio Sci 34(1):209–227

Carroll KJ, Ferraro AJ (1990) Computer simulation of ELF injection in the Earth–ionosphere waveguide. Radio Sci 25:1363–1367

Chang DC, Wait JR (1974) ELF propagation along a horizontal wire located above or buried in the Earth. IEEE Trans Commun 22(4):421–427
Chu TS (1974) The radiation of extremely low frequency (ELF) waves. IEEE Trans Commun 22(4):386–388
Cummer SA (2000) Modeling electromagnetic propagation in the Earth–ionosphere waveguide. IEEE Trans Antennas Propag 48(9):1420–1429
Fraser-Smith AC, Bannister PR (1998) Reception of ELF signals at antipodal distance. Radio Sci 33:83–88
Galejs J (1971) Excitation of the terrestrial wave guide by sources in the lower ionosphere. Radio Sci 6:41–53
Galejs J (1972) Terrestrial propagation of long electromagnetic wave. Pergamon, New York
Gradshteyn IS, Ryzhik IM (1980) Table of integrals, series, and products. Academic Press, New York
Johler JR, Berry LA (1962) Propagation of terrestrial radio waves of long wavelength: theory of zonal harmonics with improved summation techniques. J Res Natl Bur Stand 66:325–376
Larsen TR, Egeland A (1968) Fine structure of the Earth-ionosphere cavity resonances. J Geophys Res 73:4986–4989
Nickolaenko AP, Hayakawa M (2002) Resonances in the Earth–ionosphere cavity. Kluwer Academic, Dordrecht
Pan WY (2004) LF VLF ELF wave propagation. UESTC Press, Chengdu. (In Chinese)
Schumann WO (1957) Uber elektrische eigenschwingungen des hohraumes erdeluft-ionosphare angeregt durch blitzentladungen. Z Angew Phys 9(8):373–378
Simpson JJ (2009) Current and future applications of 3D global Earth–ionosphere models based on the full-vector Maxwell's equations FDTD method. Surv Geophys 30(2):105–130
Simpson JJ, Taflove A (2004) Three-dimensional FDTD modeling of impulsive ELF antipodal propagation and Schumann resonance of the Earth-sphere. IEEE Trans Antennas Propag 52(2):443–451
Simpson JJ, Taflove A (2007) A Review of progress in FDTD Maxwell's equations modeling of impulsive subionospheric propagation below 300 kHz. IEEE Trans Antennas Propag 55(6):1582–1590
Wait JR (1962) Electromagnetic wave in stratified media. Pergamon, New York
Wang YX, Fan WS, Pan WY, Zhang HQ (2007a) Spherical harmonic series solution of fields excited by vertical electric dipole in Earth–ionosphere cavity. Chin J Radio Sci 22(2):204–211. (In Chinese)
Wang YX, Peng Q, Pan WY, Zhang HQ, Zhang ZW (2007b) The fields excited by SLF/ELF horizontal electric dipole in Earth–ionosphere cavity. Chin J Radio Sci 22(5):728–734. (In Chinese)
Wang YX, Fan WS, Pan WY (2008) Spherical harmonic series solution of fields excited by vertical electric dipole in Earth–ionosphere cavity. Front Electr Electron Eng China 3(1):61–69
Yang H, Pasko VP (2005) Three-dimensional finite-difference time-domain modeling of the Earth–ionosphere cavity resonances. Geophys Res Lett 32:L03114.1–L03114.4
Yang H, Pasko VP (2006) Three-dimensional finite difference time domain modeling of the diurnal and seasonal variations in Schumann resonance parameters. Radio Sci 41:RS2S14

Chapter 4
SLF/ELF Field in Air and Ionosphere Generated by Earthquake Radiation Source

In this chapter, we treat analytically SLF/ELF field in air and ionosphere radiated by an HED in the Earth. The region of interest consists of the lithosphere, air, and ionosphere. For the planar model, the analytical solution for SLF/ELF field in the air and ionosphere generated by an underground HED is obtained readily. Then, comparison and analyses are treated for both planar and spherical models. Finally, computations are carried out and discussions are presented.

4.1 Introduction

It is well known that a series of geophysical and disturbance effects may appear in the atmosphere and the Earth's crust near the earthquake seismogenic zone. In the past two decades, it has been found that before an earthquake, there exist low-frequency electromagnetic radiation anomalies in a very wide spectrum around the earthquake source. After continuous observational studies in the past decades, it is identified that the observations on low-frequency electromagnetic precursor may become an important manner for short-term and imminent earthquake prediction. In recent ten years, the earthquake electromagnetic satellites have been launched for receiving the low-frequency electromagnetic signals from the earthquake source by several countries, especially including Russia, France, and the United States. It was reported that the low-frequency electromagnetic radiation anomalies were indeed observed before some typical earthquakes (Gokhberg et al. 1983; Parrot 1994; Serebryakova et al. 1992; Molchanov et al. 2004). Even now, the mechanism and spectrum properties for the electromagnetic radiation by earthquake are still in the dark. Because of the attenuation properties of the strata, only the components below the VLF range can penetrate the strata and can be observed on the Earth's surface. Due to the good reflective properties for both the ground and the ionosphere, VLF/SLF/ELF waves can propagate for a long distance in Earth–ionosphere waveguide or cavity. With the development of VLF/SLF communication and navigation systems, for addressing the characteristics of the electromagnetic waves radiated by

W. Pan, K. Li, *Propagation of SLF/ELF Electromagnetic Waves*,
Advanced Topics in Science and Technology in China,
DOI 10.1007/978-3-642-39050-0_4,

VLF/SLF transmitter along the Earth's surface, mature theoretical and experimental results have been obtained. It is noted that the problem of the electromagnetic field radiated by an electric dipole in a three-layered or multi-layered region was visited by many investigators and some ideas can be borrowed (Wait 1953, 1954, 1956, 1970, 1998; Dunn 1986; King 1991, 1993; King and Sandler 1994a, 1994b, 1998; Hoh et al. 1999; Tsang et al. 2000; Collin 2004a, 2004b; Zhang and Pan 2002; Li 2009). However, in the previous research, a unified physical model for an Earth–air–ionosphere region was not employed to treat quantitatively the electromagnetic field radiated by underground radiation source. Meanwhile, the case for the observation point being located on the Earth's surface was investigated widely, and the case for the observation point located in the ionosphere was seldom considered (Nikiforova et al. 1989; Morgunov and Matveev 1990; Uyeda et al. 2002).

In view of the seismic source being generally situated in the Earth with the depth of a few kilometers to several tens of kilometers, the main radiation spectrum for earthquake should be concentrated in SLF/ELF regions. Comparing with the wavelength of SLF/ELF waves, the geometric scale of the seismogenic zone is very small. Thus, the radiation source can be idealized as an electric dipole. Moreover, for the observation point on or near the Earth's surface, the radiation efficiency of VED below Earth's surface is much lower than that of the HED. Obviously, a radiation source in the Earth is usually idealized as an HED.

In this chapter, we will address SLF/ELF field in the ionosphere radiated by an underground HED in three-layered planar and spherical models, while SLF/ELF wave propagation in the anisotropic ionosphere is treated analytically (Zhang et al. 2009).

4.2 Formulation of Problem

When the propagation distance between the observation point and the dipole source is not too large, the effect of the Earth's curvature can be neglected. As a simple flat model, the three-layered region of interest consists of Earth, air, and ionosphere above. The relevant geometry and Cartesian coordinate system are illustrated in Fig. 4.1, where an HED in the $\hat{x}$ direction is located at (0, 0, $-d$). Region 0 (($z \geq h$)) is the ionosphere, characterized by the permeability μ_0, permittivity ε_i, and conductivity σ_i; Region 1 ($0 \leq z \leq h$) is the air, characterized by the permeability μ_0 and permittivity ε_a; and Region 2 ($z \leq 0$) is the Earth, characterized by the permeability μ_0, permittivity ε_g, and conductivity σ_g.

Maxwell's equations in the three-layered region are represented by

$$\nabla \times \boldsymbol{E}_j = \mathrm{i}\omega \boldsymbol{B}_j, \tag{4.1}$$

$$\nabla \times \boldsymbol{B}_j = \mu_0(-\mathrm{i}\omega\hat{\varepsilon}_j \boldsymbol{E}_j + \hat{x} J_x), \tag{4.2}$$

where $j = i, a, g$, and

$$J_x = I\,\mathrm{d}l\delta(x)\delta(y)\delta(z+d). \tag{4.3}$$

Fig. 4.1 The physical model

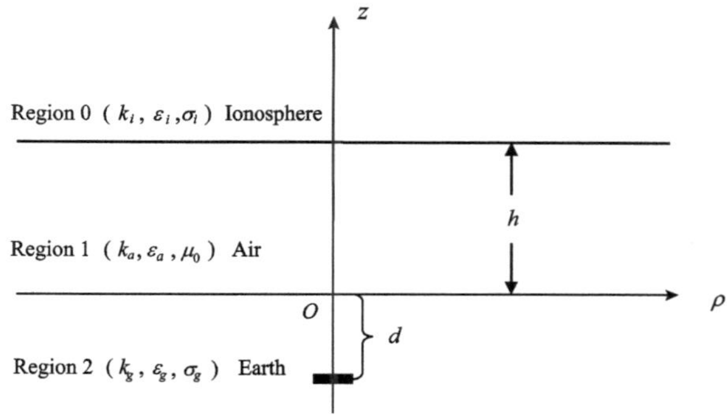

The Fourier transform of the form

$$\boldsymbol{E}(x, y, z) = \frac{1}{4\pi^2}\int_{-\infty}^{\infty}\int_{-\infty}^{\infty}\tilde{\boldsymbol{E}}(k_x, k_y, z)\mathrm{e}^{\mathrm{i}(k_x x + k_y y)}\,\mathrm{d}k_x\,\mathrm{d}k_y, \tag{4.4}$$

and similar transforms for $\boldsymbol{B}(x, y, z)$ and $J_x(x, y, z)$ are applied to Maxwell's equations. Thus,

$$\tilde{J}_x(k_x, k_y, z) = I\,\mathrm{d}l\delta(z + d). \tag{4.5}$$

The transformed Maxwell equations in Cartesian coordinate system are written as follows:

$$\mathrm{i}k_y\tilde{E}_{jz} - \frac{\partial}{\partial z}\tilde{E}_{jy} = \mathrm{i}\omega\tilde{B}_{jx}, \tag{4.6}$$

$$\frac{\partial}{\partial z}\tilde{E}_{jx} - \mathrm{i}k_x\tilde{E}_{jz} = \mathrm{i}\omega\tilde{B}_{jy}, \tag{4.7}$$

$$\mathrm{i}k_x\tilde{E}_{jy} - \mathrm{i}k_y\tilde{E}_{jx} = \mathrm{i}\omega\tilde{B}_{jz}, \tag{4.8}$$

$$\mathrm{i}k_y\tilde{B}_{jz} - \frac{\partial}{\partial z}\tilde{B}_{jy} = -\mathrm{i}\frac{k_j^2}{\omega}\tilde{E}_{jx} + \mu_0 I\,\mathrm{d}l\delta(z + d), \tag{4.9}$$

$$\frac{\partial}{\partial z}\tilde{B}_{jx} - \mathrm{i}k_x\tilde{B}_{jz} = -\frac{\mathrm{i}k_j^2}{\omega}\tilde{E}_{jy}, \tag{4.10}$$

$$\mathrm{i}k_x\tilde{B}_{jy} - \mathrm{i}k_y\tilde{B}_{jx} = -\frac{\mathrm{i}k_j^2}{\omega}\tilde{E}_{jz}. \tag{4.11}$$

In these equations,

$$k_j = \omega\sqrt{\mu_0(\varepsilon_j + \mathrm{i}\sigma_j/\omega)}; \quad j = i, a, g. \tag{4.12}$$

Equations (4.6)–(4.11) can be used to solve $\tilde{E}_y$, $\tilde{E}_z$, $\tilde{B}_y$, and $\tilde{B}_z$ in terms of $\tilde{E}_x$ and $\tilde{B}_x$. The results for the three regions can be expressed in the following forms:

$$\tilde{E}_{jy} = \frac{1}{k_j^2 - k_x^2}\left(-k_x k_y \tilde{E}_{jx} + \mathrm{i}\omega \frac{\partial}{\partial z}\tilde{B}_{jx}\right), \tag{4.13}$$

$$\tilde{E}_{jz} = \frac{1}{k_j^2 - k_x^2}\left(\mathrm{i}k_x \frac{\partial \tilde{E}_{jx}}{\partial z} + k_y \omega \tilde{B}_{jx}\right), \tag{4.14}$$

$$\tilde{B}_{jy} = \frac{1}{k_j^2 - k_x^2}\left(-\frac{\mathrm{i}k_j^2}{\omega}\frac{\partial}{\partial z}\tilde{E}_{jx} - k_x k_y \tilde{B}_{jx}\right), \tag{4.15}$$

$$\tilde{B}_{jz} = \frac{1}{k_j^2 - k_x^2}\left(-\frac{k_j^2 k_y}{\omega}\tilde{E}_{jx} + \mathrm{i}k_x \frac{\partial}{\partial z}\tilde{B}_{jx}\right). \tag{4.16}$$

With the substitutions of Eqs. (4.13)–(4.16) into (4.6) and (4.9), the following ordinary differential equations for $\tilde{E}_{jx}$ and $\tilde{B}_{jx}$ can be obtained readily. They are

$$\left(\frac{\mathrm{d}^2}{\mathrm{d}z^2} + \gamma_j^2\right)\tilde{E}_{jx} = \frac{\omega\mu_0 I\,\mathrm{d}l(k_j^2 - k_x^2)}{\mathrm{i}k_j^2}\delta(z+d), \tag{4.17}$$

$$\left(\frac{\mathrm{d}^2}{\mathrm{d}z^2} + \gamma_j^2\right)\tilde{B}_{jx} = 0, \tag{4.18}$$

where

$$\gamma_j^2 = k_j^2 - k_x^2 - k_y^2; \quad j = i, a, g. \tag{4.19}$$

In a similar manner as in Chap. 15 of the book by King et al. (1992), and considering that the transforms of tangential electric field components and those of the tangential magnetic field components are continuous on the interfaces $z = 0$ and $z = h$, the Fourier transforms of the tangential components in Region 2 (Earth) can be derived readily. They are

$$\begin{aligned}\tilde{E}_{gx} = &-\frac{\omega\mu_0 I\,\mathrm{d}l}{2k_g^2(k_g^2 - \gamma_g^2)}\left\{\left[\gamma_g k_x^2 + \frac{k_g^2}{\gamma_g}k_y^2\right]\left(\mathrm{e}^{\mathrm{i}\gamma_g|z+d|} - \mathrm{e}^{\mathrm{i}\gamma_g|z-d|}\right)\right.\\ &\left.+\left[\gamma_g(Q_2+1)k_x^2 - \frac{k_g^2}{\gamma_g}(P_3-1)k_y^2\right]\mathrm{e}^{\mathrm{i}\gamma_g|z-d|}\right\},\end{aligned} \tag{4.20}$$

$$\begin{aligned}\tilde{E}_{gy} = &-\frac{\omega\mu_0 k_x k_y I\,\mathrm{d}l}{2k_g^2(k_g^2 - \gamma_g^2)}\left\{\left(\gamma_g - \frac{k_g^2}{\gamma_g}\right)\left(\mathrm{e}^{\mathrm{i}\gamma_g|z+d|} - \mathrm{e}^{\mathrm{i}\gamma_g|z-d|}\right)\right.\\ &\left.+\left[\gamma_g(Q_2+1) + \frac{k_g^2}{\gamma_g}(P_3-1)\right]\cdot \mathrm{e}^{\mathrm{i}\gamma_g|z-d|}\right\},\end{aligned} \tag{4.21}$$

$$\tilde{B}_{gx} = \frac{\mu_0 k_x k_y I \,\mathrm{d}l}{2(k_g^2 - \gamma_g^2)}(Q_2 + P_3)\mathrm{e}^{\mathrm{i}\gamma_g|z-d|}, \tag{4.22}$$

$$\tilde{B}_{gy} = -\frac{\mu_0 I \,\mathrm{d}l}{2(k_g^2 - \gamma_g^2)}\{-(k_g^2 - \gamma_g^2)(\mathrm{e}^{\mathrm{i}\gamma_g|z+d|} + \mathrm{e}^{\mathrm{i}\gamma_g|z-d|}) + \left[(Q_2+1)k_x^2 - (P_3 - 1)k_y^2\right]\mathrm{e}^{\mathrm{i}\gamma_g|z-d|}\}. \tag{4.23}$$

In these formulas, the factors (Q_2+1) and (P_3-1) are represented in the following forms:

$$\frac{\gamma_g}{2}(Q_2+1) = \frac{\frac{k_g^2\gamma_g}{k_a^2}\left(\frac{k_a^2\gamma_i}{k_i^2} - \mathrm{i}\gamma_a \tan\gamma_a h\right)}{\gamma_g + \frac{k_g^2\gamma_i}{k_i^2} - \mathrm{i}\left(\frac{k_g^2\gamma_a}{k_a^2} + \frac{k_a^2\gamma_i\gamma_g}{k_i^2\gamma_a}\right)\tan\gamma_a h}, \tag{4.24}$$

$$\frac{k_g^2}{2\gamma_g}(P_3 - 1) = -\frac{k_g^2\left(\frac{\gamma_a}{\gamma_i} - \mathrm{i}\tan\gamma_a h\right)}{\gamma_a + \frac{\gamma_a\gamma_g}{\gamma_i} - \mathrm{i}\left(\gamma_g + \frac{\gamma_a^2}{\gamma_i}\right)\tan\gamma_a h}. \tag{4.25}$$

In Region 1 (air), the total field consists of the upward and downward waves. The transforms of the tangential components $\tilde{E}_{ax}$ and $\tilde{B}_{ax}$ can be written in the following forms:

$$\tilde{E}_{ax} = A_1\mathrm{e}^{\mathrm{i}\gamma_a z} + A_2\mathrm{e}^{-\mathrm{i}\gamma_a z}, \tag{4.26}$$

$$\tilde{B}_{ax} = B_1\mathrm{e}^{\mathrm{i}\gamma_a z} + B_2\mathrm{e}^{-\mathrm{i}\gamma_a z}. \tag{4.27}$$

In Region 0 (ionosphere), only the upward wave exists. The transforms of the tangential components $\tilde{E}_{ix}$ and $\tilde{B}_{ix}$ are obtained readily. They are

$$\tilde{E}_{ix} = C_1\mathrm{e}^{\mathrm{i}\gamma_i z}, \tag{4.28}$$

$$\tilde{B}_{ix} = C_2\mathrm{e}^{\mathrm{i}\gamma_i z}. \tag{4.29}$$

By using the boundary conditions, the coefficients A_1, A_2, B_1, B_2, C_1, and C_2 can be determined easily. They are

$$A_1 = -\frac{\omega\mu_0 I \,\mathrm{d}l\,\mathrm{e}^{\mathrm{i}\gamma_g d}}{4k_g^2(k_g^2 - \gamma_g^2)}\left[\gamma_g(Q_2+1)k_x^2 - \frac{k_g^2}{\gamma_g}(P_3-1)k_y^2\right] + \frac{\omega\mu_0 I \,\mathrm{d}l\,\mathrm{e}^{\mathrm{i}\gamma_g d}}{4k_a^2\gamma_a(k_g^2 - \gamma_g^2)} \times\left[-(Q_2+1)k_x^2\gamma_a^2 + (P_3-1)k_y^2k_a^2 + 2(k_g^2 - \gamma_g^2)(k_a^2 - k_x^2)\right], \tag{4.30}$$

$$A_2 = -\frac{\omega\mu_0 I \,\mathrm{d}l\,\mathrm{e}^{\mathrm{i}\gamma_g d}}{4k_g^2(k_g^2 - \gamma_g^2)}\left[\gamma_g(Q_2+1)k_x^2 - \frac{k_g^2}{\gamma_g}(P_3-1)k_y^2\right] - \frac{\omega\mu_0 I \,\mathrm{d}l\,\mathrm{e}^{\mathrm{i}\gamma_g d}}{4k_a^2\gamma_a(k_g^2 - \gamma_g^2)} \times\left[-(Q_2+1)k_x^2\gamma_a^2 + (P_3-1)k_y^2k_a^2 + 2(k_g^2 - \gamma_g^2)(k_a^2 - k_x^2)\right], \tag{4.31}$$

$$B_1 = \frac{\mu_0 k_x k_y I\, \mathrm{d}l \mathrm{e}^{\mathrm{i}\gamma_g d}}{4(k_g^2 - \gamma_g^2)}(Q_2 + P_3) + \frac{\mu_0 k_x k_y I\, \mathrm{d}l \mathrm{e}^{\mathrm{i}\gamma_g d}}{4k_g^2 \gamma_a (k_g^2 - \gamma_g^2)} \times \left[\gamma_g (Q_2 + 1)k_a^2 + \frac{k_g^2}{\gamma_g}(P_3 - 1)\gamma_a^2\right], \tag{4.32}$$

$$B_2 = \frac{\mu_0 k_x k_y I\, \mathrm{d}l \mathrm{e}^{\mathrm{i}\gamma_g d}}{4(k_g^2 - \gamma_g^2)}(Q_2 + P_3) - \frac{\mu_0 k_x k_y I\, \mathrm{d}l \mathrm{e}^{\mathrm{i}\gamma_g d}}{4k_g^2 \gamma_a (k_g^2 - \gamma_g^2)} \times \left[\gamma_g (Q_2 + 1)k_a^2 + \frac{k_g^2}{\gamma_g}(P_3 - 1)\gamma_a^2\right], \tag{4.33}$$

$$C_1 = -\frac{\omega \mu_0 I\, \mathrm{d}l \mathrm{e}^{\mathrm{i}\gamma_g d} \mathrm{e}^{-\mathrm{i}\gamma_l h}}{4(k_g^2 - \gamma_g^2)} \left\{\left[\frac{\gamma_g}{k_g^2}(Q_2 + 1)k_x^2 - \gamma_g^{-1}(P_3 - 1)k_y^2\right]\left(\mathrm{e}^{\mathrm{i}\gamma_a h} + \mathrm{e}^{-\mathrm{i}\gamma_a h}\right) - \frac{1}{k_a^2 \gamma_a}\left[-(Q_2 + 1)k_x^2 \gamma_a^2 + (P_3 - 1)k_y^2 k_a^2 + 2\left(k_g^2 - \gamma_g^2\right)\left(k_a^2 - k_x^2\right)\right] \times \left(\mathrm{e}^{\mathrm{i}\gamma_a h} - \mathrm{e}^{-\mathrm{i}\gamma_a h}\right)\right\}, \tag{4.34}$$

$$C_2 = \frac{\mu_0 k_x k_y I\, \mathrm{d}l \mathrm{e}^{\mathrm{i}\gamma_g d} \mathrm{e}^{-\mathrm{i}\gamma_l h}}{4(k_g^2 - \gamma_g^2)} \left\{(Q_2 + P_3)\left(\mathrm{e}^{\mathrm{i}\gamma_a h} + \mathrm{e}^{-\mathrm{i}\gamma_a h}\right) + \frac{1}{k_g^2 \gamma_a} \times \left[\gamma_g (Q_2 + 1)k_a^2 + \frac{k_g^2}{\gamma_g}(P_3 - 1)\gamma_a^2\right] \times \left(\mathrm{e}^{\mathrm{i}\gamma_a h} - \mathrm{e}^{-\mathrm{i}\gamma_a h}\right)\right\}. \tag{4.35}$$

With Eqs. (4.30)–(4.35), and the Fourier integrals like Eq. (4.4), the integrated formulas for the field components in Regions 0 and 1 can be derived readily.

In Region 1, it follows that

$$E_{ax} = \frac{\omega \mu_0 I\, \mathrm{d}l}{16\pi^2} \int_{-\infty}^{\infty} \mathrm{d}k_x \int_{-\infty}^{\infty} \mathrm{d}k_y \left\{-\frac{1}{k_g^2}\left[\gamma_g (Q_2 + 1)k_x^2 - \frac{k_g^2}{\gamma_g}(P_3 - 1)k_y^2\right] \times \left(\mathrm{e}^{\mathrm{i}\gamma_a z} + \mathrm{e}^{-\mathrm{i}\gamma_a z}\right) + \frac{1}{k_a^2 \gamma_a}\left[-(Q_2 + 1)k_x^2 \gamma_a^2 + (P_3 - 1)k_y^2 k_a^2 + 2\left(k_g^2 - \gamma_g^2\right)\left(k_a^2 - k_x^2\right)\right]\left(\mathrm{e}^{-\mathrm{i}\gamma_a z} - \mathrm{e}^{\mathrm{i}\gamma_a z}\right)\right\} \frac{\mathrm{e}^{\mathrm{i}\gamma_g d} \mathrm{e}^{\mathrm{i}(k_x x + k_y y)}}{(k_g^2 - \gamma_g^2)}, \tag{4.36}$$

$$E_{ay} = -\frac{\omega \mu_0 I\, \mathrm{d}l}{16\pi^2} \int_{-\infty}^{\infty} \mathrm{d}k_x \int_{-\infty}^{\infty} \mathrm{d}k_y \left\{\frac{1}{k_g^2}\left[\gamma_g (Q_2 + 1) + \frac{k_g^2}{\gamma_g}(P_3 - 1)\right] \times \left(\mathrm{e}^{\mathrm{i}\gamma_a z} + \mathrm{e}^{-\mathrm{i}\gamma_a z}\right) + \left[\frac{\gamma_a}{k_a^2}(Q_2 + 1) + \gamma_a^{-1}(P_3 - 1) + \frac{2(k_g^2 - \gamma_g^2)}{k_a^2 \gamma_a}\right]\left(\mathrm{e}^{-\mathrm{i}\gamma_a z} - \mathrm{e}^{\mathrm{i}\gamma_a z}\right)\right\} \times \frac{k_x k_y \mathrm{e}^{\mathrm{i}\gamma_g d} \mathrm{e}^{\mathrm{i}(k_x x + k_y y)}}{(k_g^2 - \gamma_g^2)}, \tag{4.37}$$

$$E_{az} = \frac{\omega\mu_0 I\,\mathrm{d}l}{16\pi^2}\int_{-\infty}^{\infty}\mathrm{d}k_x\int_{-\infty}^{\infty}\mathrm{d}k_y\left\{k_a^{-2}(Q_2-1)\left(\mathrm{e}^{\mathrm{i}\gamma_a z}+\mathrm{e}^{-\mathrm{i}\gamma_a z}\right)\right.$$
$$\left.+\frac{\gamma_g}{k_g^2\gamma_a}(Q_2+1)\left(\mathrm{e}^{-\mathrm{i}\gamma_a z}-\mathrm{e}^{\mathrm{i}\gamma_a z}\right)\right\}k_x\mathrm{e}^{\mathrm{i}\gamma_g d}\mathrm{e}^{\mathrm{i}(k_x x+k_y y)}, \tag{4.38}$$

$$B_{ax} = \frac{\mu_0 I\,\mathrm{d}l}{16\pi^2}\int_{-\infty}^{\infty}\mathrm{d}k_x\int_{-\infty}^{\infty}\mathrm{d}k_y\left\{(Q_2+P_3)\left(\mathrm{e}^{\mathrm{i}\gamma_a z}+\mathrm{e}^{-\mathrm{i}\gamma_a z}\right)+\frac{1}{k_g^2\gamma_a}\right.$$
$$\left.\times\left[\gamma_g(Q_2+1)k_a^2+\frac{k_g^2}{\gamma_g}(P_3-1)\gamma_a^2\right]\left(\mathrm{e}^{-\mathrm{i}\gamma_a z}-\mathrm{e}^{\mathrm{i}\gamma_a z}\right)\right\}$$
$$\times\frac{k_x k_y\mathrm{e}^{\mathrm{i}\gamma_g d}\mathrm{e}^{\mathrm{i}(k_x x+k_y y)}}{(k_g^2-\gamma_g^2)}, \tag{4.39}$$

$$B_{ay} = \frac{\mu_0 I\,\mathrm{d}l}{16\pi^2}\int_{-\infty}^{\infty}\mathrm{d}k_x\int_{-\infty}^{\infty}\mathrm{d}k_y\left\{\left[(P_3-1)k_y^2-(Q_2+1)k_x^2+2\left(k_g^2-\gamma_g^2\right)\right]\right.$$
$$\times\left(\mathrm{e}^{\mathrm{i}\gamma_a z}+\mathrm{e}^{-\mathrm{i}\gamma_a z}\right)$$
$$+\frac{1}{k_g^2}\left[(P_3-1)\frac{k_g^2\gamma_a}{\gamma_g}k_y^2-(Q_2+1)\frac{k_a^2\gamma_g}{\gamma_a}k_x^2\right]$$
$$\left.\times\left(\mathrm{e}^{-\mathrm{i}\gamma_a z}-\mathrm{e}^{\mathrm{i}\gamma_a z}\right)\right\}\frac{\mathrm{e}^{\mathrm{i}\gamma_g d}\mathrm{e}^{\mathrm{i}(k_x x+k_y y)}}{(k_g^2-\gamma_g^2)}, \tag{4.40}$$

$$B_{az} = -\frac{\mu_0 I\,\mathrm{d}l}{16\pi^2}\int_{-\infty}^{\infty}\mathrm{d}k_x\int_{-\infty}^{\infty}\mathrm{d}k_y\left\{\gamma_g^{-1}(P_3-1)\left(\mathrm{e}^{\mathrm{i}\gamma_a z}+\mathrm{e}^{-\mathrm{i}\gamma_a z}\right)+\gamma_a^{-1}(P_3+1)\right.$$
$$\left.\times\left(\mathrm{e}^{-\mathrm{i}\gamma_a z}-\mathrm{e}^{\mathrm{i}\gamma_a z}\right)\right\}k_y\mathrm{e}^{\mathrm{i}\gamma_g d}\mathrm{e}^{\mathrm{i}(k_x x+k_y y)}. \tag{4.41}$$

In Region 0, it follows that

$$E_{ix} = \frac{\omega\mu_0 I\,\mathrm{d}l}{16\pi^2}\int_{-\infty}^{\infty}\mathrm{d}k_x\int_{-\infty}^{\infty}\mathrm{d}k_y\left\{-\frac{1}{k_g^2}\left[\gamma_g(Q_2+1)k_x^2-\frac{k_g^2}{\gamma_g}(P_3-1)k_y^2\right]\right.$$
$$\times\left(\mathrm{e}^{\mathrm{i}\gamma_a h}+\mathrm{e}^{-\mathrm{i}\gamma_a h}\right)+\frac{1}{k_a^2\gamma_a}$$
$$\times\left[-(Q_2+1)k_x^2\gamma_a^2+(P_3-1)k_y^2k_a^2\right.$$
$$\left.\left.+2\left(k_g^2-\gamma_g^2\right)\left(k_a^2-k_x^2\right)\right]\left(\mathrm{e}^{-\mathrm{i}\gamma_a h}-\mathrm{e}^{\mathrm{i}\gamma_a h}\right)\right\}$$
$$\times\frac{\mathrm{e}^{\mathrm{i}\gamma_i(z-h)}\mathrm{e}^{\mathrm{i}\gamma_g d}\mathrm{e}^{\mathrm{i}(k_x x+k_y y)}}{(k_g^2-\gamma_g^2)}, \tag{4.42}$$

$$E_{iy} = -\frac{\omega\mu_0 I\,\mathrm{d}l}{16\pi^2}\int_{-\infty}^{\infty}\mathrm{d}k_x\int_{-\infty}^{\infty}\mathrm{d}k_y\left\{\frac{1}{k_g^2}\left[\gamma_g(Q_2+1)+\frac{k_g^2}{\gamma_g}(P_3-1)\right]\right.$$
$$\times\left(\mathrm{e}^{\mathrm{i}\gamma_a h}+\mathrm{e}^{-\mathrm{i}\gamma_a h}\right)$$
$$\left.+\left[\frac{\gamma_a}{k_a^2}(Q_2+1)+\gamma_a^{-1}(P_3-1)+\frac{2(k_g^2-\gamma_g^2)}{k_a^2\gamma_a}\right]\times\left(\mathrm{e}^{-\mathrm{i}\gamma_a h}-\mathrm{e}^{\mathrm{i}\gamma_a h}\right)\right\}$$
$$\times\frac{k_x k_y \mathrm{e}^{\mathrm{i}\gamma_i(z-h)}\mathrm{e}^{\mathrm{i}\gamma_g d}\mathrm{e}^{\mathrm{i}(k_x x+k_y y)}}{(k_g^2-\gamma_g^2)}, \tag{4.43}$$

$$E_{iz} = \frac{\omega\mu_0 I\,\mathrm{d}l}{16\pi^2}\int_{-\infty}^{\infty}\mathrm{d}k_x\int_{-\infty}^{\infty}\mathrm{d}k_y\left\{k_a^{-2}(Q_2-1)\left(\mathrm{e}^{\mathrm{i}\gamma_a h}+\mathrm{e}^{-\mathrm{i}\gamma_a h}\right)\right.$$
$$\left.+\frac{\gamma_g}{k_g^2\gamma_a}(Q_2+1)\left(\mathrm{e}^{-\mathrm{i}\gamma_a h}-\mathrm{e}^{\mathrm{i}\gamma_a h}\right)\right\}$$
$$\times k_x \mathrm{e}^{\mathrm{i}\gamma_i(z-h)}\mathrm{e}^{\mathrm{i}\gamma_g d}\mathrm{e}^{\mathrm{i}(k_x x+k_y y)}, \tag{4.44}$$

$$B_{ix} = \frac{\mu_0 I\,\mathrm{d}l}{16\pi^2}\int_{-\infty}^{\infty}\mathrm{d}k_x\int_{-\infty}^{\infty}\mathrm{d}k_y\left\{(Q_2+P_3)\left(\mathrm{e}^{\mathrm{i}\gamma_a h}+\mathrm{e}^{-\mathrm{i}\gamma_a h}\right)+\frac{1}{k_g^2\gamma_a}\right.$$
$$\left.\times\left[\gamma_g(Q_2+1)k_a^2+\frac{k_g^2}{\gamma_g}(P_3-1)\gamma_a^2\right]\left(\mathrm{e}^{-\mathrm{i}\gamma_a h}-\mathrm{e}^{\mathrm{i}\gamma_a h}\right)\right\}$$
$$\times\frac{k_x k_y \mathrm{e}^{\mathrm{i}\gamma_i(z-h)}\mathrm{e}^{\mathrm{i}\gamma_g d}\mathrm{e}^{\mathrm{i}(k_x x+k_y y)}}{(k_g^2-\gamma_g^2)}, \tag{4.45}$$

$$B_{iy} = \frac{\mu_0 I\,\mathrm{d}l}{16\pi^2}\int_{-\infty}^{\infty}\mathrm{d}k_x\int_{-\infty}^{\infty}\mathrm{d}k_y\left\{\left[(P_3-1)k_y^2-(Q_2+1)k_x^2+2\left(k_g^2-\gamma_g^2\right)\right]\right.$$
$$\times\left(\mathrm{e}^{\mathrm{i}\gamma_a h}+\mathrm{e}^{-\mathrm{i}\gamma_a h}\right)+\frac{1}{k_g^2}\left[(P_3-1)\frac{k_g^2\gamma_a}{\gamma_g}k_y^2-(Q_2+1)\frac{k_a^2\gamma_g}{\gamma_a}k_x^2\right]$$
$$\left.\times\left(\mathrm{e}^{-\mathrm{i}\gamma_a h}-\mathrm{e}^{\mathrm{i}\gamma_a h}\right)\right\}\frac{\mathrm{e}^{\mathrm{i}\gamma_i(z-h)}\mathrm{e}^{\mathrm{i}\gamma_g d}\mathrm{e}^{\mathrm{i}(k_x x+k_y y)}}{(k_g^2-\gamma_g^2)}, \tag{4.46}$$

$$B_{iz} = -\frac{\mu_0 I\,\mathrm{d}l}{16\pi^2}\int_{-\infty}^{\infty}\mathrm{d}k_x\int_{-\infty}^{\infty}\mathrm{d}k_y\left\{\gamma_g^{-1}(P_3-1)\left(\mathrm{e}^{\mathrm{i}\gamma_a h}+\mathrm{e}^{-\mathrm{i}\gamma_a h}\right)\right.$$
$$\left.+\gamma_a^{-1}(P_3+1)\left(\mathrm{e}^{-\mathrm{i}\gamma_a h}-\mathrm{e}^{\mathrm{i}\gamma_a h}\right)\right\}k_y \mathrm{e}^{\mathrm{i}\gamma_i(z-h)}\mathrm{e}^{\mathrm{i}\gamma_g d}\mathrm{e}^{\mathrm{i}(k_x x+k_y y)}. \tag{4.47}$$

With the relations

$$x=\rho\cos\phi;\qquad y=\rho\sin\phi;\qquad k_x=\lambda\cos\phi';\qquad k_y=\lambda\sin\phi', \tag{4.48}$$

we find

$$\gamma_j=\sqrt{k_j^2-k_x^2-k_y^2}=\sqrt{k_j^2-\lambda^2};\quad j=i,a,g \tag{4.49}$$

$$k_x x + k_y y = \lambda\rho\cos(\phi-\phi'); \qquad \mathrm{d}k_x\,\mathrm{d}k_y = \lambda\,\mathrm{d}\phi'\,\mathrm{d}\lambda. \tag{4.50}$$

In cylindrical coordinates, we have

$$E_\rho = E_x\cos\phi + E_y\sin\phi; \qquad E_\phi = -E_x\sin\phi + E_y\cos\phi. \tag{4.51}$$

Thus, the integrated formulas for the field components can be represented in the following forms:

$$E_{a\rho} = \frac{\omega\mu_0 I\,\mathrm{d}l}{16\pi}\cos\phi\cdot\left(F_{a\rho}^{(1)} + F_{a\rho}^{(2)} + F_{a\rho}^{(3)}\right), \tag{4.52}$$

$$E_{a\phi} = -\frac{\omega\mu_0 I\,\mathrm{d}l}{16\pi}\sin\phi\cdot\left(F_{a\phi}^{(1)} + F_{a\phi}^{(2)} + F_{a\phi}^{(3)}\right), \tag{4.53}$$

$$E_{az} = \frac{\mathrm{i}\omega\mu_0 I\,\mathrm{d}l}{8\pi}\cos\phi\cdot\left(F_{az}^{(1)} + F_{az}^{(3)}\right), \tag{4.54}$$

$$B_{a\rho} = \frac{\mu_0 I\,\mathrm{d}l}{16\pi}\sin\phi\cdot\left(G_{a\rho}^{(1)} + G_{a\rho}^{(2)} + G_{a\rho}^{(3)}\right), \tag{4.55}$$

$$B_{a\phi} = \frac{\mu_0 I\,\mathrm{d}l}{16\pi}\cos\phi\cdot\left(G_{a\phi}^{(1)} + G_{a\phi}^{(2)} + G_{a\phi}^{(3)}\right), \tag{4.56}$$

$$B_{az} = -\frac{\mathrm{i}\mu_0 I\,\mathrm{d}l}{8\pi}\sin\phi\cdot\left(G_{az}^{(2)} + G_{az}^{(3)}\right). \tag{4.57}$$

In these formulas, the terms with the superscript (1) designate the terms of the electric-type field, also known as the transverse magnetic (TM) field. The terms with the superscript (2) designate the terms of the magnetic-type field, also known as the transverse electric (TE) field. The terms with the superscript (3) designate the terms of the direct and reflected waves. We write

$$F_{a\rho}^{(1)} = -\int_0^\infty \left[\left(\frac{\gamma_g}{k_g^2} + \frac{\gamma_a}{k_a^2}\right)\mathrm{e}^{-\mathrm{i}\gamma_a z} + \left(\frac{\gamma_g}{k_g^2} - \frac{\gamma_a}{k_a^2}\right)\mathrm{e}^{\mathrm{i}\gamma_a z}\right](Q_2+1)$$
$$\times\left[J_0(\lambda\rho) - J_2(\lambda\rho)\right]\mathrm{e}^{\mathrm{i}\gamma_g d}\lambda\,\mathrm{d}\lambda, \tag{4.58}$$

$$F_{a\phi}^{(1)} = -\int_0^\infty \left[\left(\frac{\gamma_g}{k_g^2} + \frac{\gamma_a}{k_a^2}\right)\mathrm{e}^{-\mathrm{i}\gamma_a z} + \left(\frac{\gamma_g}{k_g^2} - \frac{\gamma_a}{k_a^2}\right)\mathrm{e}^{\mathrm{i}\gamma_a z}\right](Q_2+1)$$
$$\times\left[J_0(\lambda\rho) + J_2(\lambda\rho)\right]\mathrm{e}^{\mathrm{i}\gamma_g d}\lambda\,\mathrm{d}\lambda, \tag{4.59}$$

$$F_{az}^{(1)} = \int_0^\infty \left[\left(k_a^{-2} + \frac{\gamma_g}{k_g^2\gamma_a}\right)\mathrm{e}^{-\mathrm{i}\gamma_a z} + \left(k_a^{-2} - \frac{\gamma_g}{k_g^2\gamma_a}\right)\mathrm{e}^{\mathrm{i}\gamma_a z}\right](Q_2+1)$$
$$\times J_1(\lambda\rho)\mathrm{e}^{\mathrm{i}\gamma_g d}\lambda^2\,\mathrm{d}\lambda, \tag{4.60}$$

$$G_{a\rho}^{(1)} = -\int_0^\infty \left(\frac{k_g^2\gamma_a + k_a^2\gamma_g}{k_g^2\gamma_a}\mathrm{e}^{-\mathrm{i}\gamma_a z} + \frac{k_g^2\gamma_a - k_a^2\gamma_g}{k_g^2\gamma_a}\mathrm{e}^{\mathrm{i}\gamma_a z}\right)(Q_2+1)$$
$$\times\left[J_0(\lambda\rho) + J_2(\lambda\rho)\right]\mathrm{e}^{\mathrm{i}\gamma_g d}\lambda\,\mathrm{d}\lambda, \tag{4.61}$$

$$G_{a\phi}^{(1)} = -\int_0^\infty \left(\frac{k_g^2\gamma_a + k_a^2\gamma_g}{k_g^2\gamma_a}\mathrm{e}^{-\mathrm{i}\gamma_a z} + \frac{k_g^2\gamma_a - k_a^2\gamma_g}{k_g^2\gamma_a}\mathrm{e}^{\mathrm{i}\gamma_a z}\right)(Q_2+1) \times \left[J_0(\lambda\rho) - J_2(\lambda\rho)\right]\mathrm{e}^{\mathrm{i}\gamma_g d}\lambda\,\mathrm{d}\lambda, \tag{4.62}$$

$$F_{a\rho}^{(2)} = \int_0^\infty \left[(\gamma_g^{-1} + \gamma_a^{-1})\mathrm{e}^{-\mathrm{i}\gamma_a z} + (\gamma_g^{-1} - \gamma_a^{-1})\mathrm{e}^{\mathrm{i}\gamma_a z}\right](P_3-1) \times \left[J_0(\lambda\rho) + J_2(\lambda\rho)\right]\mathrm{e}^{\mathrm{i}\gamma_g d}\lambda\,\mathrm{d}\lambda, \tag{4.63}$$

$$F_{a\phi}^{(2)} = \int_0^\infty \left[(\gamma_g^{-1} + \gamma_a^{-1})\mathrm{e}^{-\mathrm{i}\gamma_a z} + (\gamma_g^{-1} - \gamma_a^{-1})\mathrm{e}^{\mathrm{i}\gamma_a z}\right](P_3-1) \times \left[J_0(\lambda\rho) - J_2(\lambda\rho)\right]\mathrm{e}^{\mathrm{i}\gamma_g d}\lambda\,\mathrm{d}\lambda, \tag{4.64}$$

$$G_{a\rho}^{(2)} = \int_0^\infty \left(\frac{\gamma_g + \gamma_a}{\gamma_g}\mathrm{e}^{-\mathrm{i}\gamma_a z} + \frac{\gamma_g - \gamma_a}{\gamma_g}\mathrm{e}^{\mathrm{i}\gamma_a z}\right)(P_3-1) \times \left[J_0(\lambda\rho) - J_2(\lambda\rho)\right]\mathrm{e}^{\mathrm{i}\gamma_g d}\lambda\,\mathrm{d}\lambda, \tag{4.65}$$

$$G_{a\phi}^{(2)} = \int_0^\infty \left(\frac{\gamma_g + \gamma_a}{\gamma_g}\mathrm{e}^{-\mathrm{i}\gamma_a z} + \frac{\gamma_g - \gamma_a}{\gamma_g}\mathrm{e}^{\mathrm{i}\gamma_a z}\right)(P_3-1) \times \left[J_0(\lambda\rho) + J_2(\lambda\rho)\right]\mathrm{e}^{\mathrm{i}\gamma_g d}\lambda\,\mathrm{d}\lambda, \tag{4.66}$$

$$G_{az}^{(2)} = \int_0^\infty \left[(\gamma_g^{-1} + \gamma_a^{-1})\mathrm{e}^{-\mathrm{i}\gamma_a z} + (\gamma_g^{-1} - \gamma_a^{-1})\mathrm{e}^{\mathrm{i}\gamma_a z}\right](P_3-1) \times J_1(\lambda\rho)\mathrm{e}^{\mathrm{i}\gamma_g d}\lambda^2\,\mathrm{d}\lambda, \tag{4.67}$$

$$F_{a\rho}^{(3)} = 2\int_0^\infty \left\{\gamma_a^{-1}\left[J_0(\lambda\rho) + J_2(\lambda\rho)\right] + \frac{\gamma_a}{k_a^2}\left[J_0(\lambda\rho) - J_2(\lambda\rho)\right]\right\} \times \left(\mathrm{e}^{-\mathrm{i}\gamma_a z} - \mathrm{e}^{\mathrm{i}\gamma_a z}\right)\mathrm{e}^{\mathrm{i}\gamma_g d}\lambda\,\mathrm{d}\lambda, \tag{4.68}$$

$$F_{a\phi}^{(3)} = 2\int_0^\infty \left\{\gamma_a^{-1}\left[J_0(\lambda\rho) - J_2(\lambda\rho)\right] + \frac{\gamma_a}{k_a^2}\left[J_0(\lambda\rho) + J_2(\lambda\rho)\right]\right\} \times \left(\mathrm{e}^{-\mathrm{i}\gamma_a z} - \mathrm{e}^{\mathrm{i}\gamma_a z}\right)\mathrm{e}^{\mathrm{i}\gamma_g d}\lambda\,\mathrm{d}\lambda, \tag{4.69}$$

$$F_{az}^{(3)} = -2k_a^{-2}\int_0^\infty J_1(\lambda\rho)\left(\mathrm{e}^{\mathrm{i}\gamma_a z} + \mathrm{e}^{-\mathrm{i}\gamma_a z}\right)\mathrm{e}^{\mathrm{i}\gamma_g d}\lambda^2\,\mathrm{d}\lambda, \tag{4.70}$$

$$G_{a\rho}^{(3)} = G_{a\phi}^{(3)} = 4\int_0^\infty J_0(\lambda\rho)\left(\mathrm{e}^{\mathrm{i}\gamma_a z} + \mathrm{e}^{-\mathrm{i}\gamma_a z}\right)\mathrm{e}^{\mathrm{i}\gamma_g d}\lambda\,\mathrm{d}\lambda, \tag{4.71}$$

$$G_{az}^{(3)} = 2\int_0^\infty \gamma_a^{-1}J_1(\lambda\rho)\left(\mathrm{e}^{-\mathrm{i}\gamma_a z} - \mathrm{e}^{\mathrm{i}\gamma_a z}\right)\mathrm{e}^{\mathrm{i}\gamma_g d}\lambda^2\,\mathrm{d}\lambda. \tag{4.72}$$

In Region 0, the integrated formulas for the field components can be represented in the following forms:

$$E_{i\rho} = \frac{\omega\mu_0 I\,\mathrm{d}l}{8\pi}\cos\phi\cdot\left(F_{i\rho}^{(1)} + F_{i\rho}^{(2)} + F_{i\rho}^{(3)}\right), \tag{4.73}$$

$$E_{i\phi} = -\frac{\omega\mu_0 I\,\mathrm{d}l}{8\pi}\sin\phi\cdot\left(F_{i\phi}^{(1)} + F_{i\phi}^{(2)} + F_{i\phi}^{(3)}\right), \tag{4.74}$$

$$E_{iz} = \frac{\mathrm{i}\omega\mu_0 I\,\mathrm{d}l}{4\pi}\cos\phi\cdot\left(F_{iz}^{(1)} + F_{iz}^{(3)}\right), \tag{4.75}$$

$$B_{i\rho} = \frac{\mu_0 I\,\mathrm{d}l}{8\pi}\sin\phi\cdot\left(G_{i\rho}^{(1)} + G_{i\rho}^{(2)} + G_{i\rho}^{(3)}\right), \tag{4.76}$$

$$B_{i\phi} = \frac{\mu_0 I\,\mathrm{d}l}{8\pi}\cos\phi\cdot\left(G_{i\phi}^{(1)} + G_{i\phi}^{(2)} + G_{i\phi}^{(3)}\right), \tag{4.77}$$

$$B_{iz} = -\frac{\mathrm{i}\mu_0 I\,\mathrm{d}l}{4\pi}\sin\phi\cdot\left(G_{iz}^{(2)} + G_{iz}^{(3)}\right), \tag{4.78}$$

where

$$F_{i\rho}^{(1)} = -\int_0^\infty \left[\frac{\gamma_g}{k_g^2}\cos(\gamma_a h) - \mathrm{i}\frac{\gamma_a}{k_a^2}\sin(\gamma_a h)\right]\left[J_0(\lambda\rho) - J_2(\lambda\rho)\right]$$
$$\times (Q_2+1)\mathrm{e}^{\mathrm{i}\gamma_g d}\mathrm{e}^{\mathrm{i}\gamma_i(z-h)}\lambda\,\mathrm{d}\lambda, \tag{4.79}$$

$$F_{i\phi}^{(1)} = -\int_0^\infty \left[\frac{\gamma_g}{k_g^2}\cos(\gamma_a h) - \mathrm{i}\frac{\gamma_a}{k_a^2}\sin(\gamma_a h)\right]\left[J_0(\lambda\rho) + J_2(\lambda\rho)\right]$$
$$\times (Q_2+1)\mathrm{e}^{\mathrm{i}\gamma_g d}\mathrm{e}^{\mathrm{i}\gamma_i(z-h)}\lambda\,\mathrm{d}\lambda, \tag{4.80}$$

$$F_{iz}^{(1)} = k_i^{-2}\int_0^\infty \left[\cos(\gamma_a h) - \frac{\mathrm{i}k_a^2\gamma_g}{k_g^2\gamma_a}\sin(\gamma_a h)\right]J_1(\lambda\rho)$$
$$\times (Q_2+1)\mathrm{e}^{\mathrm{i}\gamma_g d}\mathrm{e}^{\mathrm{i}\gamma_i(z-h)}\lambda^2\,\mathrm{d}\lambda, \tag{4.81}$$

$$G_{i\rho}^{(1)} = -\int_0^\infty \left[\cos(\gamma_a h) - \frac{\mathrm{i}k_a^2\gamma_g}{k_g^2\gamma_a}\sin(\gamma_a h)\right]\left[J_0(\lambda\rho) + J_2(\lambda\rho)\right]$$
$$\times (Q_2+1)\mathrm{e}^{\mathrm{i}\gamma_g d}\mathrm{e}^{\mathrm{i}\gamma_i(z-h)}\lambda\,\mathrm{d}\lambda, \tag{4.82}$$

$$G_{i\phi}^{(1)} = -\int_0^\infty \left[\cos(\gamma_a h) - \frac{\mathrm{i}k_a^2\gamma_g}{k_g^2\gamma_a}\sin(\gamma_a h)\right]\left[J_0(\lambda\rho) - J_2(\lambda\rho)\right]$$
$$\times (Q_2+1)\mathrm{e}^{\mathrm{i}\gamma_g d}\mathrm{e}^{\mathrm{i}\gamma_i(z-h)}\lambda\,\mathrm{d}\lambda, \tag{4.83}$$

$$F_{i\rho}^{(2)} = \int_0^\infty \left[\gamma_g^{-1}\cos(\gamma_a h) - \mathrm{i}\gamma_a^{-1}\sin(\gamma_a h)\right]\left[J_0(\lambda\rho) + J_2(\lambda\rho)\right]$$
$$\times (P_3-1)\mathrm{e}^{\mathrm{i}\gamma_g d}\mathrm{e}^{\mathrm{i}\gamma_i(z-h)}\lambda\,\mathrm{d}\lambda, \tag{4.84}$$

$$F_{i\phi}^{(2)} = \int_0^\infty \left[\gamma_g^{-1}\cos(\gamma_a h) - \mathrm{i}\gamma_a^{-1}\sin(\gamma_a h)\right]\left[J_0(\lambda\rho) - J_2(\lambda\rho)\right]$$
$$\times (P_3-1)\mathrm{e}^{\mathrm{i}\gamma_g d}\mathrm{e}^{\mathrm{i}\gamma_i(z-h)}\lambda\,\mathrm{d}\lambda, \tag{4.85}$$

$$G_{i\rho}^{(2)} = \int_0^\infty \left[\cos(\gamma_a h) - \mathrm{i}\frac{\gamma_a}{\gamma_g}\sin(\gamma_a h)\right]\left[J_0(\lambda\rho) - J_2(\lambda\rho)\right]$$

$$\times (P_3 - 1)\mathrm{e}^{\mathrm{i}\gamma_g d}\mathrm{e}^{\mathrm{i}\gamma_i(z-h)}\lambda\,\mathrm{d}\lambda, \tag{4.86}$$

$$G_{i\phi}^{(2)} = \int_0^\infty \left[\cos(\gamma_a h) - \mathrm{i}\frac{\gamma_a}{\gamma_g}\sin(\gamma_a h)\right]\left[J_0(\lambda\rho) + J_2(\lambda\rho)\right]$$

$$\times (P_3 - 1)\mathrm{e}^{\mathrm{i}\gamma_g d}\mathrm{e}^{\mathrm{i}\gamma_i(z-h)}\lambda\,\mathrm{d}\lambda, \tag{4.87}$$

$$G_{iz}^{(2)} = \int_0^\infty \left[\gamma_g^{-1}\cos(\gamma_a h) - \mathrm{i}\gamma_a^{-1}\sin(\gamma_a h)\right]J_1(\lambda\rho)$$

$$\times (P_3 - 1)\mathrm{e}^{\mathrm{i}\gamma_g d}\mathrm{e}^{\mathrm{i}\gamma_i(z-h)}\lambda^2\,\mathrm{d}\lambda, \tag{4.88}$$

$$F_{i\rho}^{(3)} = -2\int_0^\infty \left\{\gamma_a^{-1}\left[J_0(\lambda\rho) + J_2(\lambda\rho)\right] + \frac{\gamma_a}{k_a^2}\left[J_0(\lambda\rho) - J_2(\lambda\rho)\right]\right\}$$

$$\times \mathrm{i}\sin(\gamma_a h)\mathrm{e}^{\mathrm{i}\gamma_g d}\mathrm{e}^{\mathrm{i}\gamma_i(z-h)}\lambda\,\mathrm{d}\lambda, \tag{4.89}$$

$$F_{i\phi}^{(3)} = -2\int_0^\infty \left\{\gamma_a^{-1}\left[J_0(\lambda\rho) - J_2(\lambda\rho)\right] + \frac{\gamma_a}{k_a^2}\left[J_0(\lambda\rho) + J_2(\lambda\rho)\right]\right\}$$

$$\times \mathrm{i}\sin(\gamma_a h)\mathrm{e}^{\mathrm{i}\gamma_g d}\mathrm{e}^{\mathrm{i}\gamma_i(z-h)}\lambda\,\mathrm{d}\lambda, \tag{4.90}$$

$$F_{iz}^{(3)} = -2k_i^{-2}\int_0^\infty \cos(\gamma_a h)J_1(\lambda\rho)\mathrm{e}^{\mathrm{i}\gamma_g d}\mathrm{e}^{\mathrm{i}\gamma_i(z-h)}\lambda^2\,\mathrm{d}\lambda, \tag{4.91}$$

$$G_{i\rho}^{(3)} = G_{i\phi}^{(3)} = 4\int_0^\infty \cos(\gamma_a h)J_0(\lambda\rho)\mathrm{e}^{\mathrm{i}\gamma_g d}\mathrm{e}^{\mathrm{i}\gamma_i(z-h)}\lambda\,\mathrm{d}\lambda, \tag{4.92}$$

$$G_{iz}^{(3)} = -2\int_0^\infty \mathrm{i}\gamma_a^{-1}\sin(\gamma_a h)J_1(\lambda\rho)\mathrm{e}^{\mathrm{i}\gamma_g d}\mathrm{e}^{\mathrm{i}\gamma_i(z-h)}\lambda^2\,\mathrm{d}\lambda. \tag{4.93}$$

4.3 Analytical Formulas of the Field Components

In Sect. 4.2, the integrated formulas for the field components have been obtained. In this section, we will attempt to evaluate these Sommerfeld integrals.

4.3.1 Evaluation for the Electric-Type Field

By examining these Sommerfeld integrals, it is seen that all integrands including Bessel functions $J_j(\lambda\rho)$ $(j = 0, 1, 2)$ are highly oscillatory and the Sommerfeld integrals converge very slowly.

Taking into account that γ_i, γ_a, and γ_g are even functions of λ, and with the following relations between Bessel function and Hankel function:

$$J_n(\lambda\rho) = \frac{1}{2}\left[H_n^{(1)}(\lambda\rho) + H_n^{(2)}(\lambda\rho)\right], \tag{4.94}$$

$$H_n^{(1)}(-\lambda\rho) = (-1)^{n+1} H_n^{(2)}(\lambda\rho), \tag{4.95}$$

it follows that

$$F_{a\rho}^{(1)} = -\int_{-\infty}^{\infty} \frac{[(\frac{\gamma_g}{k_g^2} + \frac{\gamma_a}{k_a^2})\mathrm{e}^{-\mathrm{i}\gamma_a z} + (\frac{\gamma_g}{k_g^2} - \frac{\gamma_a}{k_a^2})\mathrm{e}^{\mathrm{i}\gamma_a z}](\frac{\gamma_i}{k_i^2} - \mathrm{i}\frac{\gamma_a}{k_a^2}\tan\gamma_a h)}{q(\lambda)} \times \left[H_0^{(1)}(\lambda\rho) - H_2^{(1)}(\lambda\rho)\right]\mathrm{e}^{\mathrm{i}\gamma_g d}\lambda\,\mathrm{d}\lambda, \tag{4.96}$$

$$F_{a\phi}^{(1)} = -\int_{-\infty}^{\infty} \frac{[(\frac{\gamma_g}{k_g^2} + \frac{\gamma_a}{k_a^2})\mathrm{e}^{-\mathrm{i}\gamma_a z} + (\frac{\gamma_g}{k_g^2} - \frac{\gamma_a}{k_a^2})\mathrm{e}^{\mathrm{i}\gamma_a z}](\frac{\gamma_i}{k_i^2} - \mathrm{i}\frac{\gamma_a}{k_a^2}\tan\gamma_a h)}{q(\lambda)} \times \left[H_0^{(1)}(\lambda\rho) + H_2^{(1)}(\lambda\rho)\right]\mathrm{e}^{\mathrm{i}\gamma_g d}\lambda\,\mathrm{d}\lambda, \tag{4.97}$$

$$F_{az}^{(1)} = \int_{-\infty}^{\infty} \frac{[(k_a^{-2} + \frac{\gamma_g}{k_g^2\gamma_a})\mathrm{e}^{-\mathrm{i}\gamma_a z} + (k_a^{-2} - \frac{\gamma_g}{k_g^2\gamma_a})\mathrm{e}^{\mathrm{i}\gamma_a z}](\frac{\gamma_i}{k_i^2} - \mathrm{i}\frac{\gamma_a}{k_a^2}\tan\gamma_a h)}{q(\lambda)} \times H_1^{(1)}(\lambda\rho)\mathrm{e}^{\mathrm{i}\gamma_g d}\lambda^2\,\mathrm{d}\lambda, \tag{4.98}$$

$$G_{a\rho}^{(1)} = -\int_{-\infty}^{\infty} \frac{[(k_g^2\gamma_a + k_a^2\gamma_g)\mathrm{e}^{-\mathrm{i}\gamma_a z} + (k_g^2\gamma_a - k_a^2\gamma_g)\mathrm{e}^{\mathrm{i}\gamma_a z}]}{k_g^2\gamma_a q(\lambda)} \times \left(\frac{\gamma_i}{k_i^2} - \mathrm{i}\frac{\gamma_a}{k_a^2}\tan\gamma_a h\right)\left[H_0^{(1)}(\lambda\rho) + H_2^{(1)}(\lambda\rho)\right]\mathrm{e}^{\mathrm{i}\gamma_g d}\lambda\,\mathrm{d}\lambda, \tag{4.99}$$

$$G_{a\phi}^{(1)} = -\int_{-\infty}^{\infty} \frac{[(k_g^2\gamma_a + k_a^2\gamma_g)\mathrm{e}^{-\mathrm{i}\gamma_a z} + (k_g^2\gamma_a - k_a^2\gamma_g)\mathrm{e}^{\mathrm{i}\gamma_a z}]}{k_g^2\gamma_a q(\lambda)} \times \left(\frac{\gamma_i}{k_i^2} - \mathrm{i}\frac{\gamma_a}{k_a^2}\tan\gamma_a h\right)\left[H_0^{(1)}(\lambda\rho) - H_2^{(1)}(\lambda\rho)\right]\mathrm{e}^{\mathrm{i}\gamma_g d}\lambda\,\mathrm{d}\lambda, \tag{4.100}$$

$$F_{i\rho}^{(1)} = -\int_{-\infty}^{\infty} \frac{[\frac{\gamma_g}{k_g^2}\cos(\gamma_a h) - \mathrm{i}\frac{\gamma_a}{k_a^2}\sin(\gamma_a h)](\frac{\gamma_i}{k_i^2} - \mathrm{i}\frac{\gamma_a}{k_a^2}\tan\gamma_a h)}{q(\lambda)} \times \left[H_0^{(1)}(\lambda\rho) - H_2^{(1)}(\lambda\rho)\right]\mathrm{e}^{\mathrm{i}\gamma_g d}\mathrm{e}^{\mathrm{i}\gamma_i(z-h)}\lambda\,\mathrm{d}\lambda, \tag{4.101}$$

$$F_{i\phi}^{(1)} = -\int_{-\infty}^{\infty} \frac{[\frac{\gamma_g}{k_g^2}\cos(\gamma_a h) - \mathrm{i}\frac{\gamma_a}{k_a^2}\sin(\gamma_a h)](\frac{\gamma_i}{k_i^2} - \mathrm{i}\frac{\gamma_a}{k_a^2}\tan\gamma_a h)}{q(\lambda)} \times \left[H_0^{(1)}(\lambda\rho) + H_2^{(1)}(\lambda\rho)\right]\mathrm{e}^{\mathrm{i}\gamma_g d}\mathrm{e}^{\mathrm{i}\gamma_i(z-h)}\lambda\,\mathrm{d}\lambda, \tag{4.102}$$

$$F_{iz}^{(1)} = k_i^{-2} \int_{-\infty}^{\infty} \frac{[\cos(\gamma_a h) - \frac{ik_a^2\gamma_g}{k_g^2\gamma_a}\sin(\gamma_a h)](\frac{\gamma_i}{k_i^2} - i\frac{\gamma_a}{k_a^2}\tan\gamma_a h)}{q(\lambda)} \times H_1^{(1)}(\lambda\rho)e^{i\gamma_g d}e^{i\gamma_i(z-h)}\lambda^2\,d\lambda, \tag{4.103}$$

$$G_{i\rho}^{(1)} = -\int_{-\infty}^{\infty} \frac{[\cos(\gamma_a h) - \frac{ik_a^2\gamma_g}{k_g^2\gamma_a}\sin(\gamma_a h)](\frac{\gamma_i}{k_i^2} - i\frac{\gamma_a}{k_a^2}\tan\gamma_a h)}{q(\lambda)} \times \left[H_0^{(1)}(\lambda\rho) + H_2^{(1)}(\lambda\rho)\right]e^{i\gamma_g d}e^{i\gamma_i(z-h)}\lambda\,d\lambda, \tag{4.104}$$

$$G_{i\phi}^{(1)} = -\int_{-\infty}^{\infty} \frac{[\cos(\gamma_a h) - \frac{ik_a^2\gamma_g}{k_g^2\gamma_a}\sin(\gamma_a h)](\frac{\gamma_i}{k_i^2} - i\frac{\gamma_a}{k_a^2}\tan\gamma_a h)}{q(\lambda)} \times \left[H_0^{(1)}(\lambda\rho) - H_2^{(1)}(\lambda\rho)\right]e^{i\gamma_g d}e^{i\gamma_i(z-h)}\lambda\,d\lambda, \tag{4.105}$$

where

$$q(\lambda) = \frac{\gamma_g}{k_g^2} + \frac{\gamma_i}{k_i^2} - i\left(\frac{\gamma_a}{k_a^2} + \frac{k_a^2\gamma_i\gamma_g}{k_g^2k_i^2\gamma_a}\right)\tan\gamma_a h. \tag{4.106}$$

In order to evaluate the above integrals for the electric-type (TM) waves, it is necessary to examine the poles and the branch points of all integrands. The pole equation for the electric-type (TM) waves can be represented as follows:

$$q(\lambda) = \frac{\gamma_g}{k_g^2} + \frac{\gamma_i}{k_i^2} - i\left(\frac{\gamma_a}{k_a^2} + \frac{k_a^2\gamma_i\gamma_g}{k_g^2k_i^2\gamma_a}\right)\tan\gamma_a h = 0. \tag{4.107}$$

In SLF/ELF ranges, for the TEM waves, the pole λ^* should satisfy the condition of $|\gamma_a h| \ll 1$ ($\gamma_a = \sqrt{k_a^2 - \lambda^{*2}}$). Then, we have

$$\frac{\gamma_g}{k_g^2} + \frac{\gamma_i}{k_i^2} - i\left(\frac{\gamma_a^2 h}{k_a^2} + \frac{k_a^2\gamma_i\gamma_g h}{k_g^2k_i^2}\right) = 0. \tag{4.108}$$

Taking into account $k_g \gg k_a$ and $k_i \gg k_a$, for zero-order approximations, we take $\gamma_g \approx k_g$ and $\gamma_i \approx k_i$. Thus, it follows that

$$\frac{1}{k_g} + \frac{1}{k_i} - ih\left(\frac{k_a^2 - \lambda^2}{k_a^2} + \frac{k_a^2}{k_gk_i}\right) = 0. \tag{4.109}$$

The solution of Eq. (4.109) can be obtained readily. We write

$$\lambda^* = k_a\left[1 + \Delta_i\Delta_g + i\frac{(\Delta_i + \Delta_g)}{k_a h}\right]^{1/2}. \tag{4.110}$$

With the substitution of Eq. (4.110), as an initial approximation, into Eq. (4.107), the pole for the TEM waves can be determined accurately by using Newton's iteration method.

In general, for the electric-type (TM) waves, Eq. (4.107) is simplified as follows:

$$\frac{1}{k_g} + \frac{1}{k_i} - \mathrm{i}\left(\frac{\gamma_a}{k_a^2} + \frac{k_a^2}{k_g k_i \gamma_a}\right) \tan \gamma_a h = 0. \tag{4.111}$$

For n-order TM wave, we have $\gamma_a h = n\pi + \delta_n$, where $|\delta_n| < \pi$, and n is an integer. Then, it follows that

$$\gamma_a = \frac{n\pi + \delta_n}{h}, \tag{4.112}$$

$$\tan \gamma_a h = \tan \delta_n. \tag{4.113}$$

Substituting Eqs. (4.112) and (4.113) into Eq. (4.111), it follows that

$$\frac{1}{k_g} + \frac{1}{k_i} - \mathrm{i}\left[\frac{n\pi + \delta_n}{k_a^2 h} + \frac{k_a^2 h}{k_g k_i (n\pi + \delta_n)}\right] \tan \delta_n = 0. \tag{4.114}$$

Considering the condition of $\frac{n\pi}{k_a^2 h} \gg 1$, we readily find

$$\frac{1}{k_g} + \frac{1}{k_i} - \mathrm{i}\frac{n\pi}{k_a^2 h} \tan \delta_n = 0. \tag{4.115}$$

Then we have

$$\delta_n \approx \tan^{-1}\left[-\frac{\mathrm{i}k_a h(\Delta_i + \Delta_g)}{n\pi}\right] \approx -\frac{\mathrm{i}k_a h(\Delta_i + \Delta_g)}{n\pi}. \tag{4.116}$$

The solution of Eq. (4.111) can be obtained readily. We write

$$\lambda^* = \left[k_a^2 - \left(\frac{n\pi + \delta_n}{h}\right)^2\right]^{1/2}. \tag{4.117}$$

Substituting Eq. (4.117), as an initial approximation, into Eq. (4.107), the pole for the electric-type (TM$_n$) waves can be determined accurately by using Newton's iteration method.

By examining the above Sommerfeld's integrals, it is seen that the three root functions γ_a, γ_g, and γ_i are included in the integrands. Obviously, for the above Sommerfeld's integrals, besides the poles, there are three branch points at $\lambda = k_a$, $\lambda = k_g$, and $\lambda = k_i$. As the Earth and ionosphere are lossy media, the wave numbers k_g and k_i have large imaginary parts. When the distance between the observation point and the dipole source is not too small, the integrations along the branch points $\lambda = k_g$ and $\lambda = k_i$ can be neglected. Since the integrands are even functions to γ_a, $\lambda = k_a$ is a removable singularity. Therefore, for the field components, the term contributed by the residues of poles should be included, while the term contributed by the integrations along the branch lines may be neglected. Those properties can be explained in a physical sense as follows: Considering the high loss characteristics of the Earth, SLF/ELF waves will travel upward vertically form the transmitting dipole source to the Earth's surface, then propagate radially in the Earth–ionosphere waveguide or cavity, and finally enter vertically into the ionosphere to the receiving point. It is noted that the attenuation rate and wave number of each propagation mode are determined by the pole equation.

Thus, with the residue theorem, Eq. (4.79) can be written in the form

$$F_{az}^{(1)} = 2\pi\mathrm{i}\sum_j \frac{[(k_a^{-2} + \frac{\gamma_g^*}{k_g^2\gamma_a^*})\mathrm{e}^{-\mathrm{i}\gamma_a^* z} + (k_a^{-2} - \frac{\gamma_g^*}{k_g^2\gamma_a^*})\mathrm{e}^{\mathrm{i}\gamma_a^* z}]}{q'(\lambda_j^*)} \times \left(\frac{\gamma_i^*}{k_i^2} - \mathrm{i}\frac{\gamma_a^*}{k_a^2}\tan\gamma_a^* h\right) H_1^{(1)}(\lambda_j^*\rho)\lambda_j^{*2}\mathrm{e}^{\mathrm{i}\gamma_g^* d}, \tag{4.118}$$

where

$$q'(\lambda) = -\frac{\lambda}{k_g^2\gamma_g} - \frac{\lambda}{k_i^2\gamma_i} - \mathrm{i}\lambda\tan\gamma_a h\left[-\frac{1}{k_a^2\gamma_a} + \frac{k_a^2}{k_g^2k_i^2}\left(\frac{\gamma_i\gamma_g}{\gamma_a^3} - \frac{\gamma_g}{\gamma_i\gamma_a} - \frac{\gamma_i}{\gamma_g\gamma_a}\right)\right] + \mathrm{i}\frac{\lambda h}{\gamma_a}\sec^2\gamma_a h\left(\frac{\gamma_a}{k_a^2} + \frac{k_a^2\gamma_i\gamma_g}{k_g^2k_i^2\gamma_a}\right), \tag{4.119}$$

$$\gamma_m^* = \sqrt{k_m^* - \lambda_j^{*2}}; \quad m = a, i, g. \tag{4.120}$$

In the above formulas, the value λ_j^* is the jth root of the pole equation (4.107).

With similar procedures, the remaining terms can be derived readily. We write

$$F_{a\rho}^{(1)} = -4\pi\mathrm{i}\sum_j \frac{[(\frac{\gamma_g^*}{k_g^2} + \frac{\gamma_a^*}{k_a^2})\mathrm{e}^{-\mathrm{i}\gamma_a^* z} + (\frac{\gamma_g^*}{k_g^2} - \frac{\gamma_a^*}{k_a^2})\mathrm{e}^{\mathrm{i}\gamma_a^* z}](\frac{\gamma_i^*}{k_i^2} - \mathrm{i}\frac{\gamma_a^*}{k_a^2}\tan\gamma_a^* h)}{q'(\lambda_j^*)} \times H_1^{(1)'}(\lambda_j^*\rho)\lambda_j^*\mathrm{e}^{\mathrm{i}\gamma_g^* d}, \tag{4.121}$$

$$F_{a\phi}^{(1)} = -\frac{4\pi\mathrm{i}}{\rho}\sum_j \frac{[(\frac{\gamma_g^*}{k_g^2} + \frac{\gamma_a^*}{k_a^2})\mathrm{e}^{-\mathrm{i}\gamma_a^* z} + (\frac{\gamma_g^*}{k_g^2} - \frac{\gamma_a^*}{k_a^2})\mathrm{e}^{\mathrm{i}\gamma_a^* z}](\frac{\gamma_i^*}{k_i^2} - \mathrm{i}\frac{\gamma_a^*}{k_a^2}\tan\gamma_a^* h)}{q'(\lambda_j^*)} \times H_1^{(1)}(\lambda_j^*\rho)\mathrm{e}^{\mathrm{i}\gamma_g^* d}, \tag{4.122}$$

$$G_{a\rho}^{(1)} = -\frac{4\pi\mathrm{i}}{\rho}\sum_j \frac{[(k_g^2\gamma_a^* + k_a^2\gamma_g^*)\mathrm{e}^{-\mathrm{i}\gamma_a^* z} + (k_g^2\gamma_a^* - k_a^2\gamma_g^*)\mathrm{e}^{\mathrm{i}\gamma_a^* z}]}{k_g^2\gamma_a^* q'(\lambda_j^*)} \times \left(\frac{\gamma_i^*}{k_i^2} - \mathrm{i}\frac{\gamma_a^*}{k_a^2}\tan\gamma_a^* h\right) H_1^{(1)}(\lambda_j^*\rho)\mathrm{e}^{\mathrm{i}\gamma_g^* d}, \tag{4.123}$$

$$G_{a\phi}^{(1)} = -4\pi\mathrm{i}\sum_j \frac{[(k_g^2\gamma_a^* + k_a^2\gamma_g^*)\mathrm{e}^{-\mathrm{i}\gamma_a^* z} + (k_g^2\gamma_a^* - k_a^2\gamma_g^*)\mathrm{e}^{\mathrm{i}\gamma_a^* z}]}{k_g^2\gamma_a^* q'(\lambda_j^*)}$$

$$\times \left(\frac{\gamma_i^*}{k_i^2} - \mathrm{i}\frac{\gamma_a^*}{k_a^2}\tan\gamma_a^* h\right) H_1^{(1)'}(\lambda_j^*\rho)\lambda_j^* \mathrm{e}^{\mathrm{i}\gamma_g^* d}, \tag{4.124}$$

$$F_{i\rho}^{(1)} = -4\pi\mathrm{i}\sum_j \frac{[\frac{\gamma_g^*}{k_g^2}\cos(\gamma_a^* h) - \mathrm{i}\frac{\gamma_a^*}{k_a^2}\sin(\gamma_a^* h)](\frac{\gamma_i^*}{k_i^2} - \mathrm{i}\frac{\gamma_a^*}{k_a^2}\tan\gamma_a^* h)}{q'(\lambda_j^*)}$$

$$\times H_1^{(1)'}(\lambda_j^*\rho)\lambda_j^* \mathrm{e}^{\mathrm{i}\gamma_g^* d}\mathrm{e}^{\mathrm{i}\gamma_i^*(z-h)}, \tag{4.125}$$

$$F_{i\phi}^{(1)} = -\frac{4\pi\mathrm{i}}{\rho}\sum_j \frac{[\frac{\gamma_g^*}{k_g^2}\cos(\gamma_a^* h) - \mathrm{i}\frac{\gamma_a^*}{k_a^2}\sin(\gamma_a^* h)](\frac{\gamma_i^*}{k_i^2} - \mathrm{i}\frac{\gamma_a^*}{k_a^2}\tan\gamma_a^* h)}{q'(\lambda_j^*)}$$

$$\times H_1^{(1)}(\lambda_j^*\rho)\mathrm{e}^{\mathrm{i}\gamma_g^* d}\mathrm{e}^{\mathrm{i}\gamma_i^*(z-h)}, \tag{4.126}$$

$$F_{iz}^{(1)} = 2\pi\mathrm{i}\sum_j \frac{[\cos(\gamma_a^* h) - \frac{\mathrm{i}k_a^2\gamma_g^*}{k_g^2\gamma_a^*}\sin(\gamma_a^* h)](\frac{\gamma_i^*}{k_i^2} - \mathrm{i}\frac{\gamma_a^*}{k_a^2}\tan\gamma_a^* h)}{k_i^2 q'(\lambda_j^*)}$$

$$\times H_1^{(1)}(\lambda_j^*\rho)\lambda_j^{*2}\mathrm{e}^{\mathrm{i}\gamma_i^*(z-h)}\mathrm{e}^{\mathrm{i}\gamma_g^* d}, \tag{4.127}$$

$$G_{i\rho}^{(1)} = -\frac{4\pi\mathrm{i}}{\rho}\sum_j \frac{[\cos(\gamma_a^* h) - \frac{\mathrm{i}k_a^2\gamma_g^*}{k_g^2\gamma_a^*}\sin(\gamma_a^* h)](\frac{\gamma_i^*}{k_i^2} - \mathrm{i}\frac{\gamma_a^*}{k_a^2}\tan\gamma_a^* h)}{q'(\lambda_j^*)}$$

$$\times H_1^{(1)}(\lambda_j^*\rho)\mathrm{e}^{\mathrm{i}\gamma_g^* d}\mathrm{e}^{\mathrm{i}\gamma_i^*(z-h)}, \tag{4.128}$$

$$G_{i\phi}^{(1)} = -4\pi\mathrm{i}\sum_j \frac{[\cos(\gamma_a^* h) - \frac{\mathrm{i}k_a^2\gamma_g^*}{k_g^2\gamma_a^*}\sin(\gamma_a^* h)](\frac{\gamma_i^*}{k_i^2} - \mathrm{i}\frac{\gamma_a^*}{k_a^2}\tan\gamma_a^* h)}{q'(\lambda_j^*)}$$

$$\times H_1^{(1)'}(\lambda_j^*\rho)\lambda_j^* \mathrm{e}^{\mathrm{i}\gamma_g^* d}\mathrm{e}^{\mathrm{i}\gamma_i^*(z-h)}. \tag{4.129}$$

In this subsection, all terms for the electric-type (TM) field are obtained readily.

4.3.2 Evaluation for the Magnetic-Type Field

The integrals for the magnetic-type (TE) field can be treated in a similar manner to those for the electric-type (TM) field. Then, Eqs. (4.63) and (4.64) can be rewritten as follows:

$$\begin{matrix} F_{a\rho}^{(2)} \\ F_{a\phi}^{(2)} \end{matrix} = -\int_{-\infty}^{\infty} \frac{[(\frac{1}{\gamma_g}+\frac{1}{\gamma_a})e^{-i\gamma_a z}+(\frac{1}{\gamma_g}-\frac{1}{\gamma_a})e^{i\gamma_a z}](\frac{\gamma_a}{\gamma_i}-i\tan\gamma_a h)}{p(\lambda)}$$

$$\times\left[H_0^{(1)}(\lambda\rho)\pm H_2^{(1)}(\lambda\rho)\right]e^{i\gamma_g d}\lambda\, d\lambda. \tag{4.130}$$

The poles of the integrands in Eq. (4.130) satisfy the pole equation for the magnetic-type (TE) field. We write

$$p(\lambda)=\frac{\gamma_a}{\gamma_g}+\frac{\gamma_a}{\gamma_i}-i\left(1+\frac{\gamma_a^2}{\gamma_g\gamma_i}\right)\tan\gamma_a h=0, \tag{4.131}$$

where

$$\gamma_j=\sqrt{k_j^2-\lambda^2}; \qquad m=a,i,g. \tag{4.132}$$

For zero-order approximations, the pole equation in Eq. (4.131) can be simplified as follows:

$$\gamma_a\left(1+\frac{k_g}{k_i}\right)-i\left(k_g+\frac{\gamma_a^2}{k_i}\right)\tan\gamma_a h=0, \tag{4.133}$$

For m-order TE wave, we have $\gamma_a h=m\pi+\delta_m$, where $|\delta_m|<\pi$, and m is an integer. Then, it follows that

$$\gamma_a=\frac{m\pi+\delta_m}{h}, \tag{4.134}$$

$$\tan\gamma_a h=\tan\delta_m. \tag{4.135}$$

Substituting Eqs. (4.134) and (4.135) into Eq. (4.133), we readily find

$$\frac{m\pi+\delta_m}{h}\left(1+\frac{k_g}{k_i}\right)-i\left[k_g+\left(\frac{m\pi+\delta_m}{h}\right)^2\frac{1}{k_i}\right]\tan\delta_m=0. \tag{4.136}$$

Because of $\delta_m \ll m\pi$, it follows that

$$\delta_m\approx-\tan^{-1}\frac{i\frac{m\pi}{h}(1+\frac{k_g}{k_i})}{k_g+\frac{(m\pi)^2}{k_i h^2}}. \tag{4.137}$$

The solution of Eq. (4.133) is obtained readily. We write

$$\lambda^*=\left[k_a^2-\left(\frac{m\pi+\delta_m}{h}\right)^2\right]^{1/2}. \tag{4.138}$$

Substituting Eq. (4.138), as an initial approximation, into Eq. (4.131), the pole for the magnetic-type (TE_n) waves can be determined accurately by using Newton's iteration method.

Similar to those for the electric-type (TM) field, it is found that, for the magnetic-type (TE) field, the integrands are also even functions to γ_a, and $\lambda = k_a$ is a removable cut. Thus, for the components of the magnetic-type (TE) field, the term of the resides of poles exists, while the term of the integrations along the branch lines can be neglected. With the residue theorem, the integrals for the magnetic-type (TE) field can be derived readily. We write

$$F_{a\rho}^{(2)} = -\frac{4\pi i}{\rho}\sum_j \frac{[(\gamma_g^{*-1} + \gamma_a^{*-1})e^{-i\gamma_a^* z} + (\gamma_g^{*-1} - \gamma_a^{*-1})e^{i\gamma_a^* z}]}{p'(\lambda_j^*)} \times \left(\frac{\gamma_a^*}{\gamma_i^*} - i\tan\gamma_a^* h\right) H_1^{(1)}(\lambda_j^*\rho)e^{i\gamma_g^* d}, \tag{4.139}$$

$$F_{a\phi}^{(2)} = -4\pi i\sum_j \frac{[(\gamma_g^{*-1} + \gamma_a^{*-1})e^{-i\gamma_a^* z} + (\gamma_g^{*-1} - \gamma_a^{*-1})e^{i\gamma_a^* z}]}{p'(\lambda_j^*)} \times \left(\frac{\gamma_a^*}{\gamma_i^*} - i\tan\gamma_a^* h\right) H_1^{(1)'}(\lambda_j^*\rho)\lambda_j^* e^{i\gamma_g^* d}, \tag{4.140}$$

$$G_{a\rho}^{(2)} = -4\pi i\sum_j \frac{[(\gamma_g^* + \gamma_a^*)e^{-i\gamma_a^* z} + (\gamma_g^* - \gamma_a^*)e^{i\gamma_a^* z}](\frac{\gamma_a^*}{\gamma_i^*} - i\tan\gamma_a^* h)}{\gamma_g^* p'(\lambda_j^*)} \times H_1^{(1)'}(\lambda_j^*\rho)\lambda_j^* e^{i\gamma_g^* d}, \tag{4.141}$$

$$G_{a\phi}^{(2)} = -\frac{4\pi i}{\rho}\sum_j \frac{[(\gamma_g^* + \gamma_a^*)e^{-i\gamma_a^* z} + (\gamma_g^* - \gamma_a^*)e^{i\gamma_a^* z}](\frac{\gamma_a^*}{\gamma_i^*} - i\tan\gamma_a^* h)}{\gamma_g^* p'(\lambda_j^*)} \times H_1^{(1)}(\lambda_j^*\rho)e^{i\gamma_g^* d}, \tag{4.142}$$

$$G_{az}^{(2)} = -2\pi i\sum_j \frac{[(\gamma_g^{*-1} + \gamma_a^{*-1})e^{-i\gamma_a^* z} + (\gamma_g^{*-1} - \gamma_a^{*-1})e^{i\gamma_a^* z}]}{p'(\lambda_j^*)} \times \left(\frac{\gamma_a^*}{\gamma_i^*} - i\tan\gamma_a^* h\right) H_1^{(1)}(\lambda_j^*\rho)\lambda_j^{*2} e^{i\gamma_g^* d}, \tag{4.143}$$

$$F_{i\rho}^{(2)} = -\frac{4\pi i}{\rho}\sum_j \frac{[\gamma_g^{*-1}\cos(\gamma_a^* h) - i\gamma_a^{*-1}\sin(\gamma_a^* h)](\frac{\gamma_a^*}{\gamma_i^*} - i\tan\gamma_a^* h)}{p'(\lambda_j^*)} \times H_1^{(1)}(\lambda_j^*\rho)e^{i\gamma_g^* d}e^{i\gamma_i^*(z-h)}, \tag{4.144}$$

$$F_{i\phi}^{(2)} = -4\pi i\sum_j \frac{[\gamma_g^{*-1}\cos(\gamma_a^* h) - i\gamma_a^{*-1}\sin(\gamma_a^* h)](\frac{\gamma_a^*}{\gamma_i^*} - i\tan\gamma_a^* h)}{p'(\lambda_j^*)} \times H_1^{(1)'}(\lambda_j^*\rho)\lambda_j^* e^{i\gamma_g^* d}e^{i\gamma_i^*(z-h)}, \tag{4.145}$$

$$G_{i\rho}^{(2)} = -4\pi \mathrm{i} \sum_j \frac{[\cos(\gamma_a^* h) - \mathrm{i}\frac{\gamma_a^*}{\gamma_g^*}\sin(\gamma_a^* h)](\frac{\gamma_a^*}{\gamma_i^*} - \mathrm{i}\tan\gamma_a^* h)}{p'(\lambda_j^*)}$$
$$\times H_1^{(1)\prime}(\lambda_j^*\rho)\lambda_j^* \mathrm{e}^{\mathrm{i}\gamma_g^* d}\mathrm{e}^{\mathrm{i}\gamma_i^*(z-h)}, \tag{4.146}$$

$$G_{i\phi}^{(2)} = -\frac{4\pi \mathrm{i}}{\rho} \sum_j \frac{[\cos(\gamma_a^* h) - \mathrm{i}\frac{\gamma_a^*}{\gamma_g^*}\sin(\gamma_a^* h)](\frac{\gamma_a^*}{\gamma_i^*} - \mathrm{i}\tan\gamma_a^* h)}{p'(\lambda_j^*)}$$
$$\times H_1^{(1)}(\lambda_j^*\rho)\mathrm{e}^{\mathrm{i}\gamma_g^* d}\mathrm{e}^{\mathrm{i}\gamma_i^*(z-h)}, \tag{4.147}$$

$$G_{iz}^{(2)} = -2\pi \mathrm{i} \sum_j \frac{[\gamma_g^{*-1}\cos(\gamma_a^* h) - \mathrm{i}\gamma_a^{*-1}\sin(\gamma_a^* h)](\frac{\gamma_a^*}{\gamma_i^*} - \mathrm{i}\tan\gamma_a^* h)}{p'(\lambda_j^*)}$$
$$\times H_1^{(1)}(\lambda_j^*\rho)\lambda_j^{*2} \mathrm{e}^{\mathrm{i}\gamma_g^* d}\mathrm{e}^{\mathrm{i}\gamma_i^*(z-h)}, \tag{4.148}$$

where

$$p'(\lambda) = \lambda\gamma_a\left(\frac{1}{\gamma_g^3} + \frac{1}{\gamma_i^3}\right) - \frac{\lambda}{\gamma_a}\left(\frac{1}{\gamma_g} + \frac{1}{\gamma_i}\right) - \mathrm{i}\frac{\lambda\tan\gamma_a h}{\gamma_g\gamma_i}$$
$$\times\left(\frac{\gamma_a^2}{\gamma_i^2} + \frac{\gamma_a^2}{\gamma_g^2} - 2\right) + \mathrm{i}\frac{\lambda h}{\gamma_a}\sec^2\gamma_a h\left(1 + \frac{\gamma_a^2}{\gamma_g\gamma_i}\right), \tag{4.149}$$

$$\gamma_m^* = \sqrt{k_m^2 - \lambda_j^{*2}}; \quad m = a, i, g. \tag{4.150}$$

In the above formulas, the value λ_j^* is the jth root of the pole equation (4.131) for the magnetic-type (TE) field. In this subsection, all terms for the magnetic-type (TE) field are obtained readily.

4.3.3 Final Formulas for SLF/ELF Fields in Air and Ionosphere

By now, all terms for both the electric-type (TM) and the magnetic-type (TE) fields have been evaluated. At small distance between the observation point and the dipole source, the direct wave and reflected wave can be neglected. Then, the final formulas of the components for SLF/ELF fields in air and ionosphere can be written readily.

In Region 0 (ionosphere), we write

$$E_{i\rho} = -\frac{\mathrm{i}\omega\mu_0 I\,\mathrm{d}l}{4}\cos\phi\cdot\left[-\mathrm{i}k_a^{-2}\sum_j \Lambda_j^{\mathrm{TM}}\frac{\partial G_j^{\mathrm{TM}}(z)}{\partial z}\bigg|_{z=h} H_1^{(1)\prime}(\lambda_j^*\rho)\right.$$
$$\times\mathrm{e}^{\mathrm{i}\gamma_g^* d}\mathrm{e}^{\mathrm{i}\gamma_i^*(z-h)} + \frac{1}{\rho}\sum_j \Lambda_j^{\mathrm{TE}} G_j^{\mathrm{TE}}(z)\bigg|_{z=h}$$

$$\times H_1^{(1)}\left(\lambda_j^*\rho\right) e^{i\gamma_g^* d} e^{i\gamma_i^*(z-h)}\Bigg], \tag{4.151}$$

$$E_{i\phi} = \frac{i\omega\mu_0 I\,dl}{4}\sin\phi\cdot\left[-\frac{i}{\rho k_a^2}\sum_j \Lambda_j^{\mathrm{TM}}\frac{\partial G_j^{\mathrm{TM}}(z)}{\partial z}\bigg|_{z=h} H_1^{(1)}\left(\lambda_j^*\rho\right)\lambda_j^{*-1}\right.$$

$$\times e^{i\gamma_g^* d} e^{i\gamma_i^*(z-h)} + \sum_j \Lambda_j^{\mathrm{TE}} G_j^{\mathrm{TE}}(z)\bigg|_{z=h}$$

$$\left.\times H_1^{(1)'}\left(\lambda_j^*\rho\right)\lambda_j^* e^{i\gamma_g^* d} e^{i\gamma_i^*(z-h)}\right], \tag{4.152}$$

$$E_{iz} = -\frac{\omega\mu_0 I\,dl}{4k_i^2}\cos\phi\cdot\sum_j \Lambda_j^{\mathrm{TM}} G_j^{\mathrm{TM}}(z)\bigg|_{z=h} H_1^{(1)}\left(\lambda_j^*\rho\right)\lambda_j^*$$

$$\times e^{i\gamma_g^* d} e^{i\gamma_i^*(z-h)}, \tag{4.153}$$

$$B_{i\rho} = -\frac{i\mu_0 I\,dl}{4}\sin\phi\cdot\left[\frac{1}{\rho}\sum_j \Lambda_j^{\mathrm{TM}} G_j^{\mathrm{TM}}(z)\bigg|_{z=h} H_1^{(1)}\left(\lambda_j^*\rho\right)\lambda_j^{*-1}\right.$$

$$\times e^{i\gamma_g^* d} e^{i\gamma_i^*(z-h)} - i\sum_j \Lambda_j^{\mathrm{TE}}\frac{\partial G_j^{\mathrm{TE}}(z)}{\partial z}\bigg|_{z=h}$$

$$\left.\times H_1^{(1)'}\left(\lambda_j^*\rho\right)\lambda_j^* e^{i\gamma_g^* d} e^{i\gamma_i^*(z-h)}\right], \tag{4.154}$$

$$B_{i\phi} = -\frac{i\mu_0 I\,dl}{4}\cos\phi\cdot\left[\sum_j \Lambda_j^{\mathrm{TM}} G_j^{\mathrm{TM}}(z)\bigg|_{z=h} H_1^{(1)'}\left(\lambda_j^*\rho\right)\right.$$

$$\times e^{i\gamma_g^* d} e^{i\gamma_i^*(z-h)} - i\frac{1}{\rho}\sum_j \Lambda_j^{\mathrm{TE}}\frac{\partial G_j^{\mathrm{TE}}(z)}{\partial z}\bigg|_{z=h}$$

$$\left.\times H_1^{(1)}\left(\lambda_j^*\rho\right) e^{i\gamma_g^* d} e^{i\gamma_i^*(z-h)}\right], \tag{4.155}$$

$$B_{iz} = -\frac{\mu_0 I\,dl}{4}\sin\phi\cdot\sum_j \Lambda_j^{\mathrm{TE}} G_j^{\mathrm{TE}}(z)\bigg|_{z=h} H_1^{(1)}\left(\lambda_j^*\rho\right)\lambda_j^{*2}$$

$$\times e^{i\gamma_g^* d} e^{i\gamma_i^*(z-h)}. \tag{4.156}$$

In Region 1 (air), we write

$$E_{a\rho} = -\frac{i\omega\mu_0 I\,dl}{4}\cos\phi\cdot\left[-ik_a^{-2}\sum_j \Lambda_j^{\mathrm{TM}}\frac{\partial G_j^{\mathrm{TM}}(z)}{\partial z} H_1^{(1)'}\left(\lambda_j^*\rho\right) e^{i\gamma_g^* d}\right.$$

$$\left.+\frac{1}{\rho}\sum_j \Lambda_j^{\mathrm{TE}} G_j^{\mathrm{TE}}(z) H_1^{(1)}\left(\lambda_j^*\rho\right) e^{i\gamma_g^* d}\right], \tag{4.157}$$

$$E_{a\phi} = \frac{\mathrm{i}\omega\mu_0 I\,\mathrm{d}l}{4}\sin\phi\cdot\left[-\frac{\mathrm{i}}{\rho k_a^2}\sum_j \Lambda_j^{\mathrm{TM}}\frac{\partial G_j^{\mathrm{TM}}(z)}{\partial z}H_1^{(1)}(\lambda_j^*\rho)\lambda_j^{*-1}\mathrm{e}^{\mathrm{i}\gamma_g^* d}\right.$$

$$\left.+\sum_j \Lambda_j^{\mathrm{TE}}G_j^{\mathrm{TE}}(z)H_1^{(1)\prime}(\lambda_j^*\rho)\lambda_j^*\mathrm{e}^{\mathrm{i}\gamma_g^* d}\right], \tag{4.158}$$

$$E_{az} = -\frac{\omega\mu_0 I\,\mathrm{d}l}{4k_a^2}\cos\phi\cdot\sum_j \Lambda_j^{\mathrm{TM}}G_j^{\mathrm{TM}}(z)H_1^{(1)}(\lambda_j^*\rho)\lambda_j^*\mathrm{e}^{\mathrm{i}\gamma_g^* d}, \tag{4.159}$$

$$B_{a\rho} = -\frac{\mathrm{i}\mu_0 I\,\mathrm{d}l}{4}\sin\phi\cdot\left[\frac{1}{\rho}\sum_j \Lambda_j^{\mathrm{TM}}G_j^{\mathrm{TM}}(z)H_1^{(1)}(\lambda_j^*\rho)\lambda_j^{*-1}\mathrm{e}^{\mathrm{i}\gamma_g^* d}\right.$$

$$\left.-\mathrm{i}\sum_j \Lambda_j^{\mathrm{TE}}\frac{\partial G_j^{\mathrm{TE}}(z)}{\partial z}H_1^{(1)\prime}(\lambda_j^*\rho)\lambda_j^*\mathrm{e}^{\mathrm{i}\gamma_g^* d}\right], \tag{4.160}$$

$$B_{a\phi} = -\frac{\mathrm{i}\mu_0 I\,\mathrm{d}l}{4}\cos\phi\cdot\left[\sum_j \Lambda_j^{\mathrm{TM}}G_j^{\mathrm{TM}}(z)H_1^{(1)\prime}(\lambda_j^*\rho)\mathrm{e}^{\mathrm{i}\gamma_g^* d}\right.$$

$$\left.-\mathrm{i}\frac{1}{\rho}\sum_j \Lambda_j^{\mathrm{TE}}\frac{\partial G_j^{\mathrm{TE}}(z)}{\partial z}H_1^{(1)}(\lambda_j^*\rho)\mathrm{e}^{\mathrm{i}\gamma_g^* d}\right], \tag{4.161}$$

$$B_{az} = -\frac{\mu_0 I\,\mathrm{d}l}{4}\sin\phi\cdot\sum_j \Lambda_j^{\mathrm{TE}}G_j^{\mathrm{TE}}(z)H_1^{(1)}(\lambda_j^*\rho)\lambda_j^{*2}\mathrm{e}^{\mathrm{i}\gamma_g^* d}. \tag{4.162}$$

In these formulas, the excitation factors Λ_j^{TM} and Λ_j^{TE}, and the normalized height-gain functions $G_j^{\mathrm{TM}}(z)$ and $G_j^{\mathrm{TE}}(z)$ are represented as follows:

$$\Lambda_j^{\mathrm{TM}} = \frac{\left(\frac{\gamma_i^*}{k_i^2} - \mathrm{i}\frac{\gamma_a^*}{k_a^2}\tan\gamma_a^* h\right)}{q'(\lambda_j^*)}, \tag{4.163}$$

$$\Lambda_j^{\mathrm{TE}} = \frac{\left(\frac{\gamma_a^*}{\gamma_i^*} - \mathrm{i}\tan\gamma_a^* h\right)}{p'(\lambda_j^*)}, \tag{4.164}$$

$$G_j^{\mathrm{TM}}(z) = \frac{[(k_g^2\gamma_a^* + k_a^2\gamma_g^*)\mathrm{e}^{-\mathrm{i}\gamma_a^* z} + (k_g^2\gamma_a^* - k_a^2\gamma_g^*)\mathrm{e}^{\mathrm{i}\gamma_a^* z}]}{k_g^2\gamma_a^*}\lambda_j^*, \tag{4.165}$$

$$G_j^{\mathrm{TE}}(z) = \left[(\gamma_g^{*-1} + \gamma_a^{*-1})\mathrm{e}^{-\mathrm{i}\gamma_a^* z} + (\gamma_g^{*-1} - \gamma_a^{*-1})\mathrm{e}^{\mathrm{i}\gamma_a^* z}\right]. \tag{4.166}$$

4.4 Comparison and Analysis for Planar and Spherical Models

The physical problem addressed in this section is similar to some extent to that in Sect. 2.4. The differences between them are as follows: (a) the excitation source in

Sect. 2.4 is located on the ground surface or in the air, while it is buried in the Earth and closed to the ground surface in this section; (b) the spherical model is adopted in Sect. 2.4 and the planar model in this section. Therefore, it can be predicted that the field distributions are basically similar for the planar and spherical models, but there are also some there are differences.

In this section, the analytical formula for the vertical electric field E_{az} can be written in the following form:

$$E_{az} = -\frac{\omega\mu_0 I\,\mathrm{d}l}{4k_a^2}\cos\phi\cdot\sum_j \Lambda_j^{\mathrm{TM}} G_j^{\mathrm{TM}}(z) H_1^{(1)}(\lambda_j^*\rho)\lambda_j^*\mathrm{e}^{\mathrm{i}\gamma_g^* d}. \tag{4.167}$$

Correspondingly, the formula for the radial electric field E_r is represented as follows:

$$E_r(r,\theta,\phi) = -\frac{I\,\mathrm{d}s\cdot\eta\cdot\cos\phi\cdot\Delta_{gs}}{2ha}\sum \Lambda_n^e F_n(z)\frac{\partial P_\nu(\cos(\pi-\theta))}{\sin\nu\pi\cdot\partial\theta}. \tag{4.168}$$

From Eqs. (4.167) and (4.168), it is seen that the variations with the azimuth angle of both the vertical electric component E_z for the planar model and the radial electric component E_r for the spherical model are described by the function $\cos\phi$. It is noted that the vertical electric components E_z and E_r, which only exist in TM modes, can be expressed by the superposition of all propagation modes. The changes of the components E_z and E_r as the height are described by the height-gain functions $G_j^{\mathrm{TM}}(z)$ and $F_n(z)$, respectively. Additionally, the changes of E_z and E_r as $\hat{\rho}$ and $\hat{\theta}$ can be described by the functions $H_1^{(1)}(\lambda_j^*\rho)\cdot\lambda_j^*$ and $\frac{\partial P_\nu(\cos(\pi-\theta))}{\sin\nu\pi\cdot\partial\theta}$, respectively.

In what follows, we will examine the wave numbers of all propagation modes.

For the plane model, the characterized modes should satisfy the modal equation (4.107). As addressed Sect. 4.3.1, with Eq. (4.110) being an initial approximation, the modal equation can be solved accurately by using Newton's iteration method. We rewrite Eq. (4.110)

$$\lambda_0^* = k_a\left[1+\Delta_i\Delta_g+\mathrm{i}\frac{(\Delta_i+\Delta_g)}{k_a h}\right]^{1/2}. \tag{4.169}$$

With the substitution of Eq. (4.110), as an initial approximation, into Eq. (4.107), the pole for the quasi-TEM waves can be determined accurately by using Newton's iteration method.

From Eqs. (4.110) and (4.117), we have

$$C_n = \frac{\lambda_n}{ka} \approx \frac{n\pi}{k_a h}\left[1-\frac{ik_a h}{(n\pi)^2}(\Delta_i+\Delta_g)\right]; \quad n=1,2,3,\ldots \tag{4.170}$$

For the spherical model, we take the first-order approximation

$$S_0 \approx \left[1+\frac{i(\Delta_i+\Delta_g)}{kh}\right]^{1/2}. \tag{4.171}$$

In lower frequency ranges, the normalized surface impedance Δ_g for the ground and Δ_i for the ionosphere are much less than 1. It is seen that there is no substantial difference between Eqs. (4.110) and (4.171). For high-order modes of the spherical model, it follows that

$$C_n = \frac{n\pi}{2kh} + \sqrt{\left(\frac{n\pi}{2kh}\right)^2 - \frac{i(\Delta_g + \Delta_i)}{kh}}. \tag{4.172}$$

In SLF/ELF ranges, $kh \ll 1$. Then, Eq. (4.172) can be simplified to

$$C_n \approx \frac{n\pi}{k_a h}\left[1 - \frac{\mathrm{i}(\Delta_g + \Delta_i)k_a h}{(n\pi)^2}\right]. \tag{4.173}$$

It is seen that Eq. (4.173) is exactly the same as Eq. (4.170). That is to say, in SLF/ELF ranges, the wave numbers for spherical model are in agreement with those for plane model.

Obviously, it is necessary to examine the height-gain functions $G_n^{\mathrm{TM}}(z)$ for the plane model and $F_n(z)$ for the spherical model. We rewrite

$$G_n^{\mathrm{TM}}(z) = \frac{1}{2}\left[\left(1 + \frac{k_a^2\gamma_g^*}{k_g^2\gamma_a^*}\right)\mathrm{e}^{-\mathrm{i}\gamma_a^* z} + \left(1 - \frac{k_a^2\gamma_g^*}{k_g^2\gamma_a^*}\right)\mathrm{e}^{\mathrm{i}\gamma_a^* z}\right], \tag{4.174}$$

$$F_n(z) = \frac{1}{1+R_g}\left[\exp\left(-\mathrm{i}k\int_0^z \sqrt{C_n^2 + \frac{2t}{a}S_n^2\,\mathrm{d}t}\right)\right.$$
$$\left. + R_g \exp\left(\mathrm{i}k\int_0^z \sqrt{C_n^2 + \frac{2t}{a}S_n^2\,\mathrm{d}t}\right)\right]. \tag{4.175}$$

In general, the conditions of $k_g^2 \gg k_a^2$ and $k_i^2 \gg k_a^2$ are satisfied in SLF/ELF ranges. Letting $\Delta_g = k_a/k_g^*$ and $\gamma_a^* = kC_n$, with Eq. (4.174), it follows that

$$G_n^{\mathrm{TM}}(z) = \frac{1}{1+R_g}\left(\mathrm{e}^{-\mathrm{i}kC_n z} + R_g \mathrm{e}^{\mathrm{i}kC_n z}\right), \tag{4.176}$$

where

$$R_g = \frac{C_n - \Delta_g}{C_n + \Delta_g}. \tag{4.177}$$

Considering the condition of $z/a \ll 1$ in the atmosphere, it is found that there is no substantial difference between Eqs. (4.175) and (4.176). Concisely stated, in SLF/ELF ranges, the height-gain functions for the spherical model are in agreement with those for the plane model.

For the planar model, the change in the propagation direction ($\hat{\rho}$ direction) is described by the function $H_1^{(1)}(\lambda_j^*\rho) \cdot \lambda_j^*$. Letting $\lambda_j^* = k_a \cdot S_j$, and considering

$k_a \cdot S_j \cdot \rho \gg 1$, the Hankel function can be approximated by

$$H_1^{(1)}\left(\lambda_j^* \rho\right) \cdot \lambda_j^* \approx \sqrt{\frac{2}{\pi k_a \rho}} \mathrm{e}^{\mathrm{i}\left(k_a S_j \rho - \frac{3\pi}{4}\right)} \cdot S_j^{\frac{1}{2}} \cdot k_a. \tag{4.178}$$

For the spherical model, the change in the propagation direction ($\hat{\theta}$ direction or $\hat{\rho}$ direction) is described by the function $\frac{\partial P_\nu(\cos(\pi-\theta))}{\sin \nu\pi \cdot \partial\theta}$. In SLF range, the condition of $\nu(\nu+1) = k_a^2 a^2 S_n^2 \gg 1$ is satisfied. Thus, the Legendre function can be approximated as

$$P_\nu\left(\cos(\pi-\theta)\right) \approx \left(\frac{1}{2\pi\nu \sin\theta}\right)^{\frac{1}{2}} \mathrm{e}^{-\mathrm{i}kaS_n\pi + \mathrm{i}\frac{\pi}{4}} \mathrm{e}^{\mathrm{i}kaS_n\theta}. \tag{4.179}$$

Considering that, in general, S_n has a large imaginary part, we have

$$\sin \nu\pi = \frac{\mathrm{e}^{-\mathrm{i}kaS_n\pi}}{2i}. \tag{4.180}$$

Then, it follows that

$$\frac{P_\nu(\cos(\pi-\theta))}{\sin\nu\pi} \approx \sqrt{\frac{2}{\pi k_a a S_n \cdot \sin\theta}} \cdot \mathrm{e}^{\mathrm{i}kaS_n\theta - \mathrm{i}\frac{\pi}{4}}, \tag{4.181}$$

and

$$\frac{\partial P_\nu(\cos(\pi-\theta))}{\sin\nu\pi \cdot \partial\theta} \approx \sqrt{\frac{2}{\pi k_a a \cdot \sin\theta}} \cdot \mathrm{e}^{\mathrm{i}\frac{\pi}{4}} \cdot kaS_n^{\frac{1}{2}} \mathrm{e}^{\mathrm{i}kaS_n\theta}. \tag{4.182}$$

Comparing Eq. (4.182) with (4.178), it is seen that there is only a difference of a negative sign, and the factor ρ in Eq. (4.182) is replaced by the factor $a\sin\theta$.

In a similar manner, by ignoring some high-order values, it is found that the excitation factors for the planar model and the spherical model are also consistent. Comparing Eq. (4.167) with Eq. (4.168), it is found that there is an attenuation factor $\mathrm{e}^{\mathrm{i}\gamma_g^* d}$ for all modes in Eq. (4.167) for planar case. The greater the vertical distance d below the Earth's surface is, the more the factor attenuates significantly. It is well known that the wave, which is generated by a horizontal source in the Earth, travels from the transmitting source a vertical distance d to the Earth's surface, then radially propagates in the air of the Earth–ionosphere waveguide, in which some modes are excited, and finally in the ionosphere to the receiving point. Considering the strata of the Earth to be a lossy medium, the wave in the strata attenuates as $\mathrm{e}^{\mathrm{i}\gamma_g^* d}$ in the vertical direction, thus comparing those for a radiation source being located on the Earth's surface, the formulas for the field components in the Earth–ionosphere radiated by an underground radiation source should include an attenuation factor $\mathrm{e}^{\mathrm{i}\gamma_g^* d}$.

Following the above analyses, it is seen that the expressions of the analytical solutions for the plane model are entirely consistent with those for the spherical model.

At large distance between the radiation source and observation point, for example 1,000 km or beyond, the effects by the Earth's curvature should be considered, and it is more suitable to treat the problem by employing the spherical model. From the above analyses and discussions, the formulas of the electromagnetic field in the air of the Earth–ionosphere waveguide generated by an underground HED are obtained readily. We write

$$E_r = -\frac{\eta \Delta_g I\, \mathrm{d}s}{2ha} \cos\phi \sum_n \Lambda_n^e F_n(z) \frac{\partial P_\nu(\cos(\pi - \theta))}{\sin \nu\pi \partial\theta} \cdot \mathrm{e}^{\mathrm{i}\gamma_g^* d}, \tag{4.183}$$

$$\begin{aligned} E_\theta = \frac{\mathrm{i}\eta I\, \mathrm{d}s}{2ha} \cos\phi \Bigg\{ & \frac{\mathrm{i}\Delta_g}{ka} \sum_n \Lambda_n^e S_n^{-2} \frac{\partial F_n(z)}{k\partial z} \frac{\partial^2 P_\nu(\cos(\pi - \theta))}{\sin \nu\pi \partial\theta^2} \cdot \mathrm{e}^{\mathrm{i}\gamma_g^* d} \\ & - \frac{1}{ka \sin\theta} \sum_m \Lambda_m^h G_m(z) S_m^{-2} \frac{\partial P_\mu(\cos(\pi - \theta))}{\sin \mu\pi \partial\theta} \mathrm{e}^{\mathrm{i}\gamma_g^* d} \Bigg\}, \end{aligned} \tag{4.184}$$

$$\begin{aligned} E_\phi = \frac{\eta I\, \mathrm{d}s}{2ha} \sin\phi \Bigg\{ & \frac{\Delta_g}{ka \sin\theta} \sum_n \Lambda_n^e S_n^{-2} \frac{\partial F_n(z)}{k\partial z} \frac{\partial P_\nu(\cos(\pi - \theta))}{\sin \nu\pi \partial\theta} \cdot \mathrm{e}^{\mathrm{i}\gamma_g^* d} \\ & - \frac{\mathrm{i}}{ka} \sum_m \Lambda_m^h G_m(z) S_m^{-2} \frac{\partial^2 P_\mu(\cos(\pi - \theta))}{\sin \mu\pi \partial\theta^2} \cdot \mathrm{e}^{\mathrm{i}\gamma_g^* d} \Bigg\}, \end{aligned} \tag{4.185}$$

$$H_r = \frac{I\, \mathrm{d}s}{2ha} \sin\phi \sum_m \Lambda_n^h G_m(z) \frac{\partial P_\mu(\cos(\pi - \theta))}{\sin \mu\pi \partial\theta} \cdot \mathrm{e}^{\mathrm{i}\gamma_g^* d}, \tag{4.186}$$

$$\begin{aligned} H_\theta = \frac{I\, \mathrm{d}s}{2ha} \sin\phi \Bigg\{ & \frac{-\mathrm{i}\Delta_g}{ka \sin\theta} \sum_n \Lambda_n^e S_n^{-2} F_n(z) \frac{\partial P_\nu(\cos(\pi - \theta))}{\sin \nu\pi \partial\theta} \cdot \mathrm{e}^{\mathrm{i}\gamma_g^* d} \\ & + \frac{1}{ka} \sum_m \Lambda_m^h S_m^{-2} \frac{\partial G_m(z)}{k\partial z} \frac{\partial^2 P_\mu(\cos(\pi - \theta))}{\sin \mu\pi \partial\theta^2} \cdot \mathrm{e}^{\mathrm{i}\gamma_g^* d} \Bigg\}, \end{aligned} \tag{4.187}$$

$$\begin{aligned} H_\phi = \frac{I\, \mathrm{d}s}{2ha} \cos\phi \Bigg\{ & \frac{-\mathrm{i}\Delta_g}{ka} \sum_n \Lambda_n^e S_n^{-2} F_n(z) \frac{\partial^2 P_\nu(\cos(\pi - \theta))}{\sin \nu\pi \partial\theta^2} \cdot \mathrm{e}^{\mathrm{i}\gamma_g^* d} \\ & + \frac{1}{ka \sin\theta} \sum_m \Lambda_m^h S_m^{-2} \frac{\partial G_m(z)}{k\partial z} \frac{\partial P_\mu[\cos(\pi - \theta)]}{\sin \mu\pi \partial\theta} \cdot \mathrm{e}^{\mathrm{i}\gamma_g^* d} \Bigg\}. \end{aligned} \tag{4.188}$$

In the above formulas, the excitation factors, the height-gain functions, and the modal equations are defined as in Chap. 2.

4.5 Computations and Conclusions

With Newton's iteration method and the initial value of the modal equation, the poles, which correspond to the propagation modes, can be identified. Furthermore, the attenuation rate and the normalized horizontal wave number, which correspond

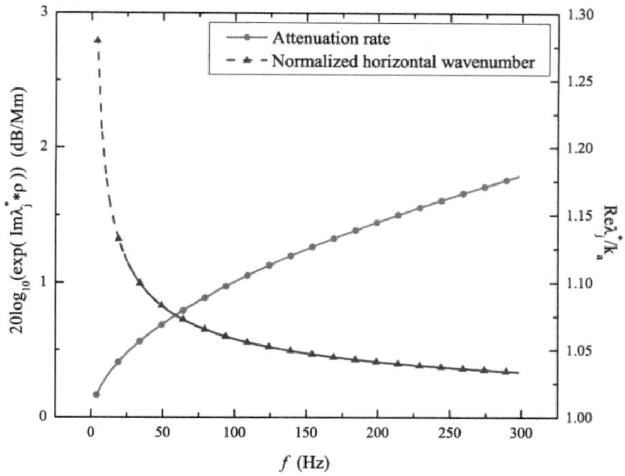

Fig. 4.2 The attenuation rate and normalized horizontal wave number versus the operating frequency for TEM mode: $h = 75$ km, $\sigma_g = 10^{-3}$ S/m, and $\sigma_g = 10^{-5}$ S/m

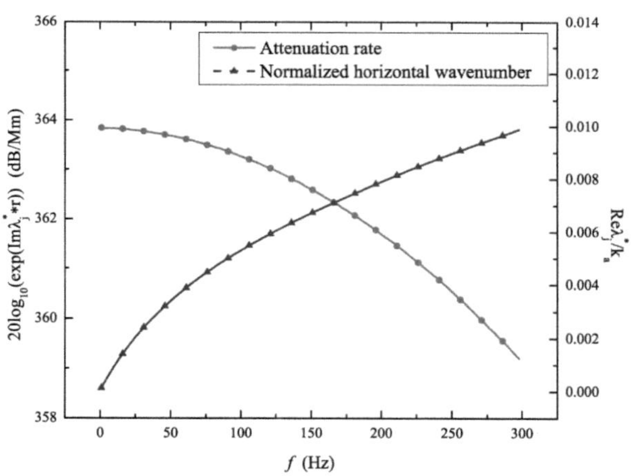

Fig. 4.3 The attenuation rate and normalized horizontal wave number versus the operating frequency for TM_1 mode: $h = 75$ km, $\sigma_g = 10^{-3}$ S/m, and $\sigma_g = 10^{-5}$ S/m

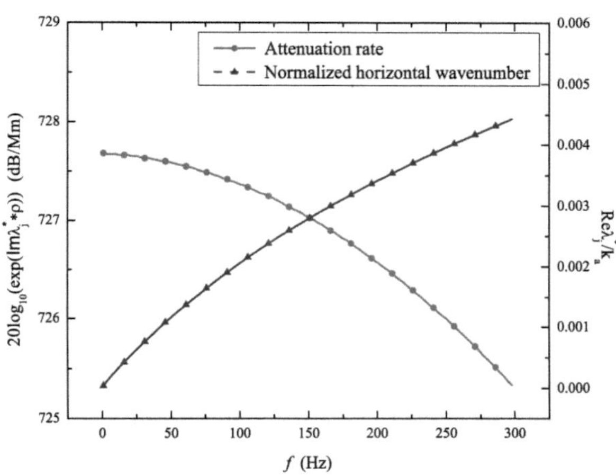

Fig. 4.4 The attenuation rate and normalized horizontal wave number versus the operating frequency for TM_2 mode: $h = 75$ km, $\sigma_g = 10^{-3}$ S/m, and $\sigma_g = 10^{-5}$ S/m

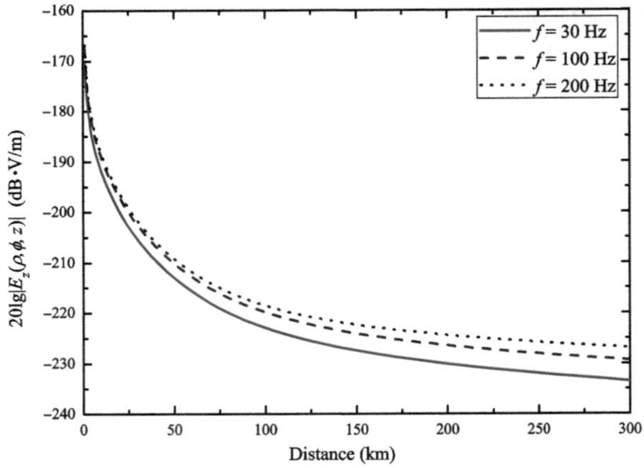

Fig. 4.5 The magnitudes of the vertical electric field E_{az} versus the propagation distance: $h = 75$ km, $\sigma_g = 10^{-3}$ S/m, $\sigma_g = 10^{-5}$ S/m, $\phi = 0$, $d = 1$ km, and $z = 0$

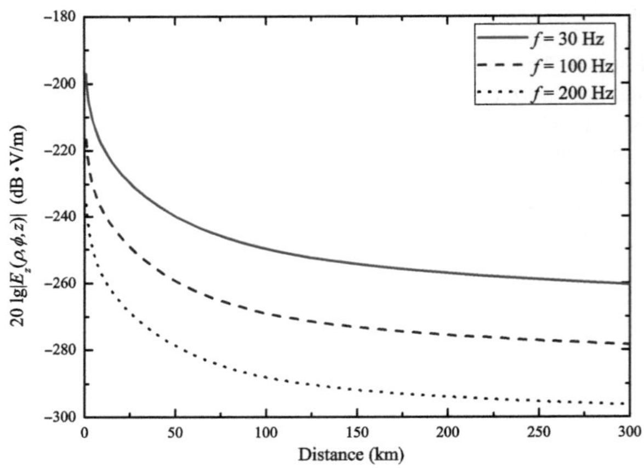

Fig. 4.6 The magnitudes of the vertical electric field E_{az} versus the propagation distance: $h = 75$ km, $\sigma_g = 10^{-3}$ S/m, $\sigma_g = 10^{-5}$ S/m, $\phi = 0$, $d = 10$ km, and $z = 0$

to each mode, are obtained readily. Graphs of the attenuation rate and the normalized wave number versus the operating frequency for TEM-, TM_1-, and TM_2-modes in SLF/ELF ranges are shown in Figs. 4.2, 4.3, 4.4, respectively. It is noted that the attenuation rate and the normalized wave number are represented by $20\log_{10}(\mathrm{e}^{\mathrm{Im}\lambda_j^*\cdot\rho})$ in dB/Mm and $(\mathrm{Re}\lambda_j^*)/k_a$, respectively. λ_j^* is the jth root of the pole Eq. (4.107). In these computations, the ionospheric height is assumed to be $h = 75$ km, and the conductivity of the Earth and that of the ionosphere are taken to be $\sigma_g = 10^{-3}$ S/m, and $\sigma_g = 10^{-5}$ S/m, respectively.

Comparing to available corresponding results in SLF range (Casey 2002), it is found that the numerical results obtained for the attenuation rate and the normalized horizontal wavenumber are in agreement with those by Casey (2002). This indicates that the calculated results in this chapter are correct.

Assuming that the unit HED is buried below the Earth's surface with a depth of $d = 1$ km, and the observation point is on the Earth's surface, $z = 0$, and with $\phi = 0$, the magnitudes of the vertical electric field E_{az} are computed at several different frequencies and shown in Fig. 4.5. With the same parameters in Fig. 4.5, in the case that the unit HED is buried below the Earth's surface with a depth of $d = 10$ km, the corresponding results are shown in Fig. 4.6. From these results, it is seen that the deeper the buried depth of the dipole source, the higher the operating frequency, the larger the absorption loss is.

References

Casey JP (2002) Extremely low frequency (ELF) propagation formulas for dipole sources radiating in a spherical earth-ionosphere waveguide. NUWC-NPT Technical Report, Naval Undersea Warfare Center Division, Newport, Rhode Island

Collin RE (2004a) Some observations about the near zone electric field of a hertzian dipole above a lossy earth. IEEE Trans Antennas Propag 52(11):3133–3137

Collin RE (2004b) Hertzian dipole radiation over a lossy earth or sea: some early and late 20th century controversies. IEEE Antennas Propag Mag 46(2):64–79

Dunn JM (1986) Lateral wave propagation in a three-layered medium. Radio Sci 21(5):787–796

Gokhberg MB, Pilipenko VA, Pokhotelov OA (1983) Observation from a satellite of electromagnetic radiation above the region of earthquake in preparation. Dokl Akad Nauk SSSR 268(1):56–58

Gradshteyn IS, Ryzhik IM (1980) Table of integrals, series, and products. Academic Press, New York

Hoh JH, Li LW, Kooi PS, Yeo TS, Leong MS (1999) Dominant lateral waves in the canopy layer of a four-layered forest. Radio Sci 34(3):681–691

King RWP (1991) The electromagnetic field of a horizontal electric dipole in the presence of a three-layered region. J Appl Phys 69(12):7987–7995

King RWP (1993) The electromagnetic field of a horizontal electric dipole in the presence of a three-layered region: supplement. J Appl Phys 74(8):4548–4845

King RWP, Sandler SS (1994a) The electromagnetic field of a vertical electric dipole over the Earth or Sea. IEEE Trans Antennas Propag 42(3):382–389

King RWP, Sandler SS (1994b) The electromagnetic field of a vertical electric dipole in the presence of a three-layered region. Radio Sci 29(1):97–113

King RWP, Sandler SS (1998) Reply. Radio Sci 33(2):255–256

King RWP, Owens M, Wu TT (1992) Lateral electromagnetic waves: theory and applications to communications, geophysical exploration, and remote sensing. Springer, New York

Li K (2009) Electromagnetic fields in stratified media. ZJU Press/Springer, Hangzhou/Berlin

Molchanov O, Fedorov E, Schekotov A (2004) Lithosphere-atmosphere-ionosphere coupling as governing mechanism for preseismic short-term events in atmosphere and ionosphere. Nat Hazards Earth Syst Sci 4:757–767

Morgunov UA, Matveev MY (1990) Electromagnetic emission on Spitak earthquake. Geophysics 6:14–191

Nikiforova NN, Yudekhiu FN, Toktosopiev AM (1989) Studies of electromagnetic emission of seismotectonic origin in the Kirghiz SSR. Phys Earth Planet Inter 57(1–2):68–75

Parrot M (1994) Statistical study of ELF/VLF emission recorded by a low-altitude satellite during seismic events. J Geophys Res 99(A12):23339–23347

Serebryakova ON, Bdiehenko SV, Chmyrev VM (1992) Electromagnetic ELF radiation from earthquake regions as observed by low-altitude satellites. Geophys Res Lett 19(2):91–94

Tsang L, Huang CC, Chan CH (2000) Surface electromagnetic fields and impedance matrix elements of stratified media. IEEE Trans Antennas Propag 48(10):1533–1543

Uyeda S et al (2002) Electric and magnetic phenomena observed before the volcano-seismic activity in 2000 in the Izu Island Region. Jpn Proc Natl Acad Sci 99(11):7352–7355

Wait JR (1953) Radiation from a vertical electric dipole over a stratified ground. IRE Trans Antennas Propag AP-1:9–12
Wait JR (1954) Radiation from a vertical electric dipole over a stratified ground. IRE Trans Antennas Propag 2:144–146
Wait JR (1956) Radiation from a vertical electric dipole over a curved stratified ground. J Res Natl Bur Stand 56:232–239
Wait JR (1970) Electromagnetic waves in stratified media, 2nd edn. Pergamon, New York
Wait JR (1998) Comment on "The electromagnetic field of a vertical electric dipole in the presence of a three-layered region" by Ronold W. P. King and Sheldon S. Sandler. Radio Sci 33(2):251–253
Zhang HQ, Pan WY (2002) Electromagnetic field of a vertical electric dipole on a perfect conductor coated with a dielectric layer. Radio Sci 37(4):1060. doi:1029/2000RS002348
Zhang HQ, Chen Y, Pan WY (2009) Fields excited by earthquake ELF/SLF radiator on the ground and in the ionosphere. Chin J Radio Sci 24(3):432–439. (In Chinese)

Chapter 5
Propagation of SLF/ELF Waves in Anisotropic Ionosphere

In this chapter, the propagation of SLF/ELF electromagnetic waves in the anisotropic ionosphere is treated analytically and numerically. Because of the fact that the lower ionosphere varies greatly with height, the lower ionosphere can be regarded as a horizontally stratified anisotropic plasma for SLF/ELF electromagnetic waves. Finally, some new numerical results and discussions are also given.

5.1 Introduction

Because of the existence of the geomagnetic field, the ionosphere behaves as remarkably anisotropic properties in LF/VLF/ELF ranges. In the past 60 years, the propagation of LF/VLF/ELF waves in a homogeneous ionosphere has been investigated widely, while the investigation on LF/VLF/ELF wave propagation in an inhomogeneous ionosphere is seldom reported.

In the past two decades, few attempts of making the experiments on the direct excitation of VLF/SLF waves in ionosphere were undertaken (Armand et al. 1988; Bannister et al. 1993). It is known that the new programmes of the active VLF/SLF wave experiments with the use of large loop antenna or line antenna onboard a spacecraft are carried out (Bannister et al. 1993). Also, the programmes in which the space borne VLF/SLF receiving experiments are used to study earthquake are carried out (Serebryakova et al. 1992). In the past two decades, the relations between the ionospheric propagation of VLF/SLF/ELF waves and earthquake have been treated widely by many investigators (Serebryakova et al. 1992; Ozaki et al. 2004). It is seen that VLF/SLF/ELF wave propagation will penetrate the lower ionosphere from the ionospheric F_2-layer to Earth's surface or from the Earth's surface to the ionospheric F_2-layer. Therefore, it is necessary to carry out VLF/SLF/ELF wave propagation in the lower ionosphere analytically.

In the past century, the theory and experiment of SLF/ELF wave propagation have been treated by many investigators and the findings are well summarized by Wait (1970) and Galejs (1972). In the available paper by Xia and Chen (2000), by

W. Pan, K. Li, *Propagation of SLF/ELF Electromagnetic Waves*,
Advanced Topics in Science and Technology in China,
DOI 10.1007/978-3-642-39050-0_5,

using the matrix-exponent formulations, the wave radiation and propagation in a stratified anisotropic plasma are analyzed and computed numerically. In the works by Li et al. (Li and Pan 1999; Li et al. 2011), the solution of the propagation of VLF/ELF electromagnetic waves penetrating the lower ionosphere is carried out, and analytical formulas are obtained for the electromagnetic field components on the sea surface generated by a space borne loop antenna. In the technical report by Chen et al. (2009), considering the case of oblique incidence, the reflection and transmission of SLF/ELF waves by the inhomogeneous ionosphere is investigated analytically and numerically by using matrix-exponent formulations.

Considering that the main parameters of the ionosphere, including electron and ion densities, and electron and ion collision frequencies, changed significantly with the height above the Earth's surface, and their changes along the horizontal direction are relatively slowly, the ionosphere is regarded as a horizontal stratified anisotropic plasma. In this chapter, based on the technical report (Chen et al. 2009), the reflection and transmission of SLF/ELF waves by the ionosphere are treated by using the matrix method and numerically. Some new numerical results and discussions are also presented.

5.2 Propagation of SLF/ELF Waves in Homogeneous Anisotropic Ionosphere

In this section, we will attempt to treat analytically SLF/ELF wave propagation in a homogeneous anisotropic ionosphere.

5.2.1 SLF/ELF Wave Propagation in Homogeneous Anisotropic Ionosphere

Considering the effect of the geomagnetic field, the ionosphere is regarded as an anisotropic plasma, which is characterized by the tensor permittivity $\bar{\bar{\varepsilon}}$. Let l, m, and n be the direction cosines of the geomagnetic field $\boldsymbol{B}_0$ in the $\hat{x}$, $\hat{y}$, and $\hat{z}$ directions, then the tensor permittivity $\bar{\bar{\varepsilon}}$ can be written in the 3×3 matrix form. It is

$$\hat{\varepsilon} = \varepsilon_0[\mathbf{I} + \mathbf{M}] \tag{5.1}$$

where ε_0 is the free-space permittivity; $\mathbf{I}$ is 3×3 unit matrix; and $\mathbf{M}$ is the susceptibility of the ionosphere, which is given by

$$\mathbf{M} = \mathbf{M}_e + \mathbf{M}_i, \tag{5.2}$$

where

$$\mathbf{M}_e = \frac{-X_e}{U_e(U_e^2 - Y_e^2)} \times \begin{bmatrix} U_e^2 - l^2Y_e^2 & inY_eU_e - lmY_e^2 & -imY_eU_e - lnY_e^2 \\ -inY_eU_e - lmY_e^2 & U_e^2 - m^2Y_e^2 & ilY_eU_e - mnY_e^2 \\ imY_eU_e - lnY_e^2 & -ilY_eU_e - mnY_e^2 & U_e^2 - n^2Y_e^2 \end{bmatrix}, \tag{5.3}$$

$$\mathbf{M}_i = \frac{-X_i}{U_i(U_i^2 - Y_i^2)} \times \begin{bmatrix} U_i^2 - l^2Y_i^2 & inY_iU_i - lmY_i^2 & -imY_iU_i - lnY_i^2 \\ -inY_iU_i - lmY_i^2 & U_i^2 - m^2Y_i^2 & ilY_iU_i - mnY_i^2 \\ imY_iU_i - lnY_i^2 & -ilY_iU_i - mnY_i^2 & U_i^2 - n^2Y_i^2 \end{bmatrix}, \tag{5.4}$$

where

$$U_e = 1 + i\frac{\nu_e}{\omega}; \qquad U_i = 1 + i\frac{\nu_i}{\omega}, \tag{5.5}$$

$$X_e = \frac{\omega_{0e}^2}{\omega^2}; \qquad X_i = \frac{\omega_{0i}^2}{\omega^2}, \tag{5.6}$$

$$\omega_{0e}^2 = \frac{N_e e^2}{\varepsilon_0 m_e}; \qquad \omega_{0i}^2 = \frac{N_i (z_i e)^2}{\varepsilon_0 m_i}, \tag{5.7}$$

$$Y_e = \frac{\omega_H^{(e)}}{\omega}; \qquad \omega_H^{(e)} = \left|\frac{eB_0}{m_e}\right|, \tag{5.8}$$

$$Y_i = \frac{\omega_H^{(i)}}{\omega}; \qquad \omega_H^{(i)} = -\left|\frac{ez_i B_0}{m_i}\right|. \tag{5.9}$$

In the above formulas, ω is the operating angular frequency; $\mathbf{M}_e$ and $\mathbf{M}_i$ are the susceptibility matrix for electrons and heavy ions respectively. ω_{0e} is the angular plasma frequency for electrons, and N_e is the electron density. ω_{0i} is the angular plasma frequency for heavy ions, and N_i is the number density of heavy ions. $\omega_H^{(e)}$ and $\omega_H^{(i)}$ are the gyro-frequency for electrons and heavy ions, respectively.

If N_i is the same order as N_e, it is clear that X_i is negligibly small compared with X_e. Namely, in the case of $N_i \sim N_e$, the effect of heavy ions can be neglected entirely. Only in the case of $N_i \gg N_e$ will the effect by X_i of the heavy ions be important. From the available monograph by Budden (1961), it is seen that for all but the very lowest frequency Y_i is very small compared with unity. That is to say, the anisotropic characteristics affected by heavy ions may be considered in ELF range, even to the lower end of SLF range. According to the analysis in the technical report by Chen et al. (2009), it is seen that when both the transmitting source and the receiving point are located under the lower boundary of the ionosphere, the effects by the heavy ions can still be neglected in SLF/ELF ranges. When the electromagnetic

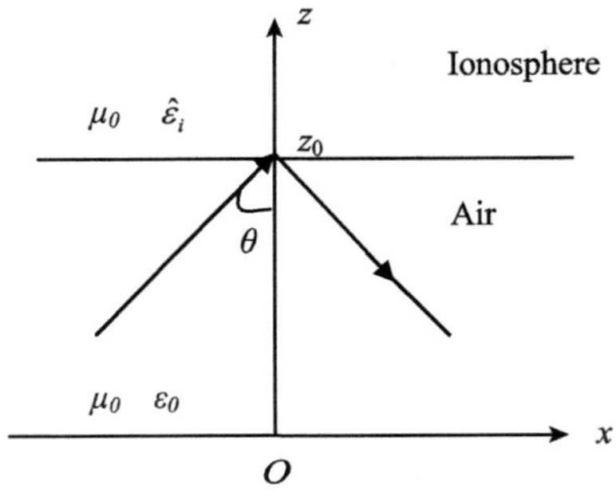

Fig. 5.1 The physical model for incidence, reflection, and refraction by the ionospheric sharp boundary

waves penetrate the ionosphere, the anisotropic characteristics affected by the heavy ions should be considered in ELF range or the lower end of SLF range.

As shown in Fig. 5.1, we take the vertical direction as $\hat{z}$ axis in a Cartesian coordinate system, and assume that the propagation direction in the horizontal plane is on $\hat{x}$ direction. For the simplicity of the analysis, the ionosphere is idealized as a homogeneous anisotropic plasma, and the sharp air–ionosphere boundary is located in the plane of $z = z_0$.

According to the waveguide propagation theory addressed in the preceding chapters, the SLF/ELF field in the Earth–ionosphere waveguide or cavity can be regarded as a superposition of multiple propagation modes and each mode includes the incident wave and its reflected wave with a certain incident angle. For the quasi-TEM mode, the sine of the incident angle can be expressed in the form

$$S_0 \approx \left[1 + \frac{\mathrm{i}(\Delta_g + \Delta_i)}{kh}\right]^{1/2}. \tag{5.10}$$

Correspondingly, for the n-order TM mode, we have

$$S_n = \sqrt{1 - C_n^2}; \quad C_n \approx \frac{n\pi}{kh}\left[1 - \frac{\mathrm{i}kh(\Delta_i + \Delta_g)}{(n\pi)^2}\right]. \tag{5.11}$$

As shown in Fig. 5.1, the changes of all propagation modes in the air and ionosphere along the horizontal directions should meet the following relations:

$$\frac{\partial}{\partial x} = \mathrm{i}kS_j; \quad j = 0, 1, 2, \ldots, n, \tag{5.12}$$

$$\frac{\partial}{\partial y} = 0. \tag{5.13}$$

From the Maxwell equation of $\nabla \times \boldsymbol{H} = -\mathrm{i}\omega\hat{\varepsilon}\boldsymbol{E}$, we have

$$\left.\begin{aligned} -\frac{\partial}{\partial z}H_y &= -\mathrm{i}\omega\varepsilon_0\left[(1+M_{xx})E_x + M_{xy}E_y + M_{xz}E_z\right] \\ \frac{\partial}{\partial z}H_x - \mathrm{i}kS_jH_z &= -\mathrm{i}\omega\varepsilon_0\left[M_{yx}E_x + (1+M_{yy})E_y + M_{yz}E_z\right] \\ \mathrm{i}kS_jH_y &= -\mathrm{i}\omega\varepsilon_0\left[M_{zx}E_x + M_{zy}E_y + (1+M_{zz})E_z\right] \end{aligned}\right\}. \tag{5.14}$$

Similarly, from the Maxwell equation of $\nabla \times \boldsymbol{E} = \mathrm{i}\omega\mu_0\boldsymbol{H}$, we have

$$\left.\begin{aligned} -\frac{\partial}{\partial z}E_y &= \mathrm{i}\omega\mu_0 H_x \\ \frac{\partial}{\partial z}E_y - \mathrm{i}kS_jE_z &= \mathrm{i}\omega\mu_0 H_y \\ \mathrm{i}kS_jE_y &= \mathrm{i}\omega\mu_0 H_z \end{aligned}\right\}. \tag{5.15}$$

With the simultaneous equations (5.14) and (5.15), and eliminating the vertical components E_z and H_z, we readily obtain

$$\left.\begin{aligned} \frac{1}{ik}\frac{\mathrm{d}E_x}{\mathrm{d}z} &= \frac{-S_jM_{zx}}{1+M_{zz}}E_x - \frac{S_jM_{zy}}{1+M_{zz}}E_y + \frac{C_j^2+M_{zz}}{1+M_{zz}}h_y \\ \frac{1}{ik}\frac{\mathrm{d}E_y}{\mathrm{d}z} &= -h_x \\ \frac{1}{ik}\frac{\mathrm{d}h_x}{\mathrm{d}z} &= \left(-M_{yx} + \frac{M_{yz}M_{zx}}{1+M_{zz}}\right)E_x - \left(C_j^2 + M_{yy} - \frac{M_{yz}M_{zy}}{1+M_{zz}}\right)E_y + \frac{S_jM_{yz}}{1+M_{zz}}h_y \\ \frac{1}{ik}\frac{\mathrm{d}h_y}{\mathrm{d}z} &= \left(1 + M_{xx} - \frac{M_{xz}M_{zx}}{1+M_{zz}}\right)E_x + \left(M_{xy} - \frac{M_{xz}M_{zy}}{1+M_{zz}}\right)E_y - \frac{S_jM_{xz}}{1+M_{zz}}h_y \end{aligned}\right\}, \tag{5.16}$$

where

$$C_j^2 = 1 - S_j^2; \qquad h = \eta H = \sqrt{\frac{\mu_0}{\varepsilon_0}}H. \tag{5.17}$$

If the column vector $\tilde{\boldsymbol{e}}$ is represented in the form of

$$\tilde{\boldsymbol{e}} = [E_x \quad -E_y \quad h_x \quad h_y]^T, \tag{5.18}$$

and the matrix $\mathbf{T}$ is written as

$$\mathbf{T} = \begin{bmatrix} \frac{-SM_{zx}}{1+M_{zz}} & \frac{SM_{zy}}{1+M_{zz}} & 0 & \frac{C^2+M_{zz}}{1+M_{zz}} \\ 0 & 0 & 1 & 0 \\ \frac{M_{yz}M_{zx}}{1+M_{zz}} - M_{yx} & C^2 + M_{yy} - \frac{M_{yz}M_{zy}}{1+M_{zz}} & 0 & \frac{SM_{yz}}{1+M_{zz}} \\ 1 + M_{xx} - \frac{M_{xz}M_{zx}}{1+M_{zz}} & \frac{M_{xz}M_{zy}}{1+M_{zz}} - M_{xy} & 0 & \frac{-SM_{xz}}{1+M_{zz}} \end{bmatrix}, \tag{5.19}$$

Equation (5.16) can be rewritten as follows:

$$\frac{\mathrm{d}\tilde{\boldsymbol{e}}}{\mathrm{d}z} = \mathrm{i}k\mathbf{T}\tilde{\boldsymbol{e}}. \tag{5.20}$$

In this section, the ionosphere is a regarded as homogeneous anisotropic plasma with a sharp boundary, it is seen that all elements in the matrix **T** in Eq. (5.19) are constants, which do not vary with in spatial position. Therefore, for the matrix equation (5.19), there should be four eigenvalues and the corresponding four eigenvectors. We write

$$\mathbf{T}\cdot\tilde{\boldsymbol{e}}_j = D_j\tilde{\boldsymbol{e}}_j; \quad j = 1, 2, 3, 4. \tag{5.21}$$

Obviously, they correspond to the four characteristic waves in the ionosphere, which the propagation spatial variation is described by an exponential function. We write

$$\tilde{\boldsymbol{e}}_j = \tilde{\boldsymbol{e}}_j|_{z=z_0}\cdot \mathrm{e}^{\mathrm{i}kD_j(z-z_0)}. \tag{5.22}$$

The eigenvalues D_j $(j = 1, 2, 3, 4)$ represents the normalized wave numbers and the attenuation rates of the characteristic waves in the ionosphere. The eigenvalues D_j of the matrix **T** are determined by the following equation:

$$\det\big[\mathbf{T} - D_j[\mathbf{I}]\big] = 0, \tag{5.23}$$

where $[\mathbf{I}]$ is the 4×4 unit matrix and "det" represents the determinant of the matrix.

Assuming that the geomagnetic field has an angular Ω with the $\hat{z}$ direction, and its projection in the horizontal plane has an angular ϕ with the $\hat{x}$ direction, the directional cosines of the geomagnetic field in the Cartesian coordinate system are expressed as $\sin\Omega\cos\phi$, $\sin\Omega\sin\phi$, and $\cos\Omega$, respectively. In the following derivations and analyses, we will neglect the contribution to the permittivity tensor by the ions in the ionosphere. We denote

$$X = \frac{\omega_{0e}^2}{\omega^2}; \qquad Z = \frac{\nu}{\omega}; \qquad t = \frac{1+\mathrm{i}Z}{X}; \qquad Y = \frac{\omega_{Te}}{\omega}, \tag{5.24}$$

$$h = \frac{Y}{X}; \qquad h_L = h\cos\Omega; \qquad h_T = h\sin\Omega, \tag{5.25}$$

then, the elements in the matrix **M**, which represents the susceptibility of the ionosphere, are denoted as follows:

$$M_{xx} = -\frac{t^2 - h_T^2\cos^2\phi}{t(t^2-h^2)}; \qquad M_{xy} = \frac{h_T^2\sin\phi\cos\phi + \mathrm{i}h_L t}{t(t^2-h^2)}, \tag{5.26}$$

$$M_{xz} = \frac{h_L h_T\cos\phi - \mathrm{i}h_T t\sin\phi}{t(t^2-h^2)}; \qquad M_{yx} = \frac{h_T^2\cos\phi\sin\phi - \mathrm{i}h_L t}{t(t^2-h^2)}, \tag{5.27}$$

$$M_{yy} = -\frac{t^2 - h_T^2\sin^2\phi}{t(t^2-h^2)}; \qquad M_{yz} = \frac{h_L h_T\sin\phi + \mathrm{i}h_T t\cos\phi}{t(t^2-h^2)}, \tag{5.28}$$

$$M_{zx} = \frac{h_L h_T \cos\phi + i h_T t \sin\phi}{t(t^2 - h^2)}; \qquad M_{zy} = \frac{h_L h_T \sin\phi - i h_T t \cos\phi}{t(t^2 - h^2)}, \tag{5.29}$$

$$M_{zz} = -\frac{t^2 - h_L^2}{t(t^2 - h^2)}. \tag{5.30}$$

Substituting Eqs. (5.26)–(5.30) into Eq. (5.23), we obtain the following Booker quartic equation:

$$b_4 D^4 + b_3 D^3 + b_2 D^2 + b_1 D + b_0 = 0, \tag{5.31}$$

where

$$b_4 = t(t^2 - h^2) - t^2 + h_L^2, \tag{5.32}$$

$$b_3 = 2S\cos\phi h_L h_T, \tag{5.33}$$

$$\begin{aligned} b_2 = {} & 2t\{(1 - S^2)h^2 - (t-1)[(1 - S^2)t - 1]\} \\ & - h_T^2(1 - S^2\cos^2\phi) - h_L^2(2 - S^2), \end{aligned} \tag{5.34}$$

$$b_1 = -2(1 - S^2)S\cos\phi h_L h_T, \tag{5.35}$$

$$\begin{aligned} b_0 = {} & (t-1)\{[(1 - S^2)t - 1]^2 - (1 - S^2)^2 h_T^2 \cos^2\phi\} \\ & - (1 - S^2)[(1 - S^2)t - 1](h_T^2 \sin^2\phi + h_L^2). \end{aligned} \tag{5.36}$$

From the complex coefficient algebraic quartic equation (5.31), it is seen that this equation has four roots, where the two roots in the first quadrant represent the upward characteristic waves, and the other two in the third quadrant represent the downward characteristic waves. When the waves from below penetrate into the ionosphere, considering that the ionosphere is a homogeneous anisotropic plasma, the waves will not encounter the reflections by the uneven surface, and there are only upward waves in the ionosphere.

Obviously, this algebraic equation can be solved readily by the traditional method. It can also be solved by the numerical method which is used to solve a transcendental equation.

When the eigenvalues of the matrix **T** are solved, the corresponding eigenvectors are obtained easily. Thus, the wave polarizations are determined uniquely. If we take the component E_y as a reference, the ratios of the rest component to E_y can be obtained readily. We write

$$\begin{aligned} P = {} & \frac{E_x}{E_y} \\ = {} & \frac{1}{M}\left[\frac{h_L^2 h_T^2 \sin^2\phi + h_T^2 t^2 \cos^2\phi}{t^2(t^2 - h^2)^2}\left(1 - S^2 - D^2 - \frac{t^2 - h_T^2\sin^2\phi}{t(t^2 - h^2)}\right)\right. \\ & \left. - \left(1 - S^2 - \frac{t^2 - h_L^2}{t(t^2 - h^2)}\right)\right], \end{aligned} \tag{5.37}$$

$$\begin{aligned} Q &= \frac{E_z}{E_y} \\ &= \frac{1}{M}\left\{\left[1 - S^2 - D^2 - \frac{t^2 - h_T^2 \sin^2\phi}{t(t^2 - h^2)}\right]\left[SD + \frac{h_L h_T \cos\phi + i h_T t \sin\phi}{t(t^2 - h^2)}\right]\right. \\ &\quad \left. - \frac{(h_T^2 \cos\phi \sin\phi - i h_L t)(h_T h_L \sin\phi - i h_T t \cos\phi)}{t^2(t^2 - h^2)^2}\right\}, \end{aligned} \tag{5.38}$$

$$\frac{\eta H_z}{E_y} = S, \tag{5.39}$$

$$\frac{\eta H_x}{E_y} = -D, \tag{5.40}$$

$$\frac{\eta H_y}{E_y} = T = DP - SQ, \tag{5.41}$$

where

$$\begin{aligned} M &= \left[1 - S^2 - \frac{t^2 - h_L^2}{t(t^2 - h^2)}\right]\frac{h_T^2 \cos\phi \sin\phi - i h_L t}{t(t^2 - h^2)} \\ &\quad - \left[SD + \frac{h_L h_T \cos\phi + i h_T t \sin\phi}{t(t^2 - h^2)}\right]\left[\frac{h_L h_T \sin\phi + i h_T t \cos\phi}{t(t^2 - h^2)}\right]. \end{aligned} \tag{5.42}$$

In Eqs. (5.37)–(5.41), corresponding to the two different characteristic waves, the normalized characteristic wave numbers D are different from each other. In general, the two characteristic waves are distinguished as the ordinary wave (O-wave) and extraordinary wave (E-wave). Then, the two upward waves in the ionosphere are expressed as follows:

$$\left.\begin{aligned} E_{ox} &= P_o \\ E_{oy} &= 1 \\ E_{oz} &= Q_o \\ \eta H_{ox} &= -D_o \\ \eta H_{py} &= T_o \\ \eta H_{oz} &= S \end{aligned}\right\} A_o \mathrm{e}^{ikD_o(z-z_0)} \mathrm{e}^{ikSx}, \tag{5.43}$$

and

$$\left.\begin{aligned} E_{ex} &= P_e \\ E_{ey} &= 1 \\ E_{ez} &= Q_e \\ \eta H_{ex} &= -D_e \\ \eta H_{ey} &= T_e \\ \eta H_{ez} &= S \end{aligned}\right\} A_e \mathrm{e}^{ikD_e(z-z_0)} \mathrm{e}^{ikSx}. \tag{5.44}$$

For SLF/ELF waves in the space between the Earth's surface and the lower boundary of the ionosphere, the main propagating mode may be understood as a vertically polarized TM_0 mode. This mode may be regarded as the superposition of the incident wave and its reflected wave with the sine of the incident angle S_0. If we take the component ηH_y as a reference, the components of the incident wave are expressed in the following forms:

$$\begin{bmatrix} E_{xi} \\ E_{zi} \\ \eta H_{yi} \end{bmatrix} = \begin{bmatrix} C \\ -S \\ 1 \end{bmatrix} \mathrm{e}^{\mathrm{i}kC(z-z_0)} \mathrm{e}^{\mathrm{i}kSx}. \tag{5.45}$$

Correspondingly, a reflected wave will be produced in the air–ionosphere interface. The propagating properties of the reflected wave should be consistent with those of the incident wave in the $\hat{x}$ direction, while the properties are opposite in the $\hat{z}$ direction. Thus, it follows that

$$\begin{bmatrix} E_{xr} \\ E_{zr} \\ \eta H_{yr} \end{bmatrix} = \begin{bmatrix} -C \\ -S \\ 1 \end{bmatrix} {}_{\parallel}R_{\parallel} \mathrm{e}^{-\mathrm{i}kC(z-z_0)} \mathrm{e}^{\mathrm{i}kSx}, \tag{5.46}$$

where ${}_{\parallel}R_{\parallel}$ represents the reflection coefficient, and the subscript $\parallel$ indicates that the polarization of the incident wave and that of the reflected wave are in the same incident plane.

Considering the effects by the geomagnetic field, the ionosphere is regarded as an anisotropic plasma. Generally, the propagating wave in the ionosphere has six components. Besides the original polarized TM mode, the new polarized TE mode is generated. Thus, another reflected wave appears, which is expressed in the following form:

$$\begin{bmatrix} E_{yr} \\ \eta H_{xr} \\ \eta H_{zr} \end{bmatrix} = \begin{bmatrix} 1 \\ C \\ S \end{bmatrix} {}_{\parallel}R_{\perp} \mathrm{e}^{-\mathrm{i}kC(z-z_0)} \mathrm{e}^{\mathrm{i}kSx}, \tag{5.47}$$

where the subscript $\perp$ in the reflection coefficient ${}_{\parallel}R_{\perp}$ indicates that the polarization plane of the reflected wave is perpendicular to the incident plane.

From the boundary condition in the air–ionosphere interface, we have

$$\left.\begin{aligned} C(1 - {}_{\parallel}R_{\parallel}) &= A_o P_o + A_e P_e \\ {}_{\parallel}R_{\perp} &= A_o + A_e \\ C\,{}_{\parallel}R_{\perp} &= -D_o A_o - D_e A_e \\ 1 + {}_{\parallel}R_{\parallel} &= T_o A_o + T_e A_e \end{aligned}\right\}. \tag{5.48}$$

By solving the above equations, it follows that

$$\left.\begin{aligned} A_o &= \frac{-2C(C+D_e)}{(C+D_o)(P_e+CT_e)-(C+D_e)(P_o+CT_o)} \\ A_e &= \frac{2C(C+D_o)}{(C+D_o)(P_e+CT_e)-(C+D_e)(P_o+CT_o)} \\ {}_{\parallel}R_{\parallel} &= \frac{(C+D_e)(P_o-CT_o)-(C+D_o)(P_e-CT_e)}{(C+D_o)(P_e+CT_e)-(C+D_e)(P_o+CT_o)} \\ {}_{\parallel}R_{\perp} &= \frac{2C(D_o-D_e)}{(C+D_o)(P_e+CT_e)-(C+D_e)(P_o+CT_o)} \end{aligned}\right\}. \tag{5.49}$$

It is noted that Eq. (5.31) will be reduced to a quasi-quadratic equation, which can be solved easily, in the following three special cases:

- Case 1: $h_L = 0$, i.e. the geomagnetic field is horizontal, which is applied to the case near the geomagnetic equator.
- Case 2: $h_T = 0$, i.e. the geomagnetic field is perpendicular to the Earth's surface, which is corresponded to the case near the two geomagnetic poles.
- Case 3: $\phi = 90°$ or $\phi = 270°$, i.e. the propagation direction is eastward or westward.

The solutions of the quasi-quadratic equation are written in the forms

$$D_o^2 = \frac{-b_2 + \sqrt{b_2^2 - 4b_0 b_4}}{2b_4}; \qquad D_e^2 = \frac{-b_2 - \sqrt{b_2^2 - 4b_0 b_4}}{2b_4}. \tag{5.50}$$

In the case of $h_L = 0$, the polarization factors are simplified as follows:

$$P = \frac{1}{M}\left\{\frac{h^2\cos^2\phi}{(t^2-h^2)^2} - \left[1 - S^2 - D^2 - \frac{t^2 - h^2\sin^2\phi}{t(t^2-h^2)}\right] \times \left(1 - S^2 - \frac{t}{t^2-h^2}\right)\right\}, \tag{5.51}$$

$$M = \left(1 - S^2 - \frac{t}{t^2-h^2}\right)\frac{h^2\cos\phi\sin\phi}{t(t^2-h^2)} - \frac{ih\cos\phi}{t^2-h^2}\left(SD + \frac{ih\sin\phi}{t^2-h^2}\right), \tag{5.52}$$

$$Q = \frac{1}{M}\left\{\left[1 - S^2 - D^2 - \frac{t^2 - h^2\sin^2\phi}{t(t^2-h^2)}\right]\left(SD + \frac{ih\sin\phi}{t^2-h^2}\right) + \frac{ih^3\cos^2\phi\sin\phi}{t(t^2-h^2)^2}\right\}. \tag{5.53}$$

In the case of $\phi = 90°$ or $\phi = 270°$, both the molecules and the denominator in Eqs. (5.51) and (5.53) are simultaneously 0. Thus, this case need to be treated in an additional manner.

When both the conditions $h_L = 0$ and $\phi = 90°$ or $\phi = 270°$ are satisfied, the solutions of the quasi-quadratic equation are obtained readily. We write

$$D_o^2 = C^2 - \frac{1}{t}; \qquad D_e^2 = C^2 - \frac{t-1}{t^2 - h^2 - t}. \tag{5.54}$$

Correspondingly, the two characteristic waves are expressed in the forms

$$\left.\begin{aligned} E_{xo} &= 0 \\ E_{yo} &= 1 \\ E_{zo} &= 0 \\ \eta H_{xo} &= -D_o \\ \eta H_{yo} &= 0 \\ \eta H_{zo} &= S \end{aligned}\right\} A_o \mathrm{e}^{ikD_o(z-z_0)}\mathrm{e}^{ikSx}, \tag{5.55}$$

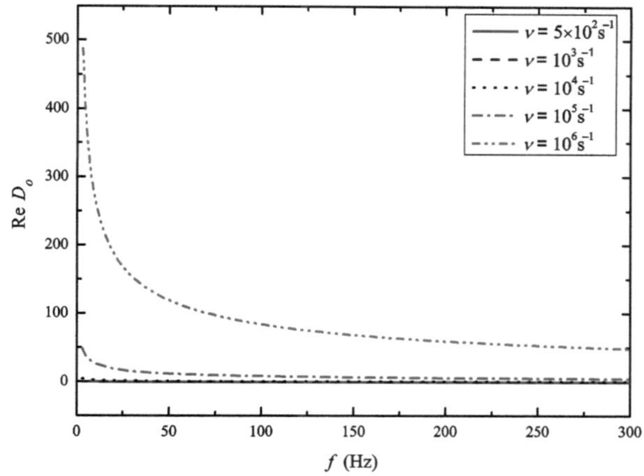

Fig. 5.2 The real part of D_o versus the operating frequency at different collision frequencies: $B_0 = 0.5 \times 10^{-4}$ T, $\omega_{0e} = 6.67 \times 10^7$ rad/s, $\Omega = \phi = 45°$

and

$$\left.\begin{aligned} E_{xe} &= 1 \\ E_{ye} &= 0 \\ E_{ze} &= L \\ \eta H_{xe} &= 0 \\ \eta H_{ye} &= D_e - SL \\ \eta H_{ze} &= 0 \end{aligned}\right\} A_e e^{ikD_e(z-z_0)} e^{ikSx}, \tag{5.56}$$

where

$$L = \frac{SD_e(t^2 - h^2) + ih \sin\phi}{t - C^2(t^2 - h^2)}. \tag{5.57}$$

With boundary condition equations, it is solved readily:

$$\left.\begin{aligned} {}_{\parallel}R_{\parallel} &= \frac{C(SL - D_e) + 1}{C(SL - D_e) - 1} \\ A_o &= {}_{\parallel}R_{\perp} = 0 \\ A_e &= -\frac{2C}{C(SL - D_e) - 1} \end{aligned}\right\}. \tag{5.58}$$

5.2.2 Computations and Discussions

With $\Omega = 45°$ and $\phi = 45°$, the real and imaginary parts of the normalized wave numbers D_0 and D_e are computed in the ranges from 3 Hz to 300 Hz and shown in Figs. 5.2, 5.3, 5.4, 5.5. In the above computations, we assume that the electron plasma angular frequency is taken as $\omega_{0e} = 6.67 \times 10^7$ rad/s, the geomagnetic field strength is $B_0 = 0.5 \times 10^{-4}$ T.

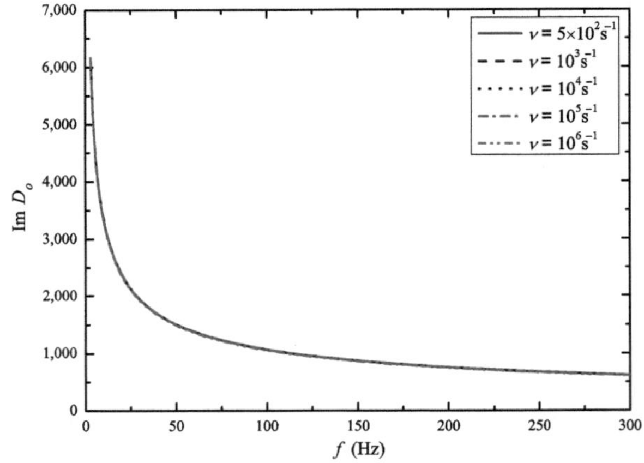

Fig. 5.3 The imaginary part of D_o versus the operating frequency at different collision frequencies: $B_0 = 0.5 \times 10^{-4}$ T, $\omega_{0e} = 6.67 \times 10^7$ rad/s, $\Omega = \phi = 45°$

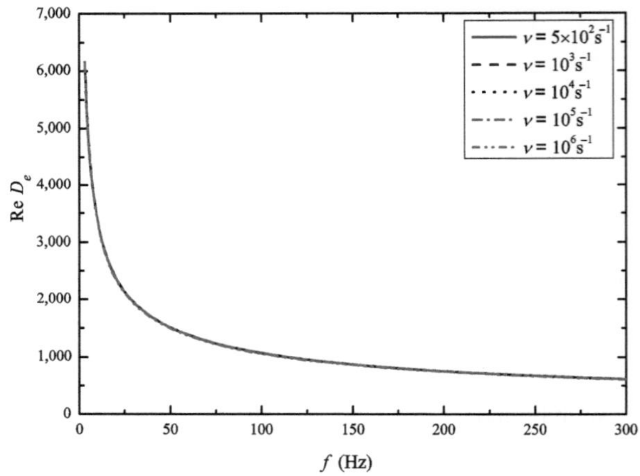

Fig. 5.4 The real part of D_e versus the operating frequency at different collision frequencies: $B_0 = 0.5 \times 10^{-4}$ T, $\omega_{0e} = 6.67 \times 10^7$ rad/s, $\Omega = \phi = 45°$

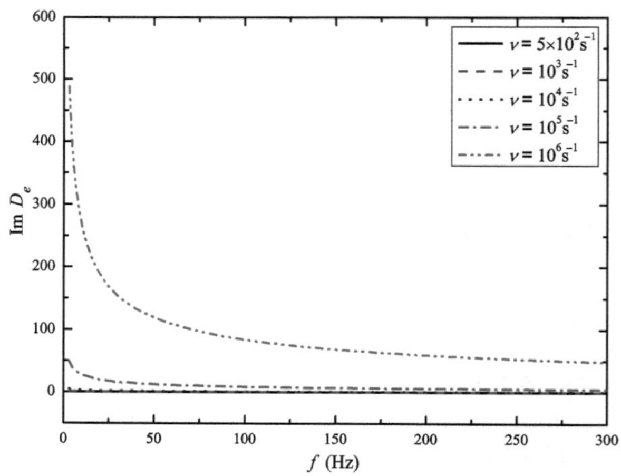

Fig. 5.5 The imaginary part of D_e versus the operating frequency at different collision frequencies: $B_0 = 0.5 \times 10^{-4}$ T, $\omega_{0e} = 6.67 \times 10^7$ rad/s, $\Omega = \phi = 45°$

Fig. 5.6 The real part of D_e versus the angle ϕ at $f = 30$ Hz: $B_0 = 0.5 \times 10^{-4}$ T, $\omega_{0e} = 6.67 \times 10^7$ rad/s, $\nu = 10^3$ s^{-1}

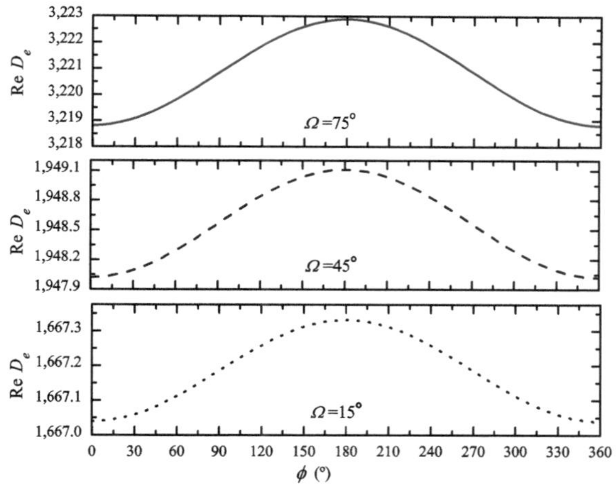

From these computations for SLF/ELF ranges, it is concluded as follows:

- The real and imaginary parts of D_0 are positive, while the real part is much smaller than the imaginary part. That is to say, when the ordinary wave propagation is upward in the ionosphere, the amplitude attenuates quickly, while the phase change is very small. Thus, the ordinary wave is a rapid evanescent wave.
- The real and imaginary parts of D_e are also positive, but the real part is much larger than the imaginary part. Then, it is seen that the extraordinary wave can propagate in the ionosphere with little attenuation.
- The imaginary part of D_e is decreased with the decrease of the collision frequency ν, the attenuation of the extraordinary wave mainly results by the collision absorption. In the ionosphere, the electron collision frequency ν will decrease with the increase of the height z above the Earth's surface. Therefore, the ionosphere attenuation occurs mainly in the lower ionosphere, for which the ionosphere height is in the range of 70–100 km. Above the lower ionosphere, the collision frequency ν becomes small, and the attenuation of the extraordinary wave will be small, too.
- The real and imaginary parts of D_o and those of D_e will decrease with the increase of the operating frequency.

We assume that the angular plasma frequency is $\omega_{0e} = 6.67 \times 10^7$ rad/s, the electron collision frequency is $\nu = 10^3$ s^{-1}, the operating frequencies are taken as $f = 30$ Hz and 300 Hz, and the angle Ω are taken as 75°, 45°, and 15°, respectively. In the case of the incident wave being TEM wave, the real and imaginary parts of D_e versus the angle ϕ are computed and shown in Figs. 5.6, 5.7, 5.8, 5.9. From the above calculated results, we conclude as follows:

- The real and imaginary parts of D_e to the azimuth are symmetrical. The minima of the real and imaginary parts are at $\phi = 0$ or 360°, while the maxima at $\phi =$ 180°.
- The real and imaginary parts of D_e at lower latitudes are larger than those at the higher latitudes, and the directional effect is more significant.

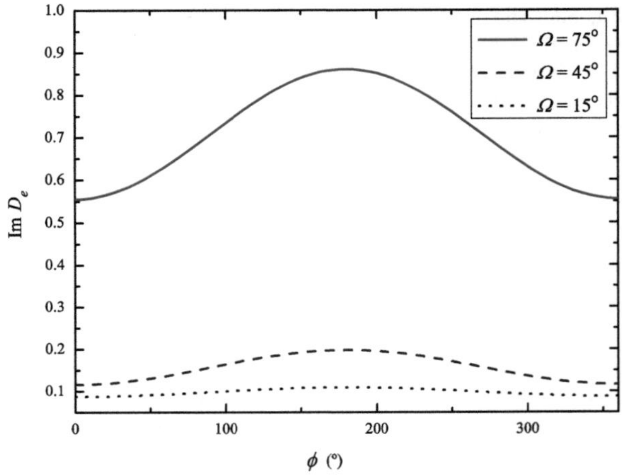

Fig. 5.7 The imaginary part of D_e versus the angle ϕ at $f = 30$ Hz: $B_0 = 0.5 \times 10^{-4}$ T, $\omega_{0e} = 6.67 \times 10^7$ rad/s, $\nu = 10^3$ s^{-1}

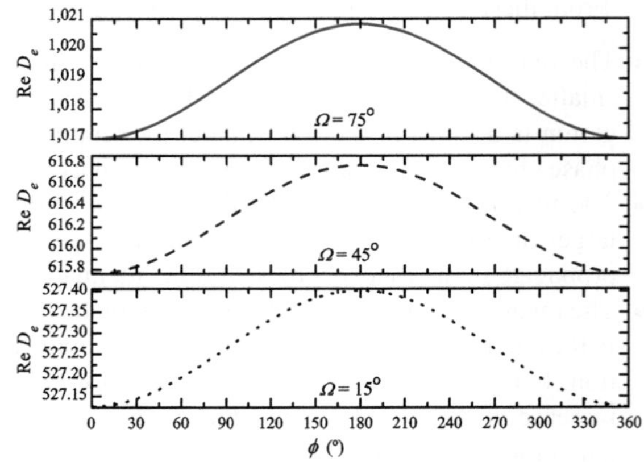

Fig. 5.8 The real part of D_e versus the angle ϕ at $f = 300$ Hz: $B_0 = 0.5 \times 10^{-4}$ T, $\omega_{0e} = 6.67 \times 10^7$ rad/s, $\nu = 10^3$ s^{-1}

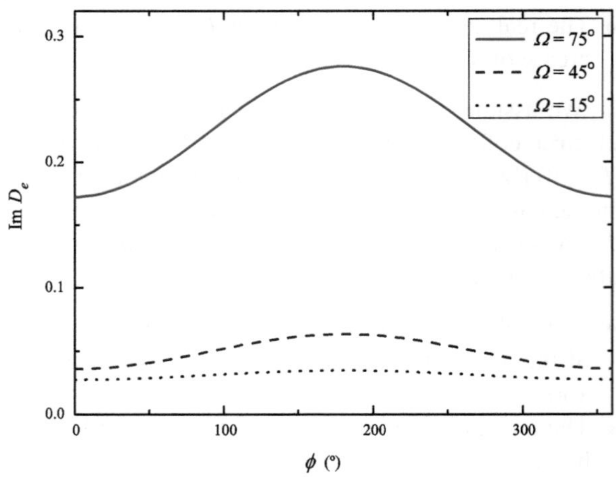

Fig. 5.9 The imaginary part of D_e versus the angle ϕ at $f = 300$ Hz: $B_0 = 0.5 \times 10^{-4}$ T, $\omega_{0e} = 6.67 \times 10^7$ rad/s, $\nu = 10^3$ s^{-1}

- The higher the operating frequency is, the smaller the real and imaginary parts of D_e are.

Assuming that the angular plasma frequency is $\omega_{0e} = 6.67 \times 10^7$ rad/s, the geomagnetic inclination angle is $\Omega = 45°$, and the electron collision frequency is $\nu = 10^3\ \mathrm{s}^{-1}$, the reflection coefficients $_{\|}R_{\|}$ and $_{\|}R_{\perp}$ and the transmission coefficients A_o and A_e versus the operating frequency are computed and shown in Figs. 5.10–5.13. From Figs. 5.10 and 5.11, it is seen that the amplitudes of $_{\|}R_{\|}$ and $_{\|}R_{\perp}$ are close to 1 and 0, respectively. It is indicated that, if the incident wave is vertically polarized, the reflected wave is primarily the vertically polarized wave, and the horizontally polarized wave is very small. From Figs. 5.12 and 5.13, it is seen that both the amplitudes of A_o and A_e are very small, and their phase difference is 90°.

With the same parameters as above except for the plasma angular frequency being taken as $\omega_{0e} = 3.33 \times 10^6$ rad/s, the reflection coefficients $_{\|}R_{\|}$ and $_{\|}R_{\perp}$ are computed and shown in Figs. 5.14 and 5.15, respectively. Comparing Fig. 5.14 with Fig. 5.10, it is found that the lower the angular plasma frequency is, the smaller the reflection coefficient $_{\|}R_{\|}$ is.

With $f = 30$ Hz, $B_0 = 0.5 \times 10^{-4}$ T, $\omega_{0e} = 6.67 \times 10^7$ rad/s, $\Omega = 45°$, and $\nu = 10^3\ \mathrm{s}^{-1}$, the reflection coefficients $_{\|}R_{\|}$ and $_{\|}R_{\perp}$ and the transmission coefficients A_o and A_e versus the angle ϕ are computed and shown in Figs. 5.16, 5.17, 5.18, 5.19, respectively.

We assume that the angular plasma frequency is $\omega_{0e} = 6.67 \times 10^7$ rad/s, the electron collision frequency is $\nu = 10^3\ \mathrm{s}^{-1}$, the operating frequencies are taken as $f = 30$ Hz, and the angle Ω is taken as 75°, 45°, and 15°, respectively. In the case of the incident wave being a TEM wave, the magnitudes of the reflection coefficients $_{\|}R_{\|}$ and $_{\|}R_{\perp}$ versus the angle ϕ are computed and shown in Figs. 5.20–5.21. From the above calculated results, it is concluded as follows:

- The minimum for the magnitude of the reflection coefficient $_{\|}R_{\|}$ is at $\phi = 90°$, while the maximum at $\phi = 270°$.
- The minimum for the magnitude of the reflection coefficient $_{\|}R_{\perp}$ is at $\phi = 180°$, while the maximum at $\phi = 0°$ or 360°.
- The magnitudes of the reflection coefficients $_{\|}R_{\|}$ at lower latitudes are larger than those at the higher latitudes, and the directional effect is more significant.

5.3 Propagation of SLF/ELF Waves in Inhomogeneous Anisotropic Ionosphere

In this section, the inhomogeneous ionosphere is idealized as a horizontal stratified anisotropic plasma. In what follows, we will attempt to treat analytically SLF/ELF wave propagation for a horizontal stratified anisotropic ionosphere.

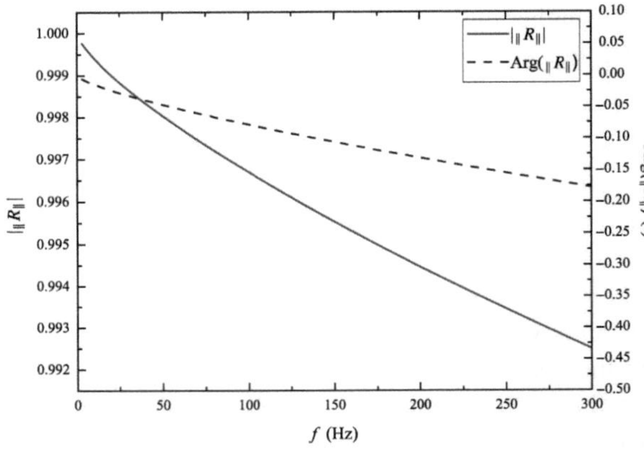

Fig. 5.10 The magnitudes and phases of the reflection coefficient $_{\parallel}R_{\parallel}$ versus the operating frequency: $B_0 = 0.5 \times 10^{-4}$ T, $\omega_{0e} = 6.67 \times 10^7$ rad/s, $\Omega = \phi = 45°$, $\nu = 10^3\ \mathrm{s}^{-1}$

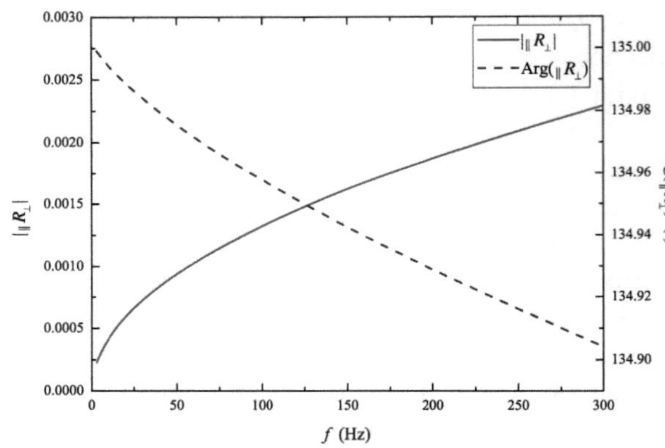

Fig. 5.11 The magnitudes and phases of the reflection coefficient $_{\parallel}R_{\perp}$ versus the operating frequency: $B_0 = 0.5 \times 10^{-4}$ T, $\omega_{0e} = 6.67 \times 10^7$ rad/s, $\Omega = \phi = 45°$, $\nu = 10^3\ \mathrm{s}^{-1}$

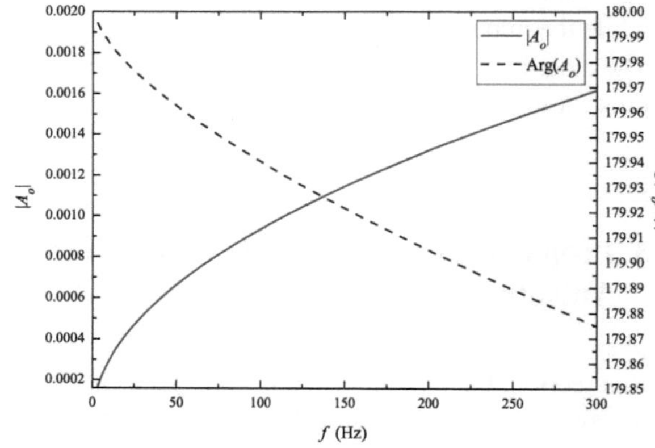

Fig. 5.12 The magnitudes and phases of the transmission coefficient A_o versus the operating frequency: $B_0 = 0.5 \times 10^{-4}$ T, $\omega_{0e} = 6.67 \times 10^7$ rad/s, $\Omega = \phi = 45°$, $\nu = 10^3\ \mathrm{s}^{-1}$

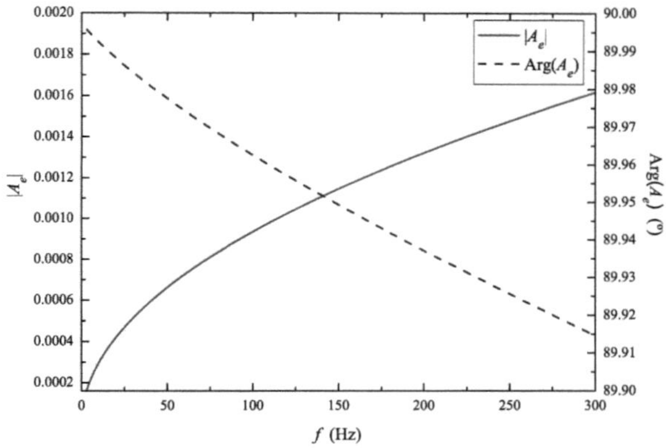

Fig. 5.13 The magnitudes and phases of the transmission coefficient A_e versus the operating frequency: $B_0 = 0.5 \times 10^{-4}$ T, $\omega_{0e} = 6.67 \times 10^7$ rad/s, $\Omega = \phi = 45°$, $\nu = 10^3$ s^{-1}

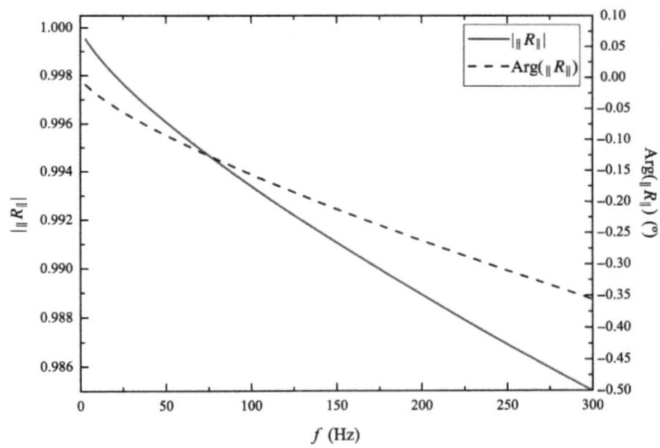

Fig. 5.14 The magnitudes and phases of the reflection coefficient $_{\parallel}R_{\parallel}$ versus the operating frequency: $B_0 = 0.5 \times 10^{-4}$ T, $\omega_{0e} = 3.33 \times 10^6$ rad/s, $\Omega = \phi = 45°$, $\nu = 10^3$ s^{-1}

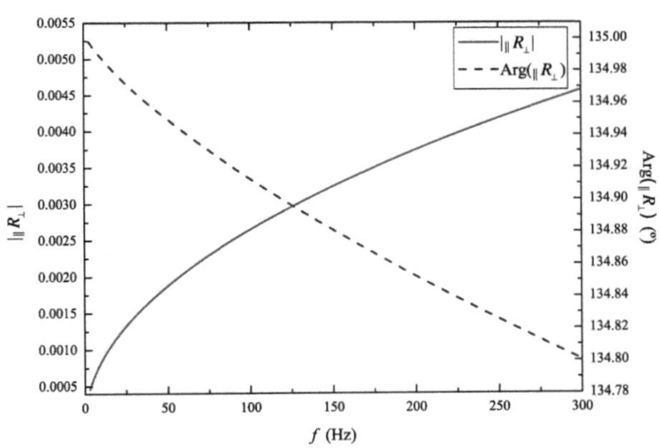

Fig. 5.15 The magnitudes and phases of the reflection coefficient $_{\parallel}R_{\perp}$ versus the operating frequency: $B_0 = 0.5 \times 10^{-4}$ T, $\omega_{0e} = 3.33 \times 10^6$ rad/s, $\Omega = \phi = 45°$, $\nu = 10^3$ s^{-1}

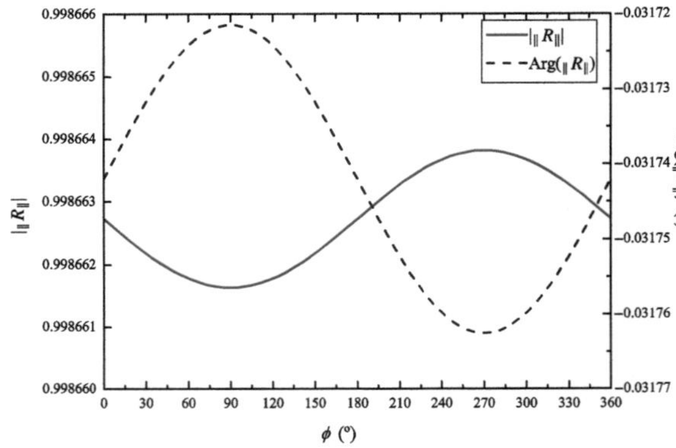

Fig. 5.16 The magnitudes and phases of the reflection coefficient $_{\parallel}R_{\parallel}$ versus the angle ϕ: $f = 30$ Hz, $B_0 = 0.5 \times 10^{-4}$ T, $\omega_{0e} = 6.67 \times 10^7$ rad/s, $\Omega = 45°$, $\nu = 10^3$ s^{-1}

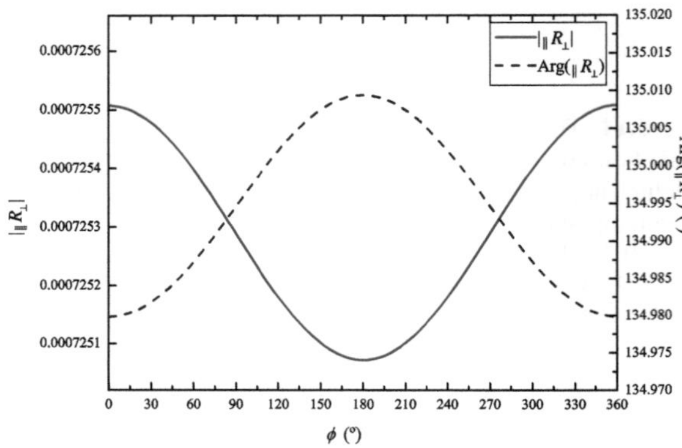

Fig. 5.17 The magnitudes and phases of the reflection coefficient $_{\parallel}R_{\perp}$ versus the angle ϕ: $f = 30$ Hz, $B_0 = 0.5 \times 10^{-4}$ T, $\omega_{0e} = 6.67 \times 10^7$ rad/s, $\Omega = 45°$, $\nu = 10^3$ s^{-1}

5.3.1 SLF/ELF Wave Propagation in Stratified Anisotropic Ionosphere

As shown in Fig. 5.22, the lower ionosphere is divided into N layers; each interface between two layers is parallel to the plane of the ground; and each layer is regarded as a homogeneous anisotropic plasma. Above the height z_n, the region is regarded as a homogeneous anisotropic plasma with a sharp boundary. Each layered plasma can be characterized by the tensor permittivity $\bar{\bar{\varepsilon}}_i$, which is determined by the parameters including the electron and ion densities, the collision frequencies of the electron and ion, the geomagnetic field strength, and so on.

Considering the ionosphere is a stratified anisotropic plasma, the wave will encounter reflection in the interface between two layers in the ionosphere. As addressed in Sect. 5.2, it is seen that there exist two upward waves and the other two downward waves. Note that the downward waves can be understood as the reflected wave caused by the interface between two layers of the ionosphere. As for the two upward waves, the first one is a rapidly evanescent wave, and the other one is a wave which can propagate. It is similar for the two downward waves. The characteristics for the two downward waves are also similar.

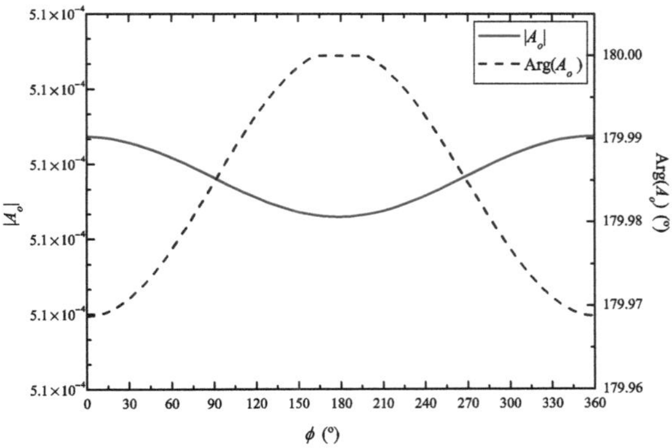

Fig. 5.18 The magnitudes and phases of the transmission coefficient A_o versus the angle ϕ: $f = 30$ Hz, $B_0 = 0.5 \times 10^{-4}$ T, $\omega_{0e} = 6.67 \times 10^7$ rad/s, $\Omega = 45°$, $\nu = 10^3$ s^{-1}

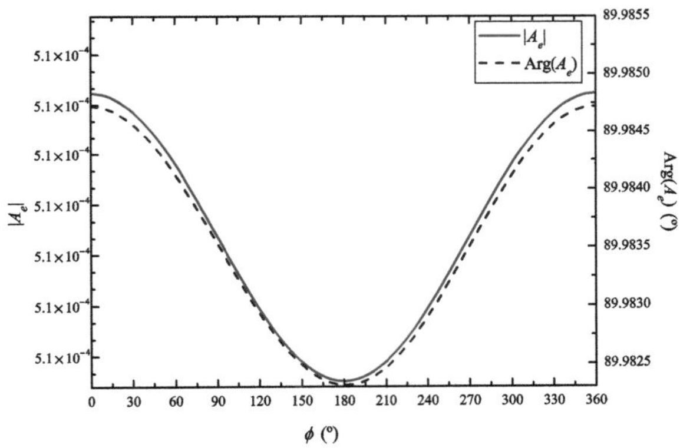

Fig. 5.19 The magnitudes and phases of the transmission coefficient A_e versus the angle ϕ: $f = 30$ Hz, $B_0 = 0.5 \times 10^{-4}$ T, $\omega_{0e} = 6.67 \times 10^7$ rad/s, $\Omega = 45°$, $\nu = 10^3$ s^{-1}

For convenience in the description of the electromagnetic field in each layer, we denote a column vector, which represents the four horizontal components in the ith layer. We write

$$\tilde{\boldsymbol{e}}_i = [e_{xi} \quad e_{yi} \quad h_{xi} \quad h_{yi}]^T. \tag{5.59}$$

We assume that $\tilde{\boldsymbol{e}}_i^1$, $\tilde{\boldsymbol{e}}_i^2$, $\tilde{\boldsymbol{e}}_i^3$, and $\tilde{\boldsymbol{e}}_i^4$ indicate the four eigenvectors in the ith layer, which correspond to the four characteristic waves. Here $\tilde{\boldsymbol{e}}_i^1$ and $\tilde{\boldsymbol{e}}_i^2$ represent the two upward waves, while $\tilde{\boldsymbol{e}}_i^3$ and $\tilde{\boldsymbol{e}}_i^4$ represent the two downward waves. The corresponding eigenvalues are expressed as D_i^1, D_i^2, D_i^3, and D_i^4, respectively. Then, the field components in the ith of the ionosphere are expressed as follows:

$$\begin{aligned}\tilde{\boldsymbol{e}}_i = {} & A_i^1\tilde{\boldsymbol{e}}_i^1 \cdot \mathrm{e}^{\mathrm{i}kD_i^1(z-z_{i-1})} + A_i^2\tilde{\boldsymbol{e}}_i^2 \cdot \mathrm{e}^{\mathrm{i}kD_i^2(z-z_{i-1})} \\ & + A_i^3\tilde{\boldsymbol{e}}_i^3\mathrm{e}^{\mathrm{i}kD_i^3(z-z_{i-1})} + A_i^4\tilde{\boldsymbol{e}}_i^4\mathrm{e}^{\mathrm{i}kD_i^4(z-z_{i-1})}.\end{aligned} \tag{5.60}$$

In the above formula, the first and second terms represent the two upward waves, while the third and fourth terms represent the two downward waves. It is noted that the reference plane is taken as the lower boundary z_{i-1} of the ith layer.

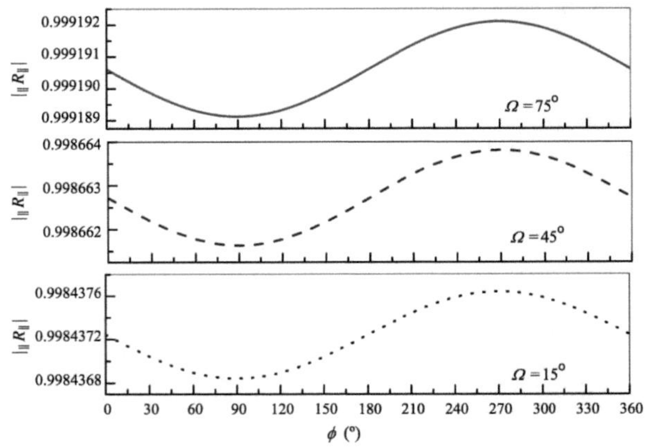

Fig. 5.20 The magnitudes of the reflection coefficient $_{\parallel}R_{\parallel}$ versus the angle ϕ: $f = 30$ Hz, $B_0 = 0.5 \times 10^{-4}$ T, $\omega_{0e} = 6.67 \times 10^7$ rad/s, and $\nu = 10^3$ s^{-1}

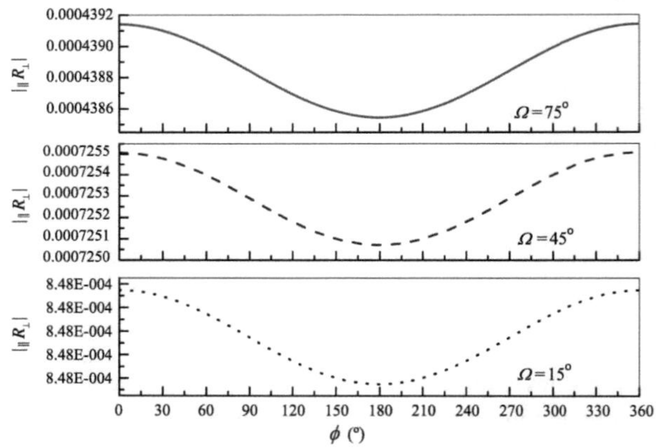

Fig. 5.21 The magnitudes of the reflection coefficient $_{\parallel}R_{\perp}$ versus the angle ϕ: $f = 30$ Hz, $B_0 = 0.5 \times 10^{-4}$ T, $\omega_{0e} = 6.67 \times 10^7$ rad/s, $\nu = 10^3$ s^{-1}

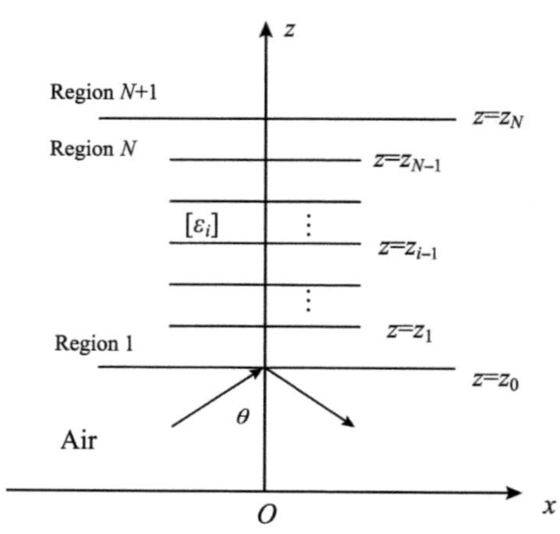

Fig. 5.22 The multi-layered ionospheric model for incidence, reflection, and refraction of plane wave

We denote that the 4 × 4 matrix $[\boldsymbol{e}_i]$ is in the form of

$$[\boldsymbol{e}_i] = \begin{bmatrix} \tilde{\boldsymbol{e}}_i^1 & \tilde{\boldsymbol{e}}_i^2 & \tilde{\boldsymbol{e}}_i^3 & \tilde{\boldsymbol{e}}_i^4 \end{bmatrix}; \tag{5.61}$$

the matrix D_i is a diagonal matrix,

$$[D_i] = \begin{bmatrix} \mathrm{e}^{\mathrm{i}kD_i^1(z-z_{i-1})} & 0 & 0 & 0 \\ 0 & \mathrm{e}^{\mathrm{i}kD_i^2(z-z_{i-1})} & 0 & 0 \\ 0 & 0 & \mathrm{e}^{\mathrm{i}kD_i^3(z-z_{i-1})} & 0 \\ 0 & 0 & 0 & \mathrm{e}^{\mathrm{i}kD_i^4(z-z_{i-1})} \end{bmatrix}; \tag{5.62}$$

and $\mathbf{A}$ is a column vector, which is written in the form of

$$\mathbf{A}_i = \begin{bmatrix} A_i^1 & A_i^2 & A_i^3 & A_i^4 \end{bmatrix}^T. \tag{5.63}$$

Then, Eq. (5.60) can be rewritten in the following form:

$$\tilde{\boldsymbol{e}}_i = [\boldsymbol{e}_i] \cdot [D_i] \cdot \mathbf{A}_i. \tag{5.64}$$

With the boundary condition at $z = z_i$, it is follows that

$$\tilde{\boldsymbol{e}}_{i+1}|_{z=z_1} = \tilde{\boldsymbol{e}}_i|_{z=z_1}. \tag{5.65}$$

It can be rewritten as follows:

$$[\boldsymbol{e}_{i+1}] \cdot \mathbf{A}_{i+1} = [\boldsymbol{e}_i] \cdot \left[D_i^*\right] \cdot \mathbf{A}_i, \tag{5.66}$$

where $[D_i^*]$ represents the value of Eq. (5.62) at $z = z_i$. It is

$$\left[D_i^*\right] = \begin{bmatrix} \mathrm{e}^{\mathrm{i}kD_i^1\Delta z_i} & 0 & 0 & 0 \\ 0 & \mathrm{e}^{\mathrm{i}kD_i^2\Delta z_i} & 0 & 0 \\ 0 & 0 & \mathrm{e}^{\mathrm{i}kD_i^3\Delta z_i} & 0 \\ 0 & 0 & 0 & \mathrm{e}^{\mathrm{i}kD_i^4\Delta z_i} \end{bmatrix}, \tag{5.67}$$

where Δz_i is the thickness of the ith layer of the ionosphere.

It is known that the four column vectors in the matrix $[\boldsymbol{e}_i]$ represent the normalized polarization factors of the four characteristic waves, and the four column vectors are linearly independent. Thus, for the matrix $[\boldsymbol{e}_i]$, there exists a corresponding inverse matrix, and the recurrence relations are obtained readily. We write

$$\mathbf{A}_i = \left[D_i^*\right]^{-1} \cdot [\boldsymbol{e}_i]^{-1} \cdot [\boldsymbol{e}_{i+1}] \cdot \mathbf{A}_{i+1}; \quad i = 1, 2, \ldots, N, \tag{5.68}$$

where

$$\left[D_i^*\right]^{-1} = \begin{bmatrix} \mathrm{e}^{-\mathrm{i}kD_i^1\Delta z_i} & 0 & 0 & 0 \\ 0 & \mathrm{e}^{-\mathrm{i}kD_i^2\Delta z_i} & 0 & 0 \\ 0 & 0 & \mathrm{e}^{-\mathrm{i}kD_i^3\Delta z_i} & 0 \\ 0 & 0 & 0 & \mathrm{e}^{-\mathrm{i}kD_i^4\Delta z_i} \end{bmatrix}. \tag{5.69}$$

If the procedure in Eq. (5.68) is continued, it follows that

$$\mathbf{A}_1 = \left[D_1^*\right]^{-1} \cdot [\boldsymbol{e}_1]^{-1} \cdot [\boldsymbol{e}_2] \cdot \left[D_2^*\right]^{-1} \cdot [\boldsymbol{e}_2]^{-1} \cdot [\boldsymbol{e}_3] \cdots \cdot \left[D_N^*\right]^{-1} \cdot [\boldsymbol{e}_N]^{-1} \cdot [\boldsymbol{e}_{N+1}] \cdot \mathbf{A}_{N+1}. \tag{5.70}$$

If the matrix $[P]$ is denoted by

$$[P] = \left[D_1^*\right]^{-1} \cdot [\boldsymbol{e}_1]^{-1} \cdot [\boldsymbol{e}_2] \cdot \left[D_2^*\right]^{-1} \cdot [\boldsymbol{e}_2]^{-1} \cdot [\boldsymbol{e}_3] \cdots \left[D_N^*\right]^{-1} \cdot [\boldsymbol{e}_N]^{-1} \cdot [\boldsymbol{e}_{N+1}]. \tag{5.71}$$

Equation (5.70) can be rewritten in the form

$$\mathbf{A}_1 = [P] \cdot \mathbf{A}_{N+1}. \tag{5.72}$$

In the Nth layer, there does not exist an uneven surface. Thus, the wave will not encounter the reflection, and there are only the upward waves in the Nth layer of the ionosphere. We write

$$\mathbf{A}_{N+1} = \left[A_{N+1}^1 \quad A_{N+1}^2 \quad 0 \quad 0\right]^T. \tag{5.73}$$

When the incident TM wave is below the $z = z_0$ plane, the components of the incident wave are expressed in the matrix form. We write

$$\begin{bmatrix} E_{xi} \\ E_{zi} \\ \eta H_{yi} \end{bmatrix} = \begin{bmatrix} C \\ -S \\ 1 \end{bmatrix} \mathrm{e}^{\mathrm{i}kC(z-z_0)} \mathrm{e}^{\mathrm{i}kSx}. \tag{5.74}$$

The horizontally and vertically polarized reflection waves are generated. They are

$$\begin{bmatrix} E_{xr} \\ E_{zr} \\ \eta H_{yr} \end{bmatrix} = \begin{bmatrix} -C \\ -S \\ 1 \end{bmatrix} {}_{\parallel}R_{\parallel} \mathrm{e}^{-\mathrm{i}kC(z-z_0)} \mathrm{e}^{\mathrm{i}kSx}, \tag{5.75}$$

and

$$\begin{bmatrix} E_{yr} \\ \eta H_{xr} \\ \eta H_{zr} \end{bmatrix} = \begin{bmatrix} 1 \\ C \\ S \end{bmatrix} {}_{\parallel}R_{\perp} \mathrm{e}^{-\mathrm{i}kC(z-z_0)} \mathrm{e}^{\mathrm{i}kSx}. \tag{5.76}$$

In the air below the plane $z = z_0$, assuming that the horizontal magnetic field component of the incident wave is normalized to 1, the four components can be expressed in the following form:

$$\mathbf{e}_0|_{z=z_0} = \left.\begin{bmatrix} E_x \\ E_y \\ \eta H_x \\ \eta H_y \end{bmatrix}\right|_{z=z_0} = \begin{bmatrix} C(1 - {}_{\parallel}R_{\parallel}) \\ {}_{\parallel}R_{\perp} \\ C\,{}_{\parallel}R_{\perp} \\ 1 + {}_{\parallel}R_{\parallel} \end{bmatrix}. \tag{5.77}$$

With the boundary conditions at the interface $z = z_0$, it is seen that the field components should be continuous. We have

$$\tilde{\boldsymbol{e}}_0|_{z=z_0} = \tilde{\boldsymbol{e}}_1|_{z=z_0} = [\boldsymbol{e}_1]\cdot\mathbf{A}_1 = [\boldsymbol{e}_1]\cdot[P]\cdot\mathbf{A}_{N+1}. \tag{5.78}$$

In Eq. (5.71), each element of the matrix $[P]$ depends on the properties of the ionosphere, the geomagnetic direction, and the incident angle, and it is independent of the amplitude and polarization of the incident wave. When the matrix $[P]$ is solved, for Eq. (5.78), there are only four unknown parameters: ${}_{\|}R_{\|}$, ${}_{\|}R_{\perp}$, A^1_{N+1}, and A^2_{N+1}. It is noted that ${}_{\|}R_{\|}$ and ${}_{\|}R_{\perp}$ represent the reflection coefficients, while A^1_{N+1} and A^2_{N+1} represent the transmission coefficients.

Let

$$[Q] = [\boldsymbol{e}_1]\cdot[P] = \begin{bmatrix} Q_{11} & Q_{12} & Q_{13} & Q_{14} \\ Q_{21} & Q_{22} & Q_{23} & Q_{24} \\ Q_{31} & Q_{32} & Q_{33} & Q_{34} \\ Q_{41} & Q_{42} & Q_{43} & Q_{44} \end{bmatrix}, \tag{5.79}$$

then the following equations are obtained readily:

$$\left.\begin{aligned} C(1 - {}_{\|}R_{\|}) &= Q_{11}A^1_{N+1} + Q_{12}A^2_{N+1} \\ {}_{\|}R_{\perp} &= Q_{21}A^1_{N+1} + Q_{22}A^2_{N+1} \\ C\,{}_{\|}R_{\perp} &= Q_{31}A^1_{N+1} + Q_{32}A^2_{N+1} \\ 1 + {}_{\|}R_{\|} &= Q_{41}A^1_{N+1} + Q_{42}A^2_{N+1} \end{aligned}\right\}. \tag{5.80}$$

Thus, the solutions are expressed as follows:

$$\left.\begin{aligned} {}_{\|}R_{\|} &= \frac{(CQ_{21} - Q_{31})(CQ_{42} - Q_{12}) - (CQ_{22} - Q_{32})(CQ_{41} - Q_{11})}{(CQ_{21} - Q_{31})(CQ_{42} + Q_{12}) - (CQ_{22} - Q_{32})(CQ_{41} + Q_{11})} \\ {}_{\|}R_{\perp} &= \frac{2C(Q_{21}Q_{32} - Q_{22}Q_{31})}{(CQ_{21} - Q_{31})(CQ_{42} + Q_{12}) - (CQ_{22} - Q_{32})(CQ_{41} + Q_{11})} \\ A^1_{N+1} &= \frac{-2C(CQ_{22} - Q_{32})}{(CQ_{21} - Q_{31})(CQ_{42} + Q_{12}) - (CQ_{22} - Q_{32})(CQ_{41} + Q_{11})} \\ A^2_{N+1} &= \frac{2C(CQ_{21} - Q_{31})}{(CQ_{21} - Q_{31})(CQ_{42} + Q_{12}) - (CQ_{22} - Q_{32})(CQ_{41} + Q_{11})} \end{aligned}\right\}. \tag{5.81}$$

5.3.2 Computations and Discussions

In general, the changes of the ionosphere's electrical parameters with the ionosphere height are dependent on the geographic latitude, season, and local time. According to the data in Tables 1.2 and 1.3 in the book *Blaunstein and Plohotniuc* (2008), the profiles for the ionospheric electron density and the electron collision frequency for daytime and nighttime can be obtained readily.

By using the method addressed in Sect. 5.3.1, the following computations are carried out. We assume that the ionospheric parameters are taken as those for nighttime model as shown in Table 2.1 in Chap. 2, the geomagnetic field strength is $B_0 = 0.5 \times 10^{-4}$ T, and the angles Ω and ϕ are taken as $\Omega = \phi = 45°$, the reflection coefficients ${}_{\parallel}R_{\parallel}$ and ${}_{\parallel}R_{\perp}$ and the transmission coefficients A_N^1 and A_N^2 versus the incident angle θ are computed at $f = 30$ Hz and $f = 300$ Hz, and shown in Figs. 5.23 and 5.24, respectively. From the above calculated results, we conclude as follows:

- The magnitudes of the reflection coefficient ${}_{\parallel}R_{\parallel}$ are close to 1, and it indicates that the ionosphere has good reflection characteristics in the SLF/ELF ranges.
- The magnitudes of the reflection coefficient ${}_{\parallel}R_{\perp}$ are not close to 0, and they are smaller than those of ${}_{\parallel}R_{\parallel}$. This indicates that, for SLF/ELF incident waves, the polarized coupling and conversion, which are resulted by the anisotropic properties of the ionosphere, are generated and should not be neglected.
- Despite that the magnitudes of the transmission coefficients A_N^1 and A_N^2 are relatively smaller, they should not be neglected. It is shown that for SLF/ELF waves there is still a small part that can penetrate the ionosphere to reach outer space. It is well known that for SLF/ELF waves the ordinary wave in the ionosphere attenuates faster than the extraordinary wave. This corresponds to that the transmission coefficient of the ordinary wave is smaller than that of the extraordinary wave.
- When the operating frequency increases from 30 Hz to 300 Hz, the characteristic value for the extraordinary wave decreases. That is to say, with the operating frequency increasing, the ionospheric refractive index in SLF/ELF regions reduces. Thus, the difference between the ionospheric and atmospheric refractive index becomes smaller. Correspondingly, at the air–ionosphere boundary, the reflection coefficient reduces, while the operating frequency increases.
- Because the electron density and collision frequency in daytime are larger than those in nighttime, the transmission coefficient in daytime is smaller than that in nighttime. That is to say, it is more difficult to penetrate into the ionosphere in daytime.
- The changes of the reflection and transmission coefficients as the incident angle are more complicated than those for the homogeneous ionosphere with uniform sharp boundary. It is resulted by the internal reflection, the conversion, and the coupling in the stratified ionosphere.

In this chapter, the propagation of SLF/ELF electromagnetic waves in the planar stratified anisotropic ionosphere is treated analytically and some new numerical results and discussions are presented specifically.

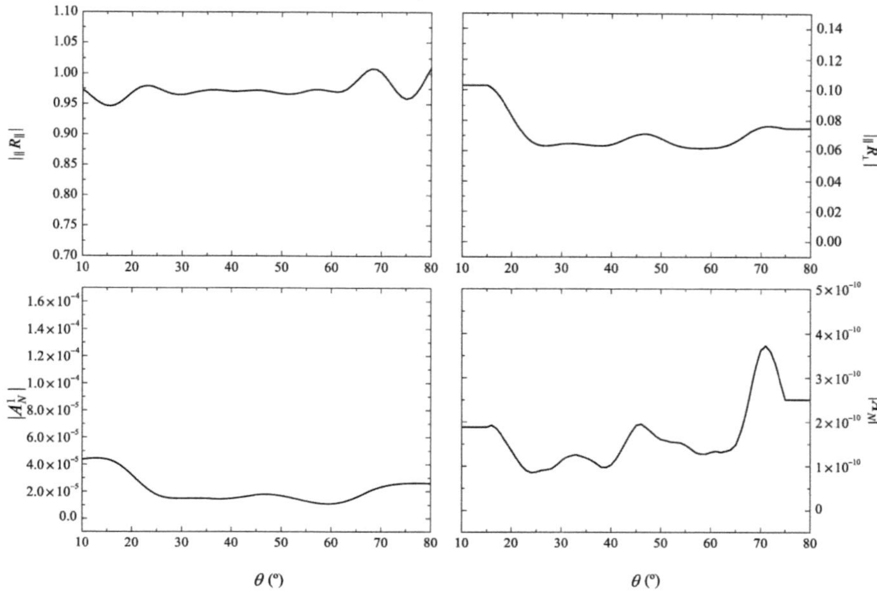

Fig. 5.23 The magnitudes of the reflection and transmission coefficients versus the incident angle θ at $f = 30$ Hz

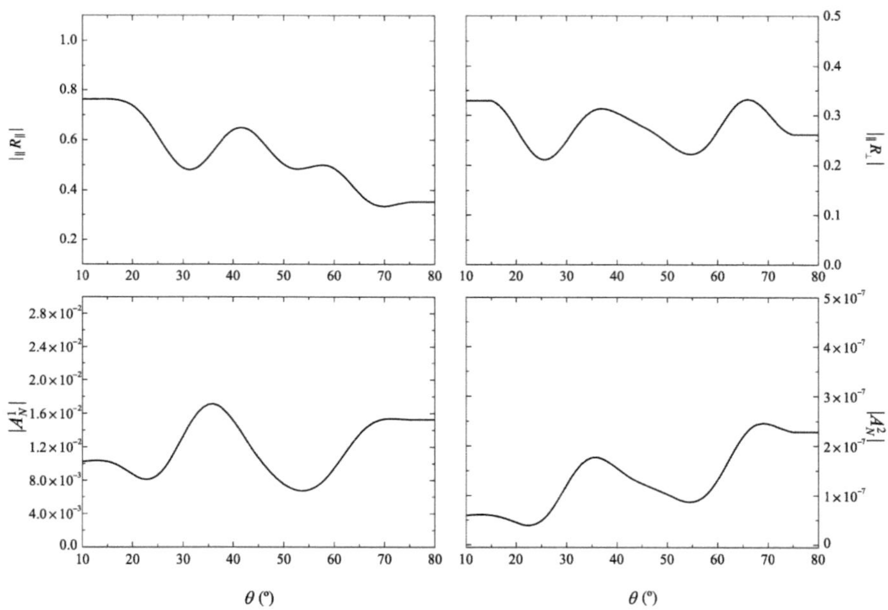

Fig. 5.24 The magnitudes of the reflection and transmission coefficients versus the incident angle θ at $f = 300$ Hz

References

Armand NA et al (1988) Experimental researches in the ionosphere of the Earth of the radiation of loop antenna in a range VLF waves, installed onboard the orbital complex "Progress-28"–"Souz TM-2". Radiotech Electron 33:2225–2233

Bannister PB et al (1993) Orbiting transmitter and antenna for space borne communications at ELF/VLF to submerged submarine. In: AGARD conference proceeding, vol 529, pp 33.1–33.14

Blaunstein N, Plohotniuc E (2008) Ionosphere and applied aspects of radio communication and radar. CRC Press, Boca Raton

Budden KG (1961) Radio waves in the ionosphere. Cambridge University Press, Cambridge

Chen Y, Wang YX, Pan WY (2009) The incidence, reflection, and transmission of SLF/ELF waves by the ionosphere. Technical Report, China Research Institute of Radiowave Propagation. Qingdao, China. (In Chinese)

Galejs J (1972) Terrestrial propagation of long electromagnetic waves. Pergamon, New York

Li K, Pan WY (1999) Propagation of VLF electromagnetic waves penetrating the low ionosphere. Indian J Radio Space Phys 33:87–94

Li K, Sun XY, Zhai HT (2011) Propagation of ELF electromagnetic waves in the lower ionosphere. IEEE Trans Antennas Propag 59(2):661–666

Lu H, Zhu XQ, Zhang HQ (2008) Influence of ions on SLF/ELF electromagnetic wave propagation in ionosphere. Chin J Radio Sci 23(6):1128–1132. (In Chinese)

Ozaki M, Nagano I, Yagitani S, Miyamura K (2004) Ionospheric propagation of ELF/VLF waves radiated from earthquake. In: Proceedings of Asia-Pacific radio science conference, Qingdao, China, Aug 24-27, pp 539–542

Serebryakova ON, Bdiehenko SV, Chmyrev VM (1992) Electromagnetic ELF radiation from earthquake regions as observed by low-altitude satellites. Geophys Res Lett 19(2):91–94

Wait JR (1970) Electromagnetic waves in stratified media, 2nd edn. Pergamon, New York

Xia MY, Chen ZY (2000) Matrix-exponent formulations for wave radiation and propagation in anisotropic stratified ionosphere. IEEE Trans Antennas Propag 48(2):339–341

Chapter 6
ELF Wave Propagation Along Sea-Rock Boundary and mCSEM Method

In this chapter, ELF wave propagation along the surface of the ocean floor or layered ocean floor is treated analytically and some computations are carried out carefully. Finally, the mCSEM method is addressed specifically.

6.1 Introduction

Over a century ago, the first analysis of the field excited by an HED near or on the boundary between two different half-spaces was treated in the dissertation of von Hörschelmann (1911), a student of Sommerfeld. As with the VED addressed by Sommerfeld (1909), the field is expressed in terms of the derivatives of general integrals, and those kinds of the integrals including Bessel function are defined as Sommerfeld's integrals. It is well known that for the Sommerfeld integrals it is difficult to obtain the analytical solutions. Lately, the subsequent developments on this problem were carried out by many investigators (Van der Pol and Niessen 1930; 1935; Norton 1937; Baños 1966; Wait 1953, 1956, 1961, 1970). Since the 1980s, the properties of lateral waves excited by an HED on the planar boundary between two different media have been treated extensively (King et al. 1980, 1984, 1986, 1989; King and Brown 1984; King 1989; Pan 1985; Wu and King 1987). The details are addressed in an excellent monograph by King et al. (1992). With the extension of the two-layered case, the electromagnetic field of an HED in three- or multi-layered region has been investigated widely (Wait 1970; Dunn 1984, 1986; Hoh et al. 1999; King 1991, 1993; Li et al. 1998, 2004; Collin 2004a, 2004b; Li and Lu 2005; Zhang et al. 2005), and the new developments on this problem were summarized in a recent book (Li 2009).

In the early 1980s, a new plan was proposed for detecting the subbottom structure by measuring the seabed conductivity with ELF wave propagation along the boundary between the sea water and the ocean floor (Chave and Cox 1982). With the great progress over 30 years, the mCSEM method has been developed as a useful method in geophysical prospecting and diagnostics with low cost and high efficiency.

W. Pan, K. Li, *Propagation of SLF/ELF Electromagnetic Waves*,
Advanced Topics in Science and Technology in China,
DOI 10.1007/978-3-642-39050-0_6,

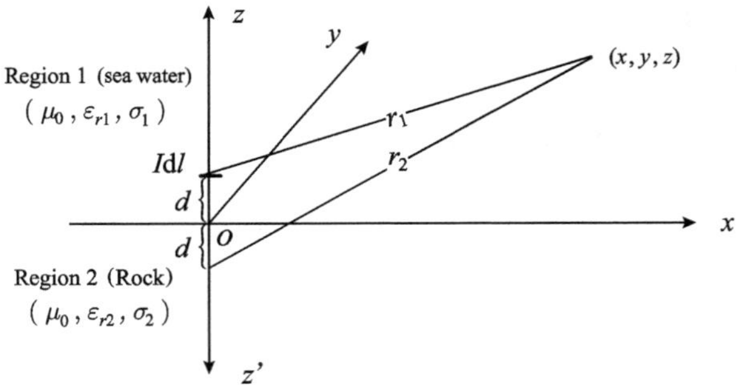

Fig. 6.1 The physical model for an HED on the surface of the ocean floor

In this chapter, we will treat analytically ELF wave propagation along the boundary between sea water and ocean floor or layered ocean floor and address the mCSEM method specifically.

6.2 ELF Field of HED on the Boundary Between Sea Water and Ocean Floor

In this section, we will treat analytically the ELF electromagnetic field radiated by an HED on or near the boundary between the sea water and ocean floor.

6.2.1 The Integrated Formulas of the Field in Sea Water and Ocean Floor

The geometry under consideration is shown in Fig. 6.1, where the HED in the $\hat{x}$ direction is located at $(0, 0, d)$. Region 1 ($z \geq 0$) is the space occupied by the sea water characterized by the permeability μ_0, relative permittivity ε_{r1}, and conductivity σ_1. Region 2 ($z \leq 0$) is the ocean floor characterized by the permeability μ_0, relative permittivity ε_{r2}, and conductivity σ_2.

In the monograph by King et al. (1992), the integrated formulas and approximated formulas are obtained for the electromagnetic field radiated by an HED on or near the boundary between the sea wave and the ocean floor. The integrated formulas in Region 1 (sea water) are written in the following forms:

$$E_{1\rho} = -\frac{\omega\mu_0 I\,\mathrm{d}l}{4\pi k_1^2}\cos\phi$$

$$\times\left(\int_0^\infty \left\{k_1^2 J_0(\lambda\rho) - \frac{\lambda^2}{2}\left[J_0(\lambda\rho) - J_2(\lambda\rho)\right]\right\}\gamma_1^{-1}\mathrm{e}^{\mathrm{i}\gamma_1|z-d|}\lambda\,\mathrm{d}\lambda\right.$$

$$+\int_0^\infty \left\{\frac{\gamma_1 Q}{2}\left[J_0(\lambda\rho)-J_2(\lambda\rho)\right]-\frac{k_1^2 P}{2\gamma_1}\left[J_0(\lambda\rho)+J_2(\lambda\rho)\right]\right\}$$

$$\left.\times\, e^{i\gamma_1(z+d)}\lambda\, d\lambda\right), \tag{6.1}$$

$$E_{1\phi}=\frac{\omega\mu_0 I\, dl}{4\pi k_1^2}\sin\phi$$

$$\times\left(\int_0^\infty \left\{k_1^2 J_0(\lambda\rho)-\frac{\lambda^2}{2}\left[J_0(\lambda\rho)+J_2(\lambda\rho)\right]\right\}\gamma_1^{-1}e^{i\gamma_1|z-d|}\lambda\, d\lambda\right.$$

$$+\int_0^\infty \left\{\frac{\gamma_1 Q}{2}\left[J_0(\lambda\rho)+J_2(\lambda\rho)\right]-\frac{k_1^2 P}{2\gamma_1}\left[J_0(\lambda\rho)-J_2(\lambda\rho)\right]\right\}$$

$$\left.\times\, e^{i\gamma_1(z+d)}\lambda\, d\lambda\right), \tag{6.2}$$

$$E_{1z}=\frac{i\omega\mu_0 I\, dl}{4\pi k_1^2}\cos\phi\left[\int_0^\infty \left(\pm e^{i\gamma_1|z-d|}+Qe^{i\gamma_1(z+d)}\right)J_1(\lambda\rho)\lambda^2\, d\lambda\right], \tag{6.3}$$

$$B_{1\rho}=-\frac{\mu_0 I\, dl}{4\pi}\sin\phi\left(\pm\int_0^\infty J_0(\lambda\rho)e^{i\gamma_1|z-d|}\lambda\, d\lambda\right.$$

$$+\int_0^\infty \left\{\frac{Q}{2}\left[J_0(\lambda\rho)+J_2(\lambda\rho)\right]-\frac{P}{2}\left[J_0(\lambda\rho)-J_2(\lambda\rho)\right]\right\}$$

$$\left.\times\, e^{i\gamma_1(z+d)}\lambda\, d\lambda\right), \tag{6.4}$$

$$B_{1\phi}=-\frac{\mu_0 I\, dl}{4\pi}\cos\phi\left(\pm\int_0^\infty J_0(\lambda\rho)e^{i\gamma_1|z-d|}\lambda\, d\lambda\right.$$

$$+\int_0^\infty \left\{\frac{Q}{2}\left[J_0(\lambda\rho)-J_2(\lambda\rho)\right]-\frac{P}{2}\left[J_0(\lambda\rho)+J_2(\lambda\rho)\right]\right\}$$

$$\left.\times\, e^{i\gamma_1(z+d)}\lambda\, d\lambda\right), \tag{6.5}$$

$$B_{1z}=\frac{i\mu_0 I\, dl}{4\pi}\sin\phi\left[\int_0^\infty \left(e^{i\gamma_1|z-d|}-Pe^{i\gamma_1(z+d)}\right)\gamma_1^{-1}J_1(\lambda\rho)\lambda^2\, d\lambda\right], \tag{6.6}$$

where

$$\gamma_i=\sqrt{k_i^2-\lambda^2};\quad i=1,2, \tag{6.7}$$

$$P=\frac{\gamma_2-\gamma_1}{\gamma_2+\gamma_1};\qquad Q=\frac{k_1^2\gamma_2-k_2^2\gamma_1}{k_1^2\gamma_2+k_2^2\gamma_1}, \tag{6.8}$$

$$M=\gamma_1+\gamma_2;\qquad N=k_1^2\gamma_2+k_2^2\gamma_1. \tag{6.9}$$

In Eqs. (6.3)–(6.5), the signs "±" are included, where the upper sign "+" and the lower sign "−" correspond to $z > d$ and $0 \le z \le d$ in Region 1, respectively.

The integrated formulas in Region 2 (ocean floor) are expressed as follows:

$$E_{2\rho} = -\frac{\omega\mu_0 I\,\mathrm{d}l}{4\pi}\cos\phi \times \left(\int_0^{\infty}\left\{M^{-1}\left[J_0(\lambda\rho)+J_2(\lambda\rho)\right]+\frac{\gamma_1\gamma_2}{N}\left[J_0(\lambda\rho)-J_2(\lambda\rho)\right]\right\} \times \mathrm{e}^{\mathrm{i}(\gamma_1 d-\gamma_2 z)}\lambda\,\mathrm{d}\lambda\right), \tag{6.10}$$

$$E_{2\phi} = \frac{\omega\mu_0 I\,\mathrm{d}l}{4\pi}\sin\phi \times \left(\int_0^{\infty}\left\{M^{-1}\left[J_0(\lambda\rho)-J_2(\lambda\rho)\right]+\frac{\gamma_1\gamma_2}{N}\left[J_0(\lambda\rho)+J_2(\lambda\rho)\right]\right\} \times \mathrm{e}^{\mathrm{i}(\gamma_1 d-\gamma_2 z)}\lambda\,\mathrm{d}\lambda\right), \tag{6.11}$$

$$E_{2z} = -\frac{\mathrm{i}\omega\mu_0 I\,\mathrm{d}l}{2\pi}\cos\phi\int_0^{\infty}\frac{\gamma_1}{N}J_1(\lambda\rho)\mathrm{e}^{\mathrm{i}(\gamma_1 d-\gamma_2 z)}\lambda^2\,\mathrm{d}\lambda, \tag{6.12}$$

$$B_{2\rho} = \frac{\mu_0 I\,\mathrm{d}l}{4\pi}\sin\phi \times \left(\int_0^{\infty}\left\{\frac{\gamma_2}{M}\left[J_0(\lambda\rho)-J_2(\lambda\rho)\right]+\frac{k_2^2\gamma_1}{N}\left[J_0(\lambda\rho)+J_2(\lambda\rho)\right]\right\} \times \mathrm{e}^{\mathrm{i}(\gamma_1 d-\gamma_2 z)}\lambda\,\mathrm{d}\lambda\right), \tag{6.13}$$

$$B_{2\phi} = \frac{\mu_0 I\,\mathrm{d}l}{4\pi}\cos\phi \times \left(\int_0^{\infty}\left\{\frac{\gamma_2}{M}\left[J_0(\lambda\rho)+J_2(\lambda\rho)\right]+\frac{k_2^2\gamma_1}{N}\left[J_0(\lambda\rho)-J_2(\lambda\rho)\right]\right\} \times \mathrm{e}^{\mathrm{i}(\gamma_1 d-\gamma_2 z)}\lambda\,\mathrm{d}\lambda\right), \tag{6.14}$$

$$B_{2z} = \frac{\mathrm{i}\mu_0 I\,\mathrm{d}l}{2\pi}\sin\phi\int_0^{\infty}M^{-1}J_1(\lambda\rho)\mathrm{e}^{\mathrm{i}(\gamma_1 d-\gamma_2 z)}\lambda^2\,\mathrm{d}\lambda. \tag{6.15}$$

In the monograph by King et al. (1992), the complete approximated formulas are obtained. We write

$$E_{1\rho}(\rho,\phi,z) = -\frac{\omega\mu_0 I\,\mathrm{d}l}{2\pi k_1^2}\cos\phi$$

$$\times \left\{ k_2 g(\rho; k_1, k_2) e^{ik_2\rho} e^{ik_1(z+d)} - \left(\frac{k_1}{r_1^2} + \frac{i}{r_1^3}\right) e^{ik_1 r_1} \right.$$

$$+ \left(\frac{z+d}{\rho}\right)\left(\frac{ik_1}{\rho^2} - \frac{3}{2\rho^3}\right) e^{ik_1 r_2}$$

$$- \frac{1}{2}\left[\left(\frac{z-d}{r_1}\right)^2 \left(\frac{ik_1^2}{r_1} - \frac{3k_1}{r_1^2} - \frac{3i}{r_1^3}\right) e^{ik_1 r_1}\right.$$

$$\left.\left. + \left(\frac{z+d}{r_2}\right)^2 \left(\frac{ik_1^2}{r_2} - \frac{3k_1}{r_2^2} - \frac{3i}{r_2^3}\right) e^{ik_1 r_2}\right]\right\}, \tag{6.16}$$

$$E_{1\phi}(\rho, \phi, z) = \frac{\omega\mu_0 I \, dl}{2\pi k_1^2} \sin\phi$$

$$\times \left\{ k_2 h(\rho; k_1, k_2) e^{ik_2\rho} e^{ik_1(z+d)} \right.$$

$$+ \frac{1}{2}\left[\left(\frac{ik_1^2}{r_2} - \frac{3k_1}{r_2^2} - \frac{3i}{r_2^3}\right) e^{ik_1 r_2} - \left(\frac{ik_1^2}{r_1} - \frac{k_1}{r_1^2} - \frac{i}{r_1^3}\right) e^{ik_1 r_1}\right]$$

$$+ i\left(\frac{z+d}{r_2}\right)\left(\frac{ik_1^2}{\rho} - \frac{3k_1}{2\rho^2} - \frac{5i}{8\rho^3}\right) e^{ik_1 r_2}$$

$$\left. - \left(\frac{z+d}{r_2}\right)^2 \left(\frac{ik_1^2}{r_2} - \frac{3k_1}{r_2^2} - \frac{3i}{r_2^3}\right) e^{ik_1 r_2}\right\}, \tag{6.17}$$

$$E_{1z}(\rho, \phi, z) = \frac{\omega\mu_0 I \, dl}{2\pi k_1^2} \cos\phi$$

$$\times \left\{ \frac{k_2^2}{k_1}\left[f(\rho; k_1, k_2) e^{ik_2\rho} e^{ik_1(z+d)} - \left(\frac{i}{\rho^2} - \frac{3}{2k_1\rho^3}\right) e^{ik_1 r_2}\right]\right.$$

$$- \frac{1}{2}\left[\left(\frac{\rho}{r_1}\right)\left(\frac{z-d}{r_1}\right)\left(\frac{ik_1^2}{r_1} - \frac{3k_1}{r_1^2} - \frac{3i}{r_1^3}\right) e^{ik_1 r_1}\right.$$

$$\left.\left. + \left(\frac{\rho}{r_2}\right)\left(\frac{z+d}{r_2}\right)\left(\frac{ik_1^2}{r_2} - \frac{3k_1}{r_2^2} - \frac{3i}{r_2^3}\right) e^{ik_1 r_2}\right]\right\}, \tag{6.18}$$

$$B_{1\rho}(\rho, \phi, z) = -\frac{\mu_0 I \, dl}{2\pi k_1} \sin\phi$$

$$\times \left\{ k_2 h(\rho; k_1, k_2) e^{ik_2\rho} e^{ik_1(z+d)} \right.$$

$$+ \left[\frac{ik_1}{\rho^2} - \frac{7}{2\rho^3} + \left(\frac{z+d}{r_2}\right)\left(\frac{ik_1^2}{r_2} - \frac{5k_1}{r_2^2} - \frac{12i}{r_2^3}\right)\right] e^{ik_1 r_2}$$

$$
\begin{aligned}
&-\frac{1}{2}\left[\left(\frac{z-d}{r_1}\right)\left(\frac{\mathrm{i}k_1^2}{r_1}-\frac{k_1}{r_1^2}\right)\mathrm{e}^{\mathrm{i}k_1 r_1}\right.\\
&\left.\left.+\left(\frac{z+d}{r_2}\right)\left(\frac{\mathrm{i}k_1^2}{r_2}-\frac{k_1}{r_2^2}\right)\mathrm{e}^{\mathrm{i}k_1 r_2}\right]\right\},
\end{aligned}
\tag{6.19}
$$

$$
\begin{aligned}
B_{1\phi}(\rho,\phi,z) = &-\frac{\mu_0 I\,\mathrm{d}l}{2\pi k_1}\cos\phi\\
&\times\left\{k_2 g(\rho;k_1,k_2)\mathrm{e}^{\mathrm{i}k_2\rho}\mathrm{e}^{\mathrm{i}k_1(z+d)}\right.\\
&+\frac{1}{2}\left[\left(\frac{2}{\rho^3}+\frac{3\mathrm{i}}{k_1\rho^4}\right)\mathrm{e}^{\mathrm{i}k_1 r_2}-\left(\frac{z-d}{r_1}\right)\left(\frac{\mathrm{i}k_1^2}{r_1}-\frac{k_1}{r_1^2}\right)\mathrm{e}^{\mathrm{i}k_1 r_1}\right.\\
&\left.\left.-\left(\frac{z+d}{r_2}\right)\left(\frac{\mathrm{i}k_1^2}{r_2}-\frac{3k_1}{r_2^2}-\frac{6\mathrm{i}}{r_2^3}\right)\mathrm{e}^{\mathrm{i}k_1 r_2}\right]\right\},
\end{aligned}
\tag{6.20}
$$

$$
\begin{aligned}
B_{1z}(\rho,\phi,z) = &\frac{\mu_0 I\,\mathrm{d}l}{2\pi}\sin\phi\\
&\times\left\{\frac{k_2^2}{k_1^2}\left(\frac{1}{\rho^2}+\frac{3\mathrm{i}}{k_2\rho^3}-\frac{3}{k_2^2\rho^4}\right)\mathrm{e}^{\mathrm{i}k_2\rho}\mathrm{e}^{\mathrm{i}k_1(z+d)}\right.\\
&-\left(\frac{\rho}{r_2}\right)\left[\frac{1}{r_2^2}+\frac{3\mathrm{i}}{k_1 r_2^3}-\frac{3}{k_1^2 r_2^4}\right.\\
&+\left(\frac{z+d}{\rho}\right)\left(\frac{k_1}{\rho}+\frac{3\mathrm{i}}{2\rho^2}-\frac{9}{8k_1\rho^3}\right)\\
&\left.\times\left(\frac{z+d}{r_2}\right)^2\left(\frac{\mathrm{i}k_1}{r_2}-\frac{6}{r_2^2}-\frac{15\mathrm{i}}{k_1 r_2^3}\right)\right]\mathrm{e}^{\mathrm{i}k_1 r_2}\\
&\left.-\frac{1}{2}\left[\left(\frac{\rho}{r_1}\right)\left(\frac{\mathrm{i}k_1}{r_1}-\frac{1}{r_1^2}\right)\mathrm{e}^{\mathrm{i}k_1 r_1}-\left(\frac{\rho}{r_2}\right)\left(\frac{\mathrm{i}k_1}{r_2}-\frac{1}{r_2^2}\right)\mathrm{e}^{\mathrm{i}k_1 r_2}\right]\right\},
\end{aligned}
\tag{6.21}
$$

where

$$
f(\rho;k_1,k_2)=\frac{\mathrm{i}k_2}{\rho}-\frac{1}{\rho^2}-\frac{k_2^3}{k_1}\sqrt{\frac{\pi}{k_2\rho}}\mathrm{e}^{-\mathrm{i}p}\mathbf{F}(p_e), \tag{6.22}
$$

$$
g(\rho;k_1,k_2)=\frac{\mathrm{i}k_2}{\rho}-\frac{1}{\rho^2}-\frac{\mathrm{i}}{k_2\rho^3}-\frac{k_2^3}{k_1}\sqrt{\frac{\pi}{k_2\rho}}\mathrm{e}^{-\mathrm{i}p}\mathbf{F}(p_e), \tag{6.23}
$$

$$
h(\rho;k_1,k_2)=\frac{2}{\rho^2}+\frac{2\mathrm{i}}{k_2\rho^3}+\frac{\mathrm{i}k_2^2}{k_1\rho}\sqrt{\frac{\pi}{k_2\rho}}\mathrm{e}^{-\mathrm{i}p}\mathbf{F}(p_e). \tag{6.24}
$$

In the above formulas, $r_1 = \sqrt{\rho^2 + (z-d)^2}$, $r_2 = \sqrt{\rho^2 + (z+d)^2}$, $p_e = \frac{k_2^3 \rho}{2k_1^2}$, and the Fresnel integral $\mathbf{F}(p_e)$ is defined by

$$\mathbf{F}(p_e) = \frac{1}{2}(1+\mathrm{i}) - C_2(p_e) - \mathrm{i}S_2(p_e), \tag{6.25}$$

where

$$C_2(p_e) + \mathrm{i}S_2(p_e) = \int_0^{p_e} \frac{\mathrm{e}^{\mathrm{i}t}}{(2\pi t)^{1/2}} \mathrm{d}t. \tag{6.26}$$

In Eqs. (6.16)–(6.21), the second term including the propagation factor $\mathrm{e}^{\mathrm{i}k_1 r_1}$ and the third term including $\mathrm{e}^{\mathrm{i}k_2 r_2}$ stand for the direct wave and the ideal reflected wave, respectively. The first term including $\mathrm{e}^{\mathrm{i}k_1(z+d)}\mathrm{e}^{\mathrm{i}k_2\rho}$ stands for the lateral wave. The lateral wave travels downward from the transmitting dipole a distance d in Region 1 (sea water) to the boundary, then radially along the boundary in Region 2 (ocean floor) a distance ρ, and finally vertically upward a distance z in Region 1 (sea water) to the receiving dipole.

At large distance from the transmitting dipole to the receiving point, we have $\rho \gg d$, $\rho \gg z$, $r_1 \sim \rho$, and $r_2 \sim \rho$. Because the attenuation in the ocean floor is much smaller than that in the sea water, namely, $\mathrm{Im}(k_2) \ll \mathrm{Im}(k_1)$, the two terms of the directed wave and ideal reflected wave are much smaller than the term of the lateral wave. In this case, there are only the terms of lateral wave in Eqs. (6.16)–(6.21). We write

$$E_{1\rho}(\rho, \phi, z) = -\frac{\omega\mu_0 I\,\mathrm{d}l}{2\pi k_1^2} \cos\phi k_2 g(\rho; k_1, k_2) \mathrm{e}^{\mathrm{i}k_2\rho}\mathrm{e}^{\mathrm{i}k_1(z+d)}, \tag{6.27}$$

$$E_{1\phi}(\rho, \phi, z) = \frac{\omega\mu_0 I\,\mathrm{d}l}{2\pi k_1^2} \sin\phi k_2 h(\rho; k_1, k_2) \mathrm{e}^{\mathrm{i}k_2\rho}\mathrm{e}^{\mathrm{i}k_1(z+d)}, \tag{6.28}$$

$$E_{1z}(\rho, \phi, z) = \frac{\omega\mu_0 I\,\mathrm{d}l}{2\pi k_1^2} \cos\phi \frac{k_2^2}{k_1} f(\rho; k_1, k_2) \mathrm{e}^{\mathrm{i}k_2\rho}\mathrm{e}^{\mathrm{i}k_1(z+d)}, \tag{6.29}$$

$$B_{1\rho}(\rho, \phi, z) = -\frac{\mu_0 I\,\mathrm{d}l}{2\pi k_1} \sin\phi k_2 h(\rho; k_1, k_2) \mathrm{e}^{\mathrm{i}k_2\rho}\mathrm{e}^{\mathrm{i}k_1(z+d)}, \tag{6.30}$$

$$B_{1\phi}(\rho, \phi, z) = -\frac{\mu_0 I\,\mathrm{d}l}{2\pi k_1} \cos\phi k_2 g(\rho; k_1, k_2) \mathrm{e}^{\mathrm{i}k_2\rho}\mathrm{e}^{\mathrm{i}k_1(z+d)}, \tag{6.31}$$

$$B_{1z}(\rho, \phi, z) = \frac{\mu_0 I\,\mathrm{d}l}{2\pi} \sin\phi \frac{k_2^2}{k_1^2} \left(\frac{1}{\rho^2} + \frac{3\mathrm{i}}{k_2\rho^3} - \frac{3}{k_2^2\rho^4} \right) \mathrm{e}^{\mathrm{i}k_2\rho}\mathrm{e}^{\mathrm{i}k_1(z+d)}. \tag{6.32}$$

Similarly, the complete approximated formulas for the electromagnetic field in Region 2 (ocean floor) are obtained readily (King et al. 1992). We write

$$E_{2\rho}(\rho, \phi, z) = -\frac{\omega\mu_0 I\,\mathrm{d}l}{2\pi k_1^2} \cos\phi$$

$$\times\left(e^{ik_1d}e^{ik_2r_0}\left\{k_2\left[\frac{ik_2}{r_0}-\frac{1}{r_0^2}-\frac{i}{k_2r_0^3}-\frac{k_2^3}{k_1}\sqrt{\frac{\pi}{k_2\rho}}e^{-ip}\mathbf{F}(P^*)\right]\right.\right.$$
$$\left.+k_1\left(\frac{z}{r_0}\right)\left(\frac{ik_2}{r_0}-\frac{1}{r_0^2}\right)\right\}$$
$$-e^{k_1z}e^{ik_1r_d}k_1\left[\frac{1}{r_d^2}+\frac{i}{k_1r_d^3}-\left(\frac{d}{\rho}\right)\left(\frac{i}{\rho^2}-\frac{3}{2k_1\rho^3}\right)\right.$$
$$\left.\left.+\left(\frac{d}{r_d}\right)^2\left(\frac{ik_1}{r_d}-\frac{3}{r_d^2}-\frac{3i}{k_1r_d^3}\right)\right]\right), \tag{6.33}$$

$$E_{2\phi}(\rho,\phi,z)=\frac{\omega\mu_0I\,dl}{2\pi k_1^2}\sin\phi$$
$$\times\left(e^{ik_1d}e^{ik_2r_0}\left\{k_2\left[\frac{2}{r_0^2}+\frac{2i}{k_2r_0^3}+\frac{ik_2^3}{k_1\rho}\sqrt{\frac{\pi}{k_2\rho}}e^{-ip}\mathbf{F}(P^*)\right.\right.\right.$$
$$\left.+\left(\frac{z}{r_0}\right)^2\left(\frac{ik_2}{r_0}-\frac{3}{r_0^2}-\frac{3i}{k_2r_0^3}\right)\right]$$
$$\left.+k_1\left(\frac{z}{r_0}\right)\left(\frac{ik_2}{r_0}-\frac{1}{r_0^2}\right)\right\}$$
$$-e^{k_1z}e^{ik_1r_d}k_1\left[\frac{1}{r_d^2}+\frac{i}{k_1r_d^3}-i\left(\frac{d}{r_d}\right)\left(\frac{ik_1}{\rho}-\frac{3}{2\rho^2}-\frac{5i}{8k_1\rho^3}\right)\right.$$
$$\left.\left.+\left(\frac{d}{r_d}\right)^2\left(\frac{ik_1}{r_d}-\frac{3}{r_d^2}-\frac{3i}{k_1r_d^3}\right)\right]\right), \tag{6.34}$$

$$E_{2z}(\rho,\phi,z)=\frac{\omega\mu_0I\,dl}{2\pi k_1}\cos\phi$$
$$\times\left\{e^{ik_1d}e^{ik_2r_0}\left[\left(\frac{\rho}{r_0}\right)\left(\frac{ik_2}{r_0}-\frac{1}{r_0^2}\right)-\frac{k_2^3}{k_1}\sqrt{\frac{\pi}{k_2\rho}}e^{-ip}\mathbf{F}(P^*)\right]\right.$$
$$\left.-e^{k_1z}e^{ik_1r_d}\left(\frac{i}{\rho^2}-\frac{3}{2k_1\rho^3}\right)\right\}, \tag{6.35}$$

$$B_{2\rho}(\rho,\phi,z)=\frac{-\mu_0I\,dl}{2\pi k_1}\sin\phi$$
$$\times\left\{e^{ik_1d}e^{ik_2r_0}k_2\left[\frac{ik_2z^2}{r_0^3}+\left(\frac{2\rho^2-z^2}{r_0^2}\right)\left(\frac{1}{r_0^2}+\frac{i}{k_2r_0^3}\right)\right.\right.$$
$$\left.+\frac{ik_2^2}{k_1\rho}\sqrt{\frac{\pi}{k_2\rho}}e^{-ip}\mathbf{F}(P^*)+\left(\frac{k_2z}{k_1r_0}\right)\left(\frac{4}{r_0^2}-\frac{1}{\rho^2}+\frac{12i}{k_2r_0^3}\right)\right]$$

$$
+ \mathrm{e}^{k_1 z} \mathrm{e}^{\mathrm{i}k_1 r_d} k_1 \left[\frac{\mathrm{i}}{\rho^2} - \frac{7}{2k_1\rho^3} + \left(\frac{d}{r_d} \right) \times \left(\frac{\mathrm{i}k_1}{r_d} - \frac{5}{r_d^2} - \frac{12\mathrm{i}}{k_1 r_d^3} \right) \right] \Bigg\}, \tag{6.36}
$$

$$
\begin{aligned}
B_{2\phi}(\rho,\phi,z) = {} & -\frac{\mu_0 I\,\mathrm{d}l}{2\pi k_1} \cos\phi \\
& \times \Bigg\{ \mathrm{e}^{\mathrm{i}k_1 d} \mathrm{e}^{\mathrm{i}k_2 r_0} k_2 \left[\frac{\mathrm{i}k_2}{r_0} - \frac{1}{r_0^2} - \frac{\mathrm{i}}{k_2 r_0^3} - \frac{k_2^2}{k_1} \sqrt{\frac{\pi}{k_2\rho}} \mathrm{e}^{-\mathrm{i}p} \mathbf{F}(P^*) \right] \\
& + \mathrm{e}^{k_1 z} \mathrm{e}^{\mathrm{i}k_1 r_d} k_1 \left[\frac{1}{k_1\rho^3} + \frac{3\mathrm{i}}{2k_1\rho^4} + \left(\frac{d}{r_d} \right) \left(\frac{1}{r_d^2} + \frac{3\mathrm{i}}{k_1 r_d^3} \right) \right] \Bigg\},
\end{aligned} \tag{6.37}
$$

$$
\begin{aligned}
B_{2z}(\rho,\phi,z) = {} & \frac{\mu_0 k_2 I\,\mathrm{d}l}{2\pi k_1} \sin\phi \\
& \times \Bigg(\mathrm{e}^{\mathrm{i}k_1 d} \mathrm{e}^{\mathrm{i}k_2 r_0} \Bigg\{ \left(\frac{\rho z}{r_0^2} \right) \left(\frac{\mathrm{i}k_2}{r_0} - \frac{3}{r_0^2} - \frac{3\mathrm{i}}{k_2 r_0^3} \right) \\
& + \left(\frac{k_2\rho}{k_1 r_0} \right) \Bigg[\frac{1}{r_0^2} + \frac{3\mathrm{i}}{k_2 r_0^3} - \frac{3}{k_2^2 r_0^4} \\
& + \left(\frac{z^2}{r_0^2} \right) \left(\frac{k_2}{r_0} + \frac{6\mathrm{i}}{r_0^2} - \frac{15}{k_2 r_0^3} \right) \Bigg] \Bigg\} \\
& - \mathrm{e}^{k_1 z} \mathrm{e}^{\mathrm{i}k_1 r_d} \left(\frac{k_1\rho}{k_2 r_0} \right) \Bigg[\frac{1}{r_d^2} + \frac{3\mathrm{i}}{k_1 r_d^3} - \frac{3}{k_1 r_d^4} \\
& + \left(\frac{d}{\rho} \right) \left(\frac{k_1}{\rho} + \frac{3\mathrm{i}}{2\rho^2} - \frac{9}{8k_1\rho^3} \right) \\
& + \left(\frac{d}{r_d} \right)^2 \left(\frac{\mathrm{i}k_1}{r_d} - \frac{6}{r_d^2} - \frac{15\mathrm{i}}{k_1 r_d^3} \right) \Bigg] \Bigg),
\end{aligned} \tag{6.38}
$$

where

$$
P^* = \frac{(R-Z)^2}{R}; \qquad R = \frac{k_2^3 \rho}{2k_1^2}, \tag{6.39}
$$

$$
Z = \frac{k_2^2 z}{2k_1} \le 0; \qquad z < 0, \tag{6.40}
$$

$$
r_0 = \sqrt{\rho^2 + z^2}; \qquad r_d = \sqrt{\rho^2 + d^2}. \tag{6.41}
$$

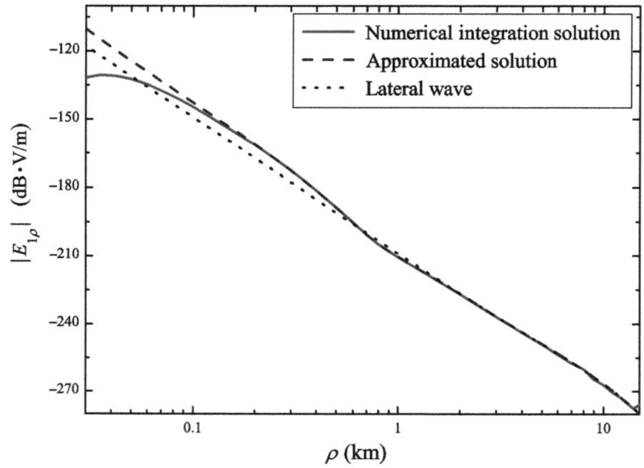

Fig. 6.2 The magnitudes of the component $E_{1\rho}(\rho, 0, 1)$ versus the propagation distance ρ for the two-layered model

6.2.2 Computations and Discussions

In order to verify the accuracy of the approximated formulas, for the actual propagation conditions, the computations by the approximated formulas and the directional numerical calculations on the integrated formula are carried out, respectively. It is noted that three waves, the directed wave, the ideal reflected wave, and the lateral wave, are included in the approximated formulas in Eqs. (6.16)–(6.21), while only the lateral wave is included in the approximated formulas in Eqs. (6.27)–(6.32).

We assume that the operating frequency is $f = 1$ Hz, the sea water is characterized by the permeability μ_0, relative permittivity $\varepsilon_{r1} = 80$, and conductivity $\sigma_1 = 4$ S/m, and the rock is characterized by the permeability μ_0, relative permittivity $\varepsilon_{r2} = 10$, and conductivity $\sigma_2 = 0.01$ S/m. With $d = 30$ m and $z = 1$ m, the magnitudes of the three electric field components, which are radiated by an HED in sea water, are computed and shown in Figs. 6.2, 6.3, 6.4, respectively.

From these computations, it is seen that the results by using the directional numerical calculation method are in agreement with those by using the approximated formulas and those by using only the lateral-wave term at the regions of $\rho > 600$ m, and the differences between the three computed results are within 0.7 dB. At the regions of $\rho < 100$ m, the differences are enlarged over 5 dB.

We assume that the transmitting source is located at the height of $d = 30$ m, and z is taken as -1 km, -2 km, and -3 km, respectively. The components $E_{2\rho}(\rho, 0, z')$, $E_{2\phi}(\rho, \frac{\pi}{2}, z')$, and $E_{2z}(\rho, 0, z')$ in the ocean floor are computed and shown in Figs. 6.5, 6.6, 6.7, respectively. It is noted that the computations are obtained by the directional numerical calculation method. And the differences of the computed results by using the approximated formulas and those by the directional numerical calculation method are similar to those of the field components in the sea water.

From Figs. 6.5–6.7, it is seen that at $\rho < 1$ km, the deeper the depth of the receiving point, the smaller the field strength. This is resulted by the fact that the deeper the depth of the receiving point, the larger the distance from the transmitting source

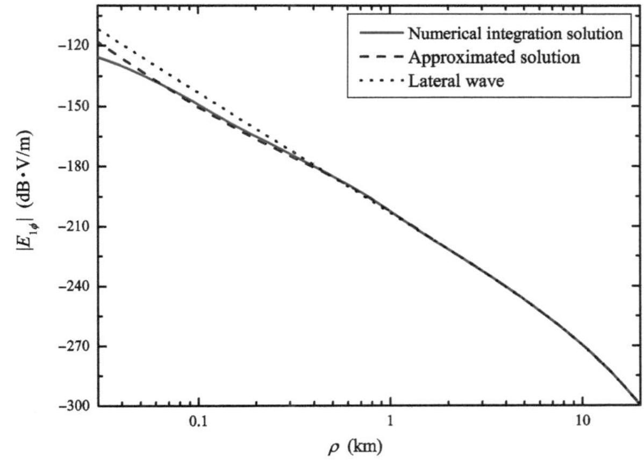

Fig. 6.3 The magnitudes of the component $E_{1\phi}(\rho, \frac{\pi}{2}, 1)$ versus the propagation distance ρ for the two-layered model

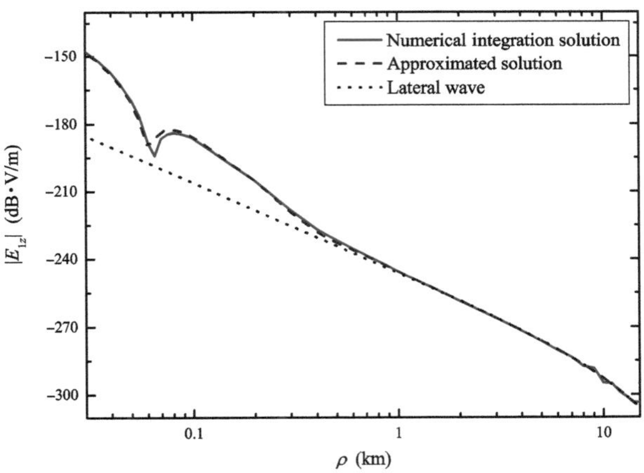

Fig. 6.4 The magnitudes of the component $E_{1z}(\rho, 0, 1)$ versus the propagation distance ρ for the two-layered model

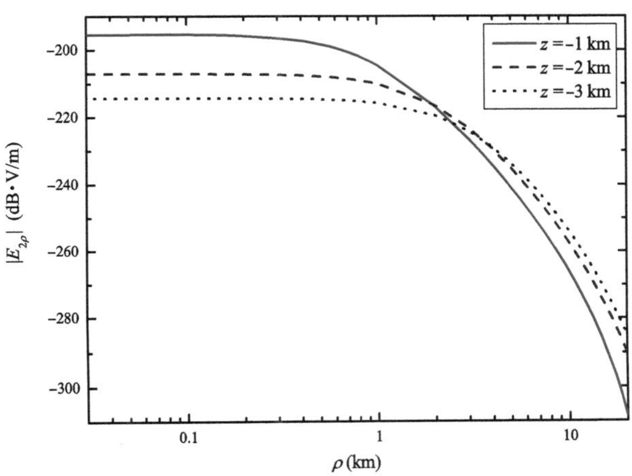

Fig. 6.5 The magnitudes of the component $E_{2\rho}(\rho, 0, z')$ versus the propagation distance ρ for the two-layered model

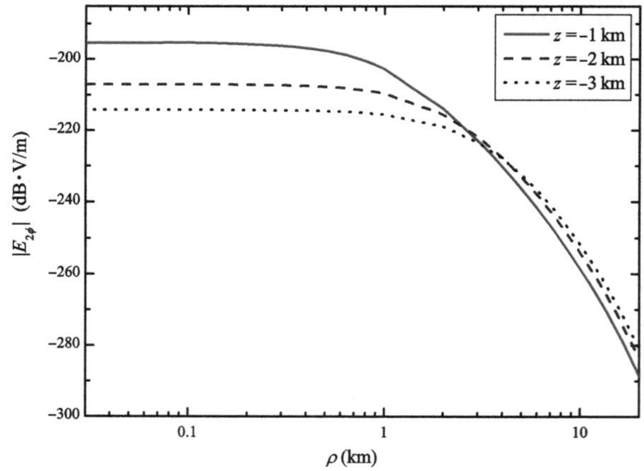

Fig. 6.6 The magnitudes of the component $E_{2\phi}(\rho, \frac{\pi}{2}, z')$ versus the propagation distance ρ for the two-layered model

and the receiving point, and correspondingly the greater the propagation loss. At considerable large distance between the transmitting source and the receiving point, when the depth of the receiving point increases, the total propagation distance does not increase significantly. Meanwhile, the directions of the energy flow in the ocean floor, which are not completely in the vertical and horizontal directions, show a complex trajectory. Therefore, at a large distance of $\rho > 5$–10 km, when the depth of the receiving point is deeper, the field strength may be stronger. We assume that the distance ρ is taken as 1 km, 5 km, and 10 km, respectively. The magnitudes of the field components versus the depth z' are shown in Figs. 6.8, 6.9, 6.10, respectively.

6.2.3 Poynting Vector and Its Trajectory in Ocean Floor

The ELF wave, which is radiated by an HED near the boundary between the sea water and ocean floor, propagates mainly in the lateral-wave form. In order to illustrate the penetrating depth of the lateral wave, it is necessary to analyze the direction and the trajectory of Poynting vector in the ocean floor. By classic electromagnetic theory, it is well known that the energy-flow direction in a medium is mainly described by the Poynting vector. The electromagnetic field radiated by an HED in the ocean floor can be divided into the electric-type (TM) components and magnetic-type (TE) components, where the orientation changes of the electric-type components E_2 and those of the magnetic-type components are described by using the functions $\cos\phi$ and $\sin\phi$, respectively. The Poynting vector of the electric-type components and that of the magnetic-type components are defined, respectively, as follows:

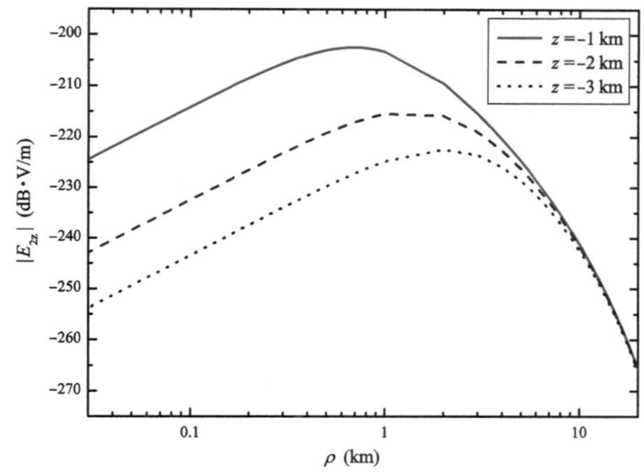

Fig. 6.7 The magnitudes of the component $E_{2z}(\rho, 0, z')$ versus the propagation distance ρ for the two-layered model

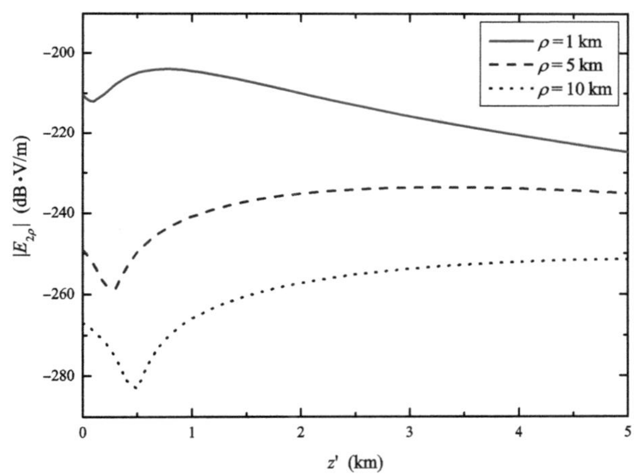

Fig. 6.8 The magnitudes of the component $E_{2\rho}(\rho, 0, z')$ versus the depth z' of the receiving point for the two-layered model

$$\boldsymbol{S}_{2E} = \frac{1}{2\mu_0}\left(\boldsymbol{E}_{2E} \times \boldsymbol{B}_{2E}^*\right) = \frac{1}{2\mu_0}\left(\hat{z}E_{2\rho}B_{2\phi}^* - \hat{\rho}E_{2z}B_{2\phi}^*\right), \tag{6.42}$$

$$\boldsymbol{S}_{2M} = \frac{1}{2\mu_0}\left(\boldsymbol{E}_{2M} \times \boldsymbol{B}_{2M}^*\right) = \frac{1}{2\mu_0}\left(-\hat{z}E_{2\phi}B_{2\rho}^* + \hat{\rho}E_{2\phi}B_{2z}^*\right). \tag{6.43}$$

Evidently, there exists the $\hat{\phi}$ component of the Poynting vector in the ocean floor, i.e.

$$\boldsymbol{S}_{2EM} = \frac{\hat{\phi}}{2\mu_0}\left(E_{2z}B_{2\rho}^* - E_{2\rho}B_{2z}^*\right). \tag{6.44}$$

The $\hat{\phi}$ component, which is in the azimuth direction, does not play a role for the outward diffusion and the downward penetration of the wave energy. Thus, it

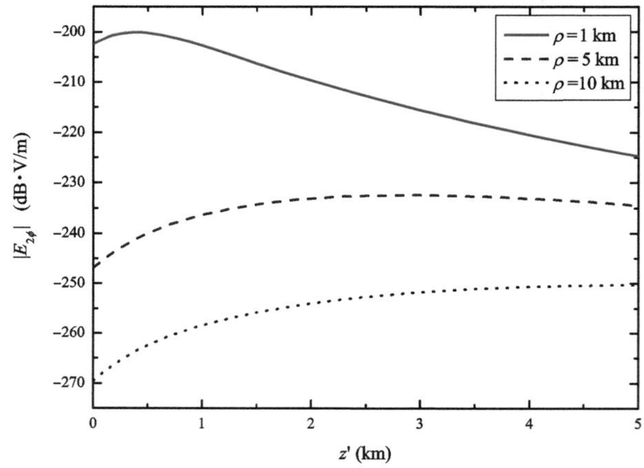

Fig. 6.9 The magnitudes of the component $E_{2\phi}(\rho, \frac{\pi}{2}, z')$ versus the depth z' of the receiving point for the two-layered model

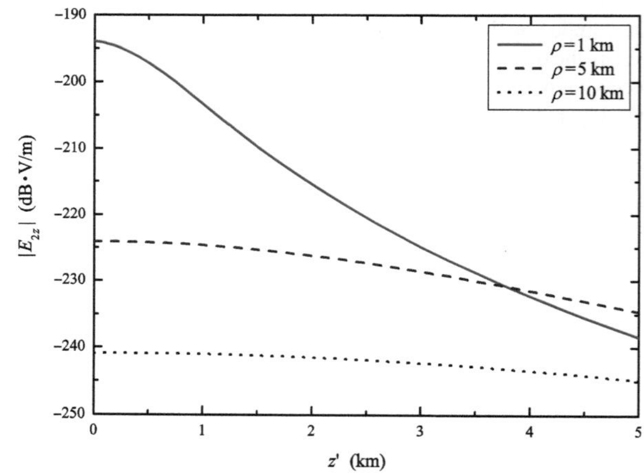

Fig. 6.10 The magnitudes of the component $E_{2z}(\rho, 0, z')$ versus the depth z' of the receiving point for the two-layered model

is not focused on the analyses. It is well known that the direction of the Poynting vector represents the direction of the energy flow of the electromagnetic wave. In the stratified media, the direction of the Poynting vector will be changed with the propagation path, so that its trajectory is certainly not a straight line. The slope of the trajectory of the wave vector should be expressed as follows:

$$\frac{\mathrm{d}z}{\mathrm{d}\rho} = -\frac{\mathrm{d}z'}{\mathrm{d}\rho} = \frac{\mathrm{Re}(S_{2z})}{\mathrm{Re}(S_{2\rho})}. \tag{6.45}$$

For the electric-type (TM) wave, it is written in the form of

$$\frac{\mathrm{d}z}{\mathrm{d}\rho} = -\frac{\mathrm{d}z'}{\mathrm{d}\rho} = -\frac{\mathrm{Re}(E_{2\rho}B_{2\phi}^*)}{\mathrm{Re}(E_{2z}B_{2\phi}^*)}. \tag{6.46}$$

For the magnetic-type (TE) wave, it is written in the form of

$$\frac{dz}{d\rho} = -\frac{dz'}{d\rho} = -\frac{\mathrm{Re}(E_{2\phi}B_{2\rho}^*)}{\mathrm{Re}(E_{2\phi}B_{2z}^*)}. \tag{6.47}$$

In the ocean floor, the components of the electromagnetic can be expressed by Eqs. (6.33)–(6.38). Considering the distances between the transmitting source and the receiving point are generally in the range of 20–30 km in practical applications of mCSEM method, it is seen that the parameter P^* in Eq. (6.39) is actually close to 0, and the term of Fresnel integral in Eqs. (6.33)–(6.38) can be neglected. In order to obtain the analytical formula for the energy-flow trajectory in the region near the transmitting source, only the most fundamental-mode term of the formulas of the electromagnetic field is retained, while the rest terms are neglected. Thus, we write

$$E_{2\rho} \approx -\frac{\omega\mu_0 I\,dl}{2\pi k_1} e^{ik_1 d}\cos\phi\left(-\frac{1}{r_0^2}\right)\left(\frac{z}{r_0}+\frac{k_2}{k_1}\right)e^{ik_2 r_0}, \tag{6.48}$$

$$E_{2\phi} \approx \frac{\omega\mu_0 k_2 I\,dl}{2\pi k_1} e^{ik_1 d}\sin\phi\left(\frac{z}{k_1}+iz\right)\left(\frac{i}{k_2 r_0^3}\right)e^{ik_2 r_0}, \tag{6.49}$$

$$E_{2z} \approx \frac{\omega\mu_0 I\,dl}{2\pi k_1} e^{ik_1 d}\cos\phi\left(\frac{\rho}{r_0}\right)\left(-\frac{1}{r_0^2}\right)e^{ik_2 r_0}, \tag{6.50}$$

$$B_{2\rho} \approx -\frac{\mu_0 k_2 I\,dl}{2\pi k_1}\sin\phi e^{ik_1 d}e^{ik_2 r_0} \times\left[\left(\frac{2\rho^2-z^2}{r_0^2}\right)\left(\frac{i}{k_2 r_0^3}\right)-\left(\frac{3k_2 z}{k_1 r_0}\right)\left(\frac{2i}{k_2 r_0^3}\right)\right], \tag{6.51}$$

$$B_{2\phi} \approx -\frac{\mu_0 k_2 I\,dl}{2\pi k_1} e^{ik_1 d}\cos\phi\left(-\frac{i}{k_2 r_0^3}\right)e^{ik_2 r_0}, \tag{6.52}$$

$$B_{2z} \approx -\frac{\mu_0 k_2 I\,dl}{2\pi k_1} e^{ik_1 d}\sin\phi\left(\frac{3\rho}{r_0^2}\right)\left(\frac{i}{k_2 r_0^3}\right)\left(z-\frac{i}{k_1}\right)e^{ik_2 r_0}. \tag{6.53}$$

For the electric-type (TM) wave, with the substitutions of Eqs. (6.48)–(6.53) into Eq. (6.46), the dip angle of Poynting vector is expressed in Eq. (6.54):

$$\frac{dz}{d\rho} = \frac{\mathrm{Re}[(\frac{z}{r_0}+\frac{k_2}{k_1})(\frac{1}{r_0^2})]}{\mathrm{Re}[\frac{\rho}{r_0}(-\frac{1}{r_0^2})]}, \tag{6.54}$$

where

$$r_0 = \sqrt{(\rho^2+z^2)}; \qquad \frac{k_2}{k_1} = \sqrt{\frac{\sigma_2}{\sigma_1}} = \delta, \tag{6.55}$$

$$k_i = \beta_i + i\alpha_i = e^{i\frac{\pi}{4}}\sqrt{\omega\mu_0\sigma_i}; \quad i = 1, 2. \tag{6.56}$$

Thus, Eq. (6.54) can be rewritten in the form

$$\begin{aligned}\frac{\mathrm{d}z}{\mathrm{d}\rho} &= \mathrm{Re}\left\{\left(\frac{z}{r_0}+\delta\right)\left[\frac{r_0}{\rho}-\mathrm{i}\left(\frac{1}{\beta_2\rho}+\frac{\alpha_2 r_0}{\beta_2\rho}\right)\right]\right\}\\ &= \frac{z}{\rho}+\frac{\delta r_0}{\rho}\approx\frac{z}{\rho}+\delta.\end{aligned} \tag{6.57}$$

With the change of variable, $z' = -z = s\rho$, we have

$$\frac{\mathrm{d}z'}{\mathrm{d}\rho} = s + \rho\frac{\mathrm{d}s}{\mathrm{d}\rho}. \tag{6.58}$$

With the substitution of Eq. (6.57) into Eq. (6.58), it follows that

$$\rho\frac{\mathrm{d}s}{\mathrm{d}\rho} = -\delta. \tag{6.59}$$

The solution of the above differential equation can be expressed as follows:

$$s = -\delta\ln\left(\frac{\rho}{\rho_0}\right). \tag{6.60}$$

Then, we write

$$z' = -\delta\rho\ln\left(\frac{\rho}{\rho_0}\right). \tag{6.61}$$

It is noted that when $\rho = 0$, $z' \approx 0$, and when $\rho = \rho_0$, $z' = 0$.

When ρ is in the interval from 0 to ρ_0, the maximum $z' = z'_m$ is reached at $\rho = \rho_m$. We write

$$z'_m = \frac{\delta\rho_0}{\mathrm{e}}. \tag{6.62}$$

When $\rho = \rho_m$ and $z' = z'_m$, there only exist the horizontal components for the Poynting vector. Evidently, there exists the outward diffusion rather than the downward penetration.

In Fig. 6.11, a schematic diagram of the trajectory of Poynting vector for the electric-type (TM) wave in the ocean floor is shown. In the computations, the operating frequency $f = 0.5$ Hz, and the conductivity σ_1 is 4 S/m, while the conductivity σ_2 is taken as 0.001 S/m and 0.0001 S/m, respectively.

From this figure, it is seen that the farther the receiving point from the transmitting source, the deeper the wave penetrates in the ocean floor. The maximum of the penetration depth is approximated as $z'_m = \sqrt{\frac{\sigma_2}{\sigma_1}}\cdot\frac{\rho_0}{\mathrm{e}}$, where ρ_0 is the horizontal distance between the receiving point and the transmitting source. In this section, in order to obtain the simplified analytical solution of the flow trajectory, we take the two assumptions of $k_2 r \ll 1$ and $P^* = \frac{(R-Z)^2}{R} \ll 1$.

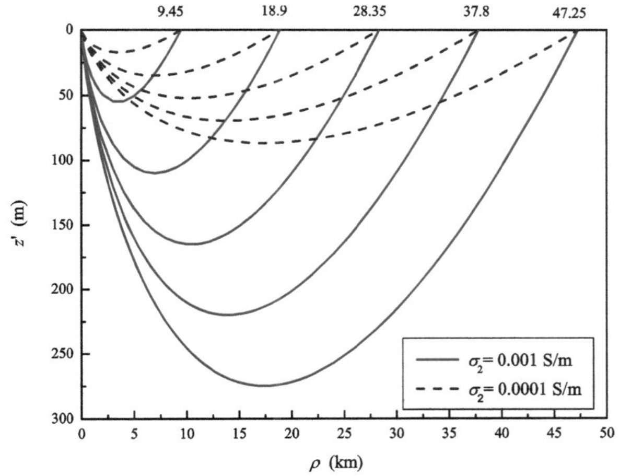

Fig. 6.11 Schematic diagram of the trajectory of Poynting vector for the electric-type (TM) wave in the ocean floor

For instance, $f = 0.5$ Hz, $\sigma_2 = 0.001$ S/m, if $r_0 \geq 10$ km, then $k_2 r_0 \geq 0.628$. Obviously, in practical cases, the above two assumptions are not strictly satisfied, so that the energy-flow trajectory in the actual ocean floor is more complex than that in Fig. 6.11. However, the direction of the energy flow still can be described qualitatively by the trajectory in Fig. 6.11.

For the magnetic-type (TE) wave, with the substitutions of Eqs. (6.48)–(6.53) into Eq. (6.47), the dip angle of Poynting vector is expressed in Eq. (6.63):

$$\frac{\mathrm{d}z}{\mathrm{d}\rho} = -\frac{\mathrm{Re}(E_{2\phi}B_{2\rho}^*)}{\mathrm{Re}(E_{2\phi}B_{2z}^*)} = -\frac{\mathrm{Re}(\frac{2\rho^2 - z^2}{r_0^2} - \frac{3k_2 z}{k_1 r_0})}{\mathrm{Re}[(\frac{3\rho}{r_0^2})(z + \frac{\mathrm{i}}{k_1})]}. \tag{6.63}$$

With $k_i = \sqrt{\mathrm{i}\omega\mu_0\sigma_i}$, Eq. (6.63) is simplified as

$$\frac{\mathrm{d}z}{\mathrm{d}\rho} = -2\left[\frac{2\rho^2 - z^2}{3\rho z} - \delta\sqrt{\left(1 + \frac{z^2}{\rho^2}\right)}\right] \approx -2\left(\frac{2\rho^2 - z^2}{3\rho z} - \delta\right), \tag{6.64}$$

where $\delta = \sqrt{\frac{\sigma_2}{\sigma_1}}$ is a small constant. For mathematical convenience, the small constant δ is temporarily omitted. With the change of variable, $z' = -z = s\rho$, we have

$$\frac{\mathrm{d}z'}{\mathrm{d}\rho} = s + \rho\left(\frac{\mathrm{d}s}{\mathrm{d}\rho}\right). \tag{6.65}$$

It follows that

$$\rho\frac{\mathrm{d}s}{\mathrm{d}\rho} = -\frac{4 + s^2}{3s}. \tag{6.66}$$

Then, we rewrite

$$\frac{2s\,\mathrm{d}s}{s^2+4} = -\frac{2\,\mathrm{d}\rho}{3\rho}. \tag{6.67}$$

Integrating the two sides in Eq. (6.67), we have

$$\ln\left(s^2+4\right) = \ln\left(\frac{\rho_1}{\rho}\right)^{\frac{2}{3}}, \tag{6.68}$$

where ρ_1 is the integration constant. With Eq. (6.68), the solution is obtained readily. We have

$$z' = \rho^{\frac{2}{3}}\left(\rho_1^{\frac{2}{3}} - 4\rho^{\frac{2}{3}}\right)^{\frac{1}{2}}. \tag{6.69}$$

Here, we denote that ρ_0 is the distance from the energy-flow trajectory to the boundary of the sea water and the ocean floor. When $\rho = \rho_0$, $z' = 0$, we obtain $\rho_1^{\frac{2}{3}} = 4\rho_0^{\frac{2}{3}}$ or $\rho_1^{\frac{2}{3}} = 4^{\frac{2}{3}}\rho_0$. Thus, Eq. (6.69) can be rewritten as follows:

$$z' = 2\rho^{\frac{2}{3}}\left(\rho_0^{\frac{2}{3}} - \rho^{\frac{2}{3}}\right)^{\frac{1}{2}}. \tag{6.70}$$

From Eq. (6.70), it can be resolved that the depth of the energy-flow trajectory penetrating into the ocean floor is maximum at $\rho = \rho_m$. We write

$$\rho_m = \left(\frac{2}{3}\right)^{\frac{3}{2}}\rho_0 \approx 0.544\rho_0, \tag{6.71}$$

$$z'_m = \sqrt{2}\rho_m \approx 0.77\rho_0. \tag{6.72}$$

Comparing Eqs. (6.72) with (6.62), it is seen that the energy-flow trajectory for the magnetic-type (TE) wave penetrates into the sea floor deeper than that of the electric-type (TM) wave.

In the above analyses and derivations, the small constant δ is omitted in Eq. (6.65), and when the constant δ is added, the maximum of the penetration depth should be satisfied to the following equation:

$$2\rho_m^2 - z'^2_m - 3\rho_m z'_m\delta = 0. \tag{6.73}$$

Considering that generally δ is significantly small, we readily obtain

$$z'_m \approx (\sqrt{2} - 1.5\delta)\rho_m, \tag{6.74}$$

where

$$\rho_m = \left[\frac{3}{2}\left(1+\frac{\delta}{2}\right)\right]^{-\frac{3}{2}}\rho_0. \tag{6.75}$$

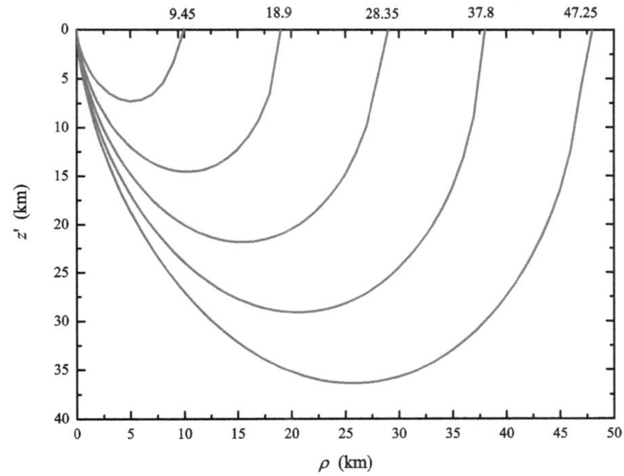

Fig. 6.12 Schematic diagram of the trajectory of Poynting vector for the magnetic-type (TE) wave in the ocean floor

In Fig. 6.12, the schematic diagram of the trajectory of Poynting vector for the magnetic-type (TE) wave in the ocean floor is shown. In the computations, the operating frequency is $f = 0.5$ Hz, and the conductivity σ_1 and σ_2 are taken as 4 S/m and 0.001 S/m, respectively.

In the process of the derivations of Eqs. (6.69) and (6.70), in order to obtain conveniently the analytical formulas of the energy-flow trajectory, the field components in the ocean floor are expressed in the simplified forms. Therefore, the plotted trajectory indicates approximately the propagation direction of the energy flow.

6.3 ELF Wave Excited by HED in the Three-Layered Region

The three-layered region of interest consists of sea water, sedimentary rock layer, and seabed rock. It is well known that the sea water and rock are media with high losses. In this case, the electromagnetic field radiated by an HED in the three-layered region has been treated analytically in the work by Dunn (1986). In this section, based on the work by Dunn, we will attempt to give the corresponding numerical calculated results in ELF range.

In another case, the intermeddle layer is lossless or with small loss, the trapped surface wave can be excited efficiently by a dipole source in the presence of the three-layered region. This problem has been treated analytically by several investigators (King 1991, 1993; Collin 2004a, 2004b; Li and Lu 2005; Zhang et al. 2005) and summarized in the book (Li 2009). It should be pointed out that the above theory and computations are not suitable to the case in this section.

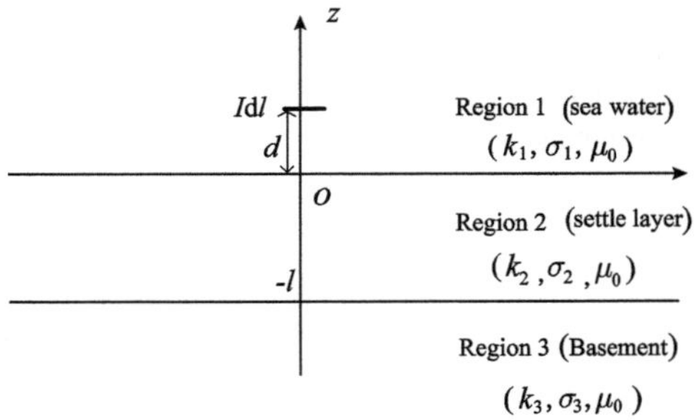

Fig. 6.13 The physical model for the electromagnetic field radiated by an HED in the presence of three-layered region

6.3.1 Analytical Formulas for EM Field in the Three-Layered Region

The relevant geometry and Cartesian coordinate system are shown in Fig. 6.13, where an HED in the $\hat{x}$ direction is located at $(0, 0, d)$ in Region 1 (sea water). Region 1 ($z \geq 0$) is occupied by the sea water characterized by the permeability μ_0 and relative permittivity ε_{r1}, and conductivity σ_1. Region 2 ($-l \leq z \leq 0$) is the sedimentary layer characterized by the permeability μ_0, relative permittivity ε_{r2}, and conductivity σ_2. Region 3 ($z \leq -l$) is occupied by the bedrock characterized by the permeability μ_0, relative permittivity ε_{r3}, and conductivity σ_3. It is noted that the condition of $\sigma_1 \gg \sigma_2 \gg \sigma_3$ should be satisfied. The wave numbers in the three-layered region are

$$k_j = \omega\sqrt{\mu_0(\varepsilon_0\varepsilon_{rj} + \mathrm{i}\sigma_j/\omega)}; \quad j = 1, 2, 3. \tag{6.76}$$

From Maxwell's equations, and with the boundary conditions, the integrated formulas for the six components in the three-layered region can be obtained readily. In this section, we will not repeat the derivation process. In cylindrical coordinate system, the integrated formulas for the field components at (ρ, ϕ, z) in Region 1 are expressed in the following forms (Dunn 1986; King et al. 1992):

$$\begin{aligned} E_{1\rho} = &-\frac{\omega\mu_0 I\,\mathrm{d}l}{4\pi k_1^2}\cos\phi \\ &\times\left(\int_0^\infty \left\{k_1^2 J_0(\lambda\rho) - \frac{\lambda^2}{2}\left[J_0(\lambda\rho) - J_2(\lambda\rho)\right]\right\}\gamma_1^{-1}\mathrm{e}^{\mathrm{i}\gamma_1|z-d|}\lambda\,\mathrm{d}\lambda\right. \\ &+\int_0^\infty \left\{\frac{\gamma_1 Q_3}{2}\left[J_0(\lambda\rho) - J_2(\lambda\rho)\right]\right. \\ &\left.\left.-\frac{k_1^2 P_3}{2\gamma_1}\left[J_0(\lambda\rho) + J_2(\lambda\rho)\right]\right\}\mathrm{e}^{\mathrm{i}\gamma_1(z+d)}\lambda\,\mathrm{d}\lambda\right), \end{aligned} \tag{6.77}$$

$$E_{1\phi} = \frac{\omega\mu_0 I\,\mathrm{d}l}{4\pi k_1^2}\sin\phi \times \left(\int_0^\infty \left\{k_1^2 J_0(\lambda\rho) - \frac{\lambda^2}{2}\left[J_0(\lambda\rho) + J_2(\lambda\rho)\right]\right\}\gamma_1^{-1}\mathrm{e}^{\mathrm{i}\gamma_1|z-d|}\lambda\,\mathrm{d}\lambda + \int_0^\infty \left\{\frac{\gamma_1 Q_3}{2}\left[J_0(\lambda\rho) + J_2(\lambda\rho)\right] - \frac{k_1^2 P_3}{2\gamma_1}\left[J_0(\lambda\rho) - J_2(\lambda\rho)\right]\right\}\mathrm{e}^{\mathrm{i}\gamma_1(z+d)}\lambda\,\mathrm{d}\lambda\right), \tag{6.78}$$

$$E_{1z} = \frac{i\omega\mu_0 I\,\mathrm{d}l}{4\pi k_1^2}\cos\phi\left[\int_0^\infty \left(\pm\mathrm{e}^{\mathrm{i}\gamma_1|z-d|} + Q_3\mathrm{e}^{\mathrm{i}\gamma_1(z+d)}\right)J_1(\lambda\rho)\lambda^2\,\mathrm{d}\lambda\right], \tag{6.79}$$

$$B_{1\rho} = -\frac{\mu_0 I\,\mathrm{d}l}{4\pi}\sin\phi\left(\pm\int_0^\infty J_0(\lambda\rho)\mathrm{e}^{\mathrm{i}\gamma_1|z-d|}\lambda\,\mathrm{d}\lambda + \int_0^\infty \left\{\frac{Q_3}{2}\left[J_0(\lambda\rho) + J_2(\lambda\rho)\right] - \frac{P_3}{2}\left[J_0(\lambda\rho) - J_2(\lambda\rho)\right]\right\} \times \mathrm{e}^{\mathrm{i}\gamma_1(z+d)}\lambda\,\mathrm{d}\lambda\right), \tag{6.80}$$

$$B_{1\phi} = -\frac{\mu_0 I\,\mathrm{d}l}{4\pi}\cos\phi\left(\pm\int_0^\infty J_0(\lambda\rho)\mathrm{e}^{\mathrm{i}\gamma_1|z-d|}\lambda\,\mathrm{d}\lambda + \int_0^\infty \left\{\frac{Q_3}{2}\left[J_0(\lambda\rho) - J_2(\lambda\rho)\right] - \frac{P_3}{2}\left[J_0(\lambda\rho) + J_2(\lambda\rho)\right]\right\} \times \mathrm{e}^{\mathrm{i}\gamma_1(z+d)}\lambda\,\mathrm{d}\lambda\right), \tag{6.81}$$

$$B_{1z} = \frac{\mathrm{i}\mu_0 I\,\mathrm{d}l}{4\pi}\sin\phi\left[\int_0^\infty \left(\mathrm{e}^{\mathrm{i}\gamma_1|z-d|} - P_3\mathrm{e}^{\mathrm{i}\gamma_1(z+d)}\right)\gamma_1^{-1}J_1(\lambda\rho)\lambda^2\,\mathrm{d}\lambda\right], \tag{6.82}$$

where

$$P_3 = \frac{\gamma_3 - \gamma_1 - \mathrm{i}(\gamma_2 - \frac{\gamma_1\gamma_3}{\gamma_2})\tan\gamma_2 l}{\gamma_3 + \gamma_1 - \mathrm{i}(\gamma_2 + \frac{\gamma_1\gamma_3}{\gamma_2})\tan\gamma_2 l}, \tag{6.83}$$

$$Q_3 = \frac{\gamma_3 k_1^2 - \gamma_1 k_3^2 - \mathrm{i}(\gamma_2\frac{k_1^2 k_3^2}{k_2^2} - k_2^2\frac{\gamma_1\gamma_3}{\gamma_2})\tan\gamma_2 l}{\gamma_3 k_1^2 + \gamma_1 k_3^2 - \mathrm{i}(\gamma_2\frac{k_1^2 k_3^2}{k_2^2} + k_2^2\frac{\gamma_1\gamma_3}{\gamma_2})\tan\gamma_2 l}. \tag{6.84}$$

When $l = 0$, namely, the thickness of the sedimentary layer is 0, P_3 represented by Eq. (6.83) and Q_3 by Eq. (6.84) are reduced to

$$P_3 = \frac{\gamma_3 - \gamma_1}{\gamma_3 + \gamma_1}; \qquad Q_3 = \frac{\gamma_3 k_1^2 - \gamma_1 k_3^2}{\gamma_3 k_1^2 + \gamma_1 k_3^2}. \tag{6.85}$$

If the parameters k_3 and γ_3 in this section are replaced by k_2 and γ_2 in Sect. 6.2, it is seen that the formulas in Eq. (6.85) are completely the same as the corresponding formulas in Eq. (6.18).

When $l \to \infty$, namely, the thickness of the sedimentary layer is infinite. Because the sedimentary layer is a high-loss medium, k_2 has a positive imaginary part. When $l \to \infty$, we have $\tan \gamma_2 l \to \mathrm{i}$. Then, Eqs. (6.83) and (6.84) are reduced to

$$P_3 = \frac{\gamma_2 - \gamma_1}{\gamma_2 + \gamma_1}; \qquad Q_3 = \frac{k_1^2 \gamma_2 - k_2^2 \gamma_1}{k_1^2 \gamma_2 + k_2^2 \gamma_1}. \tag{6.86}$$

Obviously, they are completely the same as the corresponding formulas for the two-layered case.

When the condition of $\sigma_1 \gg \sigma_2 \gg \sigma_3$ is satisfied, the complete analytical formulas have been obtained for the electromagnetic field in the three-layered region (Dunn 1986). We write

$$\begin{aligned} E_{1\rho}(\rho, \phi, z) = &-\frac{\omega\mu_0}{2\pi k_1^2} \cos\phi \cdot \left\{ A k_3 \mathrm{e}^{\mathrm{i}k_1(z+d)} \mathrm{e}^{\mathrm{i}k_3\rho} g(\nu, \rho, k_3) \right. \\ &- \left(\frac{k_1}{r_1^2} + \frac{\mathrm{i}}{r_1^3} \right) \mathrm{e}^{\mathrm{i}k_1 r_1} + \left(\frac{z+d}{\rho} \right) \left(\frac{\mathrm{i}k_1}{\rho^2} - \frac{3}{2\rho^3} \right) \mathrm{e}^{\mathrm{i}k_1 r_2} \\ &- \frac{1}{2} \left[\left(\frac{z-d}{r_1} \right)^2 \left(\frac{\mathrm{i}k_1^2}{r_1} - \frac{3k_1}{r_1^2} - \frac{3i}{r_1^3} \right) \mathrm{e}^{\mathrm{i}k_1 r_1} \right. \\ &\left. \left. + \left(\frac{z+d}{r_2} \right)^2 \left(\frac{\mathrm{i}k_1^2}{r_2} - \frac{3k_1}{r_2^2} - \frac{3\mathrm{i}}{r_2^3} \right) \mathrm{e}^{\mathrm{i}k_1 r_2} \right] \right\}, \end{aligned} \tag{6.87}$$

$$\begin{aligned} E_{1\phi}(\rho, \phi, z) = &\frac{\omega\mu_0}{2\pi k_1^2} \sin\phi \cdot \left\{ A k_3 \mathrm{e}^{\mathrm{i}k_1(z+d)} \mathrm{e}^{\mathrm{i}k_3\rho} h(\nu, \rho, k_3) \right. \\ &+ \frac{1}{2} \left[\mathrm{e}^{\mathrm{i}k_1 r_2} \left(\frac{\mathrm{i}k_1^2}{r_2} - \frac{3k_1}{r_2^2} - \frac{3\mathrm{i}}{r_2^3} \right) - \mathrm{e}^{\mathrm{i}k_1 r_1} \left(\frac{\mathrm{i}k_1^2}{r_1} - \frac{k_1}{r_1^2} - \frac{\mathrm{i}}{r_1^3} \right) \right] \\ &+ \mathrm{i}\mathrm{e}^{\mathrm{i}k_1 r_2} \left(\frac{z+d}{r_2} \right) \left(\frac{\mathrm{i}k_1^2}{\rho} - \frac{3k_1}{2\rho^2} - \frac{5\mathrm{i}}{8\rho^3} \right) \\ &\left. - \mathrm{e}^{\mathrm{i}k_1 r_2} \left(\frac{z+d}{r_2} \right)^2 \left(\frac{\mathrm{i}k_1^2}{r_2} - \frac{3k_1}{r_2^2} - \frac{3\mathrm{i}}{r_2^3} \right) \right\}, \end{aligned} \tag{6.88}$$

$$E_{1z}(\rho, \phi, z) = \frac{\omega\mu_0}{2\pi k_1^2} \cos\phi \cdot \left\{ \frac{k_3^2}{k_1} \left[A \mathrm{e}^{\mathrm{i}k_1(z+d)} \mathrm{e}^{\mathrm{i}k_3\rho} f(\nu, \rho, k_3) \right. \right.$$

$$
\begin{aligned}
&- e^{ik_1 r_2}\left(\frac{i}{\rho^2} - \frac{3}{2k_1\rho^3}\right)\Bigg] + \frac{ik_2^2}{k_1^2} T_z\left(k_2 l, e^{ik_1\rho}\right) \\
&- \frac{1}{2}\Bigg[e^{ik_1 r_1}\left(\frac{\rho}{r_1}\right)\left(\frac{z-d}{r_1}\right)\left(\frac{ik_1^2}{r_1} - \frac{3k_1}{r_1^2} - \frac{3i}{r_1^3}\right) \\
&+ e^{ik_1 r_2}\left(\frac{\rho}{r_2}\right)\left(\frac{z+d}{r_2}\right)\left(\frac{ik_1^2}{r_2} - \frac{3k_1}{r_2^2} - \frac{3i}{r_2^3}\right)\Bigg]\Bigg\},
\end{aligned} \tag{6.89}
$$

$$
\begin{aligned}
B_{1\rho}(\rho,\phi,z) = &-\frac{\mu_0}{2\pi k_1}\sin\phi\cdot\Bigg\{Ae^{ik_1(z+d)}e^{ik_3\rho}k_3 h(\nu,\rho,k_3) \\
&+ e^{ik_1 r_2}\left[\frac{ik_1}{\rho^2} - \frac{7}{2\rho^3} + \left(\frac{z+d}{r_2}\right)\left(\frac{ik_1^2}{r_2} - \frac{5k_1}{r_2^2} - \frac{12i}{r_2^3}\right)\right] \\
&- \frac{1}{2}\Bigg[e^{ik_1 r_1}\left(\frac{z-d}{r_1}\right)\left(\frac{ik_1^2}{r_1} - \frac{k_1}{r_1^2}\right) \\
&+ e^{ik_1 r_2}\left(\frac{z+d}{r_2}\right)\left(\frac{ik_1^2}{r_2} - \frac{k_1}{r_2^2}\right)\Bigg]\Bigg\},
\end{aligned} \tag{6.90}
$$

$$
\begin{aligned}
B_{1\phi}(\rho,\phi,z) = &-\frac{\mu_0}{2\pi k_1}\cos\phi\cdot\Bigg\{Ae^{ik_1(z+d)}e^{ik_3\rho}k_3 g(\nu,\rho,k_3) \\
&+ e^{ik_1 r_2}\left[\frac{1}{\rho^3} + \frac{3i}{2k_1\rho^4} + \left(\frac{z+d}{r_2}\right)\left(\frac{k_1}{r_2^2} + \frac{3i}{r_2^3} - \frac{3}{k_1 r_2^4}\right)\right] \\
&- \frac{1}{2}\Bigg[e^{ik_1 r_1}\left(\frac{z-d}{r_1}\right)\left(\frac{ik_1^2}{r_1} - \frac{k_1}{r_1^2}\right) \\
&+ e^{ik_1 r_2}\left(\frac{z+d}{r_2}\right)\left(\frac{ik_1^2}{r_2} - \frac{k_1}{r_2^2}\right)\Bigg]\Bigg\},
\end{aligned} \tag{6.91}
$$

$$
\begin{aligned}
B_{1z}(\rho,\phi,z) = &\frac{\mu_0}{2\pi}\sin\phi\cdot\Bigg\{Ae^{ik_1(z+d)}e^{ik_3\rho}\frac{k_3^2}{k_1^2}\left(\frac{1}{\rho^2} + \frac{3i}{k_3\rho^3} - \frac{3}{k_3^2\rho^4}\right) \\
&- e^{ik_1 r_2}\left(\frac{\rho}{r_2}\right)\Bigg[\frac{1}{r_2^2} + \frac{3i}{k_1 r_2^3} - \frac{3}{k_1^2 r_2^4} \\
&+ \left(\frac{z+d}{r_2}\right)^2\left(\frac{ik_1}{r_2} - \frac{6}{r_2^2} - \frac{15i}{kr_2^3}\right)\Bigg] \\
&- \frac{1}{2}\Bigg[e^{ik_1 r_1}\left(\frac{\rho}{r_1}\right)\left(\frac{ik_1}{r_1} - \frac{1}{r_1^2}\right) - e^{ik_1 r_2}\left(\frac{\rho}{r_2}\right)\left(\frac{ik_1}{r_2} - \frac{1}{r_2^2}\right)\Bigg]\Bigg\},
\end{aligned} \tag{6.92}
$$

where

$$g(\nu,\rho,k_3) = \frac{\mathrm{i}k_3}{\rho} - \frac{1}{\rho^2} - \frac{\mathrm{i}}{k_3\rho^3} - \nu k_3^2\left(\frac{\pi}{k_3\rho}\right)^{\frac{1}{2}} \mathrm{e}^{-\frac{\mathrm{i}k_3\rho\nu^2}{2}} F\left(\frac{k_3\rho\nu^2}{2}\right), \tag{6.93}$$

$$h(\nu,\rho,k_3) = \frac{2}{\rho^2} + \frac{2\mathrm{i}}{k_3\rho^3} + \frac{\mathrm{i}\nu k_3}{\rho}\left(\frac{\pi}{k_3\rho}\right)^{\frac{1}{2}} \mathrm{e}^{-\frac{\mathrm{i}k_3\rho\nu^2}{2}} F\left(\frac{k_3\rho\nu^2}{2}\right), \tag{6.94}$$

$$A = \left(\cos k_2 l - \frac{\mathrm{i}k_2}{k_1}\sin k_2 l\right)^{-2}, \tag{6.95}$$

$$f(\nu,\rho,k_3) = \frac{\mathrm{i}k_3}{\rho} - \frac{1}{\rho^2} - \nu k_3^2\left(\frac{\pi}{k_3\rho}\right)^{\frac{1}{2}} \mathrm{e}^{-\frac{\mathrm{i}k_3\rho\nu^2}{2}} F\left(\frac{k_3\rho\nu^2}{2}\right), \tag{6.96}$$

$$\nu = \frac{k_3 b}{k_1}; \quad b = \frac{1 - \mathrm{i}(\frac{k_1}{k_2})\tan k_2 l}{1 - \mathrm{i}(\frac{k_1}{k_2})\tan k_2 l}, \tag{6.97}$$

$$r_1 = \sqrt{[\rho^2 + (z-d)^2]}; \qquad r_2 = \sqrt{[\rho^2 + (z+d)^2]}. \tag{6.98}$$

The Fesnel integral $F(\frac{k_3\rho\nu^2}{2})$ is defined by

$$F\left(\frac{k_3\rho\nu^2}{2}\right) = \frac{1}{2}(1+\mathrm{i}) - C_2\left(\frac{k_3\rho\nu^2}{2}\right) - \mathrm{i}S_2\left(\frac{k_3\rho\nu^2}{2}\right), \tag{6.99}$$

where

$$\begin{aligned} C_2\left(\frac{k_3\rho\nu^2}{2}\right) + \mathrm{i}S_2\left(\frac{k_3\rho\nu^2}{2}\right) &= \int_0^{\frac{k_3\rho\nu^2}{2}} \frac{\mathrm{e}^{\mathrm{i}t}}{(2\pi t)^{\frac{1}{2}}}\,\mathrm{d}t \\ &= \int_0^{\frac{k_3\rho\nu^2}{2}} \frac{\cos t}{(2\pi t)^{\frac{1}{2}}}\,\mathrm{d}t + \mathrm{i}\int_0^{\frac{k_3\rho\nu^2}{2}} \frac{\sin t}{(2\pi t)^{\frac{1}{2}}}\,\mathrm{d}t. \end{aligned} \tag{6.100}$$

In the analytical formulas in Eqs. (6.87)–(6.92), the first term including the factor $\mathrm{e}^{\mathrm{i}k_1(z+d)}\mathrm{e}^{\mathrm{i}k_3\rho}$ represents the lateral wave, which travels from the transmitting source in Region 1 (sea water), and across Region 2 (sedimentary layer) and to Region 3 (seabed rock), then radially along the boundary in Region 3 a distance ρ, and finally vertically across Region 2 and to Region 1 to the receiving point. In the case of $\rho \gg |z|$ and $\rho \gg d$, the total field is determined primarily by the lateral wave. The term including the factor $\mathrm{e}^{\mathrm{i}k_1 r_1}$ can be understood as the directed wave, while the term including the factor $\mathrm{e}^{\mathrm{i}k_1 r_2}$ can be understood as the ideal reflected wave. Due to the large loss in the sea water, the actual contributions to the total field by the term

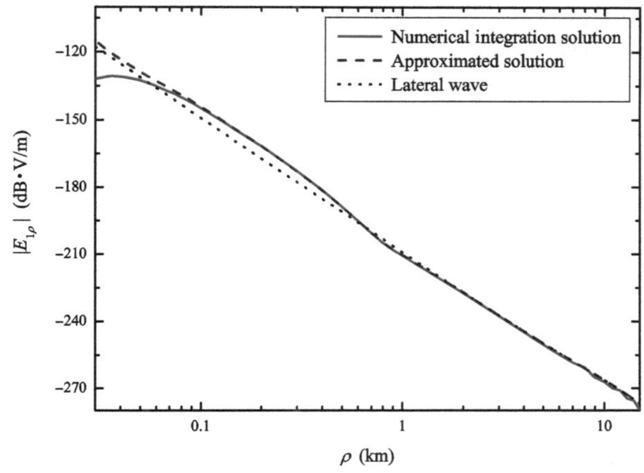

Fig. 6.14 The magnitudes of the component $E_{1\rho}(\rho, 0, 1)$ versus the distance ρ

of the directed wave and that of the ideal reflected wave are not large in the case of $\rho \gg |z|$ and $\rho \gg d$.

6.3.2 Computations and Discussions

When there exists an intermediate sedimentary layer, the factor A in Eqs. (6.90)–(6.95) appears in the approximated formulas of the field component in the sea water, which is determined mainly by the conductivity σ_2 and the thickness l of the intermediate sedimentary layer. It is indicated that the factor A represents the propagation loss of the electromagnetic wave, which penetrates the intermediate layer.

With $f = 1$ Hz, $I\,dl = 1$ A·m, $d = 30$ m, $z = 1$ m, $l = 100$ m, $\sigma_1 = 4$ S/m, $\sigma_2 = 0.1$ S/m, and $\sigma_3 = 0.004$ S/m, the computations for the components $E_{1\rho}(\rho, 0, 1)$, $E_{1\phi}(\rho, \frac{\pi}{2}, 1)$, $H_{1\rho}(\rho, \frac{\pi}{2}, 1)$, and $H_{1\phi}(\rho, 0, 1)$ are carried out and shown in Figs. 6.14, 6.15, 6.16, 6.17, respectively.

From Figs. 6.14–6.17, it is seen that the results by using the numerical integration method are in agreement with those by using the approximated formulas, and those by the lateral-wave term at $\rho > 600$ km. It is noted that the differences between the results by using the numerical integration method and those by using the approximated formula are within 0.7 dB at $\rho > 600$ m, while the differences are enlarged over 5 dB at $\rho < 100$ m.

The components $E_{1\rho}(\rho, 0, 1)$ and $E_{1\phi}(\rho, \frac{\pi}{2}, 1)$, in the sea water for two- and three-layered models are computed and shown in Figs. 6.18–6.21. In the computations, the parameters are taken as $f = 1$ Hz, $d = 30$ m, $z = 1$, $\sigma_1 = 4$ S/m for the three-layered model, while $\sigma_2 = 0.1$ S/m, $\sigma_3 = 0.004$ S/m for the two-layered model. In Figs. 6.18 and 6.19, $l = 100$ m, while $l = 2{,}000$ m in Figs. 6.20 and 6.21.

From Figs. 6.18 and 6.19, it is seen that the calculated results for the two-layered model with $\sigma_2 = 0.004$ S/m are in agreement with those for the three-layered model

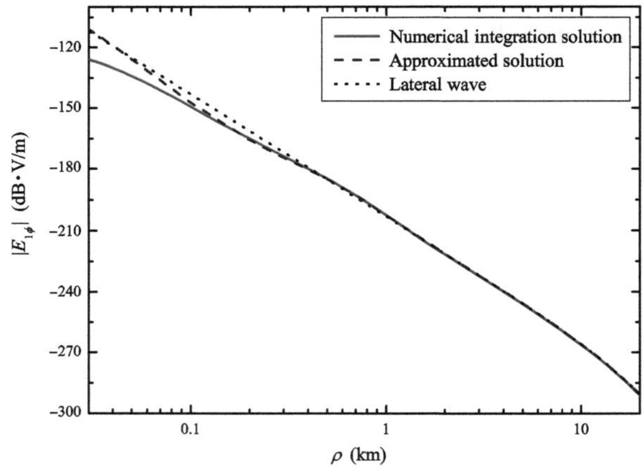

Fig. 6.15 The magnitudes of the component $E_{1\phi}(\rho, \frac{\pi}{2}, 1)$ versus the distance ρ

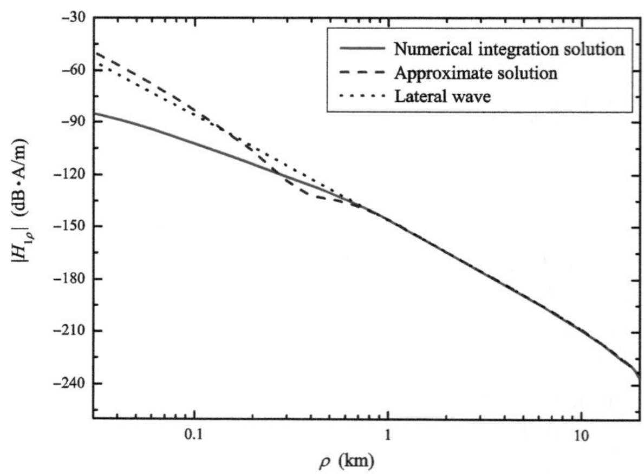

Fig. 6.16 The magnitudes of the component $H_{1\rho}(\rho, \frac{\pi}{2}, 1)$ versus the distance ρ

$l = 100$ m, $\sigma_2 = 0.1$ S/m, and $\sigma_3 = 0.004$ S/m. It is concluded that the contributions of the total field by the intermediate sedimentary layer with $l = 100$ m and $\sigma_2 = 0.1$ S/m are very small. This is resulted by the fact that ELF wave propagates mainly in Region 3 (seabed rock) in the lateral-wave form. For the two-layered model, the attenuations with $\sigma_2 = 0.1$ S/m are obviously larger than those with $\sigma_3 = 0.004$ S/m. From Figs. 6.20 and 6.21, it is seen that there exists a difference between the two- and three-layered models. Obviously, the effects of the total field by the intermediate sedimentary layer with $l = 2{,}000$ m should be considered.

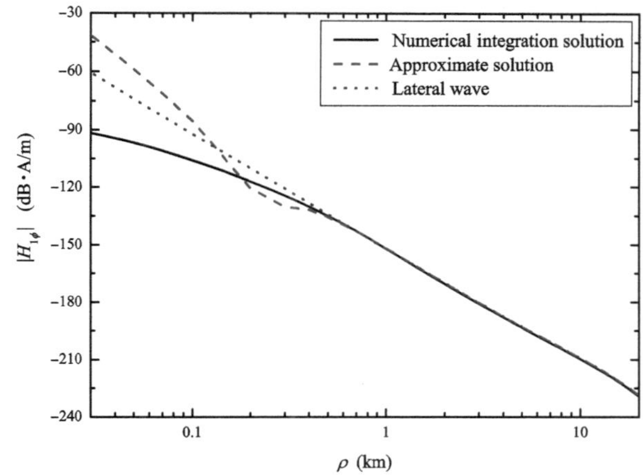

Fig. 6.17 The magnitudes of the component $H_{1\phi}(\rho, 0, 1)$ versus the distance ρ

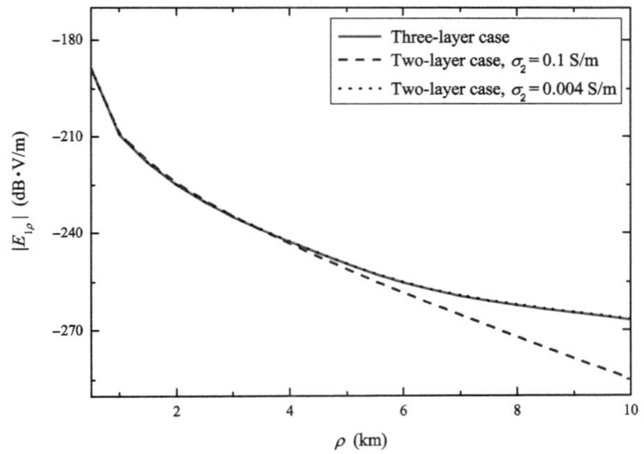

Fig. 6.18 The component $E_{1\rho}(\rho, 0, 1)$ versus the propagation distance for two- and three-layered models: $l = 100$ m

6.4 ELF Wave Propagation Along the Boundary Between Sea Water and One-Dimensionally Anisotropic Rock

Anisotropy in conductivity is found in some stratified rock which consists of alternating layers of dense rock and less dense rock. It is seen that the conductivity transverse to the bedding surfaces is always smaller than the conductivity parallel to the surfaces (Parkhomenko 1967). The problem for the ELF wave propagation along the boundary between sea water and one-dimensionally anisotropic rock was first investigated by Pan in the mid-1980s. Based on Pan's work (Pan 1985), the results on this problem are outlined specifically.

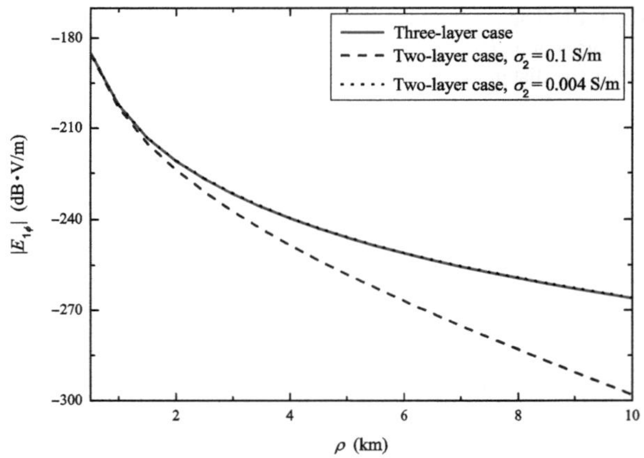

Fig. 6.19 The component $E_{1\phi}(\rho, \frac{\pi}{2}, 1)$ versus the propagation distance for two- and three-layered models: $l = 100$ m

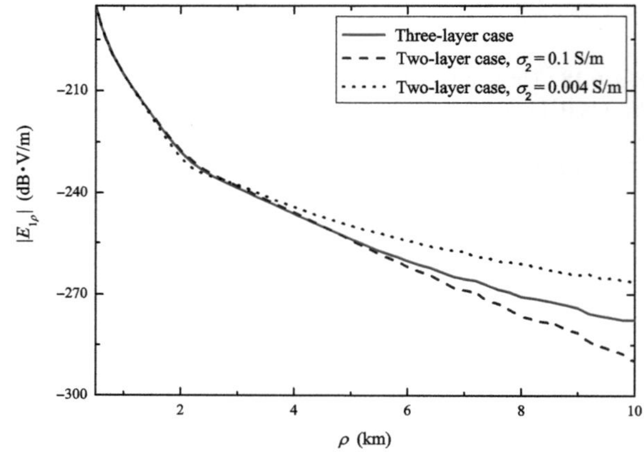

Fig. 6.20 The component $E_{1\rho}(\rho, 0, 1)$ versus the propagation distance for two- and three-layered models: $l = 2{,}000$ m

6.4.1 Formulations of the Problem

The geometry and coordinate system are shown in Fig. 6.22, where the HED in the $\hat{x}$ direction is located at $(0, 0, d)$. Region 1 ($z \geq 0$) is the upper half-space filled with the sea water characterized by the permeability μ_0, permittivity ε_1, and conductivity σ_1. Region 2 ($z \leq 0$) is the ocean floor characterized by the permeability μ_0, permittivity $\varepsilon_{2x} = \varepsilon_{2y}$ and ε_{2z}, and conductivity $\sigma_{2x} = \sigma_{2y}$ and σ_{2z}. Then, the complex permittivities $\tilde{\varepsilon}_1$ and $\tilde{\varepsilon}_2$ are represented as follows:

$$\tilde{\varepsilon}_1 = \varepsilon_1 + \mathrm{i}\frac{\sigma_1}{\omega}, \tag{6.101}$$

$$\tilde{\varepsilon}_2 = \begin{bmatrix} \varepsilon_T - \mathrm{i}\frac{\sigma_T}{\omega} & 0 & 0 \\ 0 & \varepsilon_T - \mathrm{i}\frac{\sigma_T}{\omega} & 0 \\ 0 & 0 & \varepsilon_L - \mathrm{i}\frac{\sigma_L}{\omega} \end{bmatrix}. \tag{6.102}$$

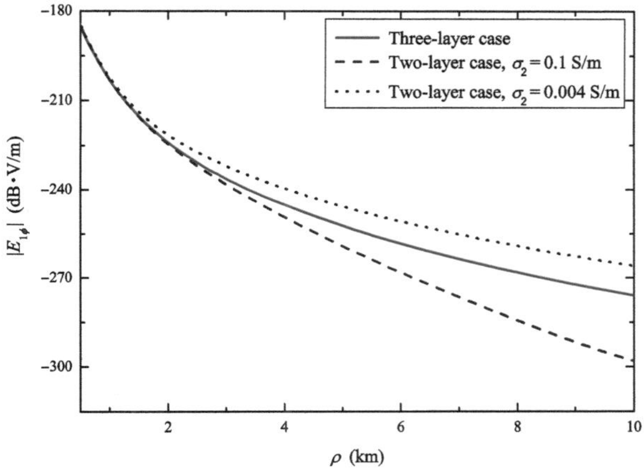

Fig. 6.21 The component $E_{1\phi}(\rho, \frac{\pi}{2}, 1)$ versus the propagation distance for two- and three-layered models: $l = 2{,}000$ m

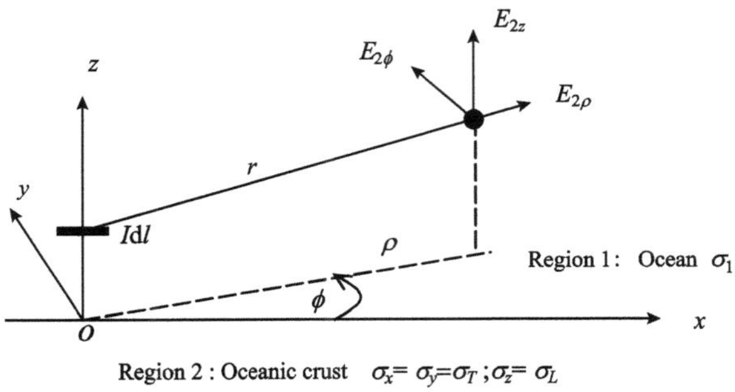

Fig. 6.22 The physical model for an HED on the boundary between sea water and one-dimensionally anisotropic rock

Maxwell's equations in the two half-spaces are expressed by

$$\nabla \times \boldsymbol{E}_j = \mathrm{i}\omega \boldsymbol{B}_j, \tag{6.103}$$

$$\nabla \times \boldsymbol{B}_j = \mu_0\big(-i\omega\tilde{\varepsilon}_j \boldsymbol{E}_j + \hat{x} J_x^e\big), \tag{6.104}$$

where $j = 1, 2$, and

$$J_x^e = I\,\mathrm{d}l\delta(x)\delta(y)\delta(z-d). \tag{6.105}$$

The Fourier transform is in the form of

$$\boldsymbol{E}(x, y, z) = \frac{1}{(2\pi)^2}\int_{-\infty}^{+\infty}\mathrm{d}\xi\int_{-\infty}^{+\infty}\mathrm{d}\eta \mathrm{e}^{\mathrm{i}(\xi x+\eta y)}\tilde{E}(\xi, \eta, z), \tag{6.106}$$

and the similar transform for $\boldsymbol{B}(x, y, z)$ and J_x^e are applied to Maxwell's equation. Then, we have

$$\tilde{J}_x^e = I\,\mathrm{d}l\delta(z-d). \tag{6.107}$$

With Maxwell's equations, and after considerable algebraic manipulation, the transforms of the components $\tilde{E}_y$, $\tilde{E}_z$, $\tilde{B}_y$, and $\tilde{B}_z$ can be expressed in terms of the transforms of the components $\tilde{E}_x$ and $\tilde{B}_x$:

$$\tilde{E}_{1y} = \frac{1}{k_1^2 - \xi^2}\left(-\xi\eta\tilde{E}_{1x} + \mathrm{i}\omega\frac{\partial}{\partial z}\tilde{B}_{1x}\right), \tag{6.108}$$

$$\tilde{E}_{1z} = \frac{1}{k_1^2 - \xi^2}\left(\mathrm{i}\xi\frac{\partial}{\partial z}\tilde{E}_{1x} + \omega\eta\tilde{B}_{1x}\right), \tag{6.109}$$

$$\tilde{B}_{1y} = \frac{1}{k_1^2 - \xi^2}\left(-\mathrm{i}\frac{k_1^2}{\omega}\frac{\partial}{\partial z}\tilde{E}_{1x} - \xi\eta\tilde{B}_{1x}\right), \tag{6.110}$$

$$\tilde{B}_{1z} = \frac{1}{k_1^2 - \xi^2}\left(-\frac{k_1^2}{\omega}\eta\tilde{E}_{1x} + \mathrm{i}\xi\frac{\partial}{\partial z}\tilde{B}_{1x}\right), \tag{6.111}$$

and

$$\tilde{E}_{2y} = \frac{1}{k_T^2 - \xi^2}\left(-\xi\eta\tilde{E}_{2x} + \mathrm{i}\omega\frac{\partial}{\partial z}\tilde{B}_{2x}\right), \tag{6.112}$$

$$\tilde{E}_{2z} = \frac{1}{k_L^2 - \xi^2}\left(\mathrm{i}\xi\frac{\partial}{\partial z}\tilde{E}_{2x} + \omega\eta\tilde{B}_{2x}\right), \tag{6.113}$$

$$\tilde{B}_{2y} = \frac{1}{k_L^2 - \xi^2}\left(-\mathrm{i}\frac{k_L^2}{\omega}\frac{\partial}{\partial z}\tilde{E}_{2x} - \xi\eta\tilde{B}_{2x}\right), \tag{6.114}$$

$$\tilde{B}_{2z} = \frac{1}{k_T^2 - \xi^2}\left(-\frac{k_T^2}{\omega}\eta\tilde{E}_{2x} + \mathrm{i}\xi\frac{\partial}{\partial z}\tilde{B}_{2x}\right), \tag{6.115}$$

where

$$k_j^2 = \omega^2\mu_0\tilde{\varepsilon}_j; \quad j = 1,\ L,\ T. \tag{6.116}$$

In Region 1, the following differential equations are obtained readily:

$$\frac{\partial^2}{\partial z^2}\tilde{B}_{1x} + \gamma_1^2\tilde{B}_{1x} = 0, \tag{6.117}$$

$$\frac{\partial^2}{\partial z^2}\tilde{E}_{1x} + \gamma_1^2\tilde{E}_{1x} = \frac{k_1^2 - \xi^2}{\mathrm{i}\omega\hat{\varepsilon}_1} I\,\mathrm{d}l\delta(z - d), \tag{6.118}$$

where

$$\gamma_1^2 = \left(k_1^2 - \xi^2 - \eta^2\right)^{\frac{1}{2}}; \qquad \mathrm{Im}\,\gamma_1 > 0. \tag{6.119}$$

In Region 2, the equations for $\tilde{B}_{2x}$ and $\tilde{E}_{2x}$ are obtained readily. We write

$$\frac{1}{k_T^2 - \xi^2}\frac{\partial^2}{\partial z^2}\tilde{B}_{2x} + \left(1 - \frac{\eta^2}{k_L^2 - \xi^2}\right)\tilde{B}_{2x}$$

$$+\frac{\mathrm{i}\xi\eta}{\omega}\left(\frac{1}{k_T^2-\xi^2}-\frac{1}{k_L^2-\xi^2}\right)\frac{\partial}{\partial z}\tilde{E}_{2x}=0, \tag{6.120}$$

$$\frac{k_L^2}{k_L^2-\xi^2}\frac{\partial^2}{\partial z^2}\tilde{E}_{2x}+k_T^2\left(1-\frac{\eta^2}{k_T^2-\xi^2}\right)\tilde{E}_{2x}$$

$$+\mathrm{i}\omega\xi\eta\left(\frac{1}{k_T^2-\xi^2}-\frac{1}{k_L^2-\xi^2}\right)\frac{\partial}{\partial z}\tilde{B}_{2x}=0. \tag{6.121}$$

With the Sommerfeld radiation condition for $z\to\infty$, the solutions of Eqs. (6.117) and (6.118) for $\tilde{B}_{1x}$ and $\tilde{E}_{1x}$ can be obtained readily. We write

$$\tilde{B}_{1x}=\mu_0 C_1 \mathrm{e}^{\mathrm{i}\gamma_1 z}; \qquad z\geq 0, \tag{6.122}$$

$$\tilde{E}_{1x}=C_2\mathrm{e}^{\mathrm{i}\gamma_1 z}-\frac{k_1^2-\xi^2}{\mathrm{i}\omega\hat{\varepsilon}_1\gamma_1}I\,\mathrm{d}l\mathrm{e}^{\mathrm{i}\gamma_1 d}\sin\gamma_1 z; \quad 0\leq z\leq d, \tag{6.123}$$

$$\tilde{E}_{1x}=\left(C_2-\frac{k_1^2-\xi^2}{\mathrm{i}\omega\hat{\varepsilon}_1\gamma_1}I\,\mathrm{d}l\sin\gamma_1 d\right)\mathrm{e}^{\mathrm{i}\gamma_1 z}; \quad z>d, \tag{6.124}$$

where C_1 and C_2 are constants of integration.

We assume that the solutions for $\tilde{B}_{2x}$ and $\tilde{E}_{2x}$ are expressed in the forms

$$\tilde{B}_{2x}=\mu_0 A\mathrm{e}^{-\mathrm{i}\gamma z}; \qquad \tilde{E}_{2x}=B\mathrm{e}^{-\mathrm{i}\gamma z}. \tag{6.125}$$

Substituting these expressions into Eqs. (6.120) and (6.121), the algebraic equations are obtained as follows:

$$\left(1-\frac{\gamma^2}{k_T^2-\xi^2}-\frac{\eta^2}{k_L^2-\xi^2}\right)\mu_0 A+\frac{\xi\eta\gamma}{\omega}\left(\frac{1}{k_T^2-\xi^2}-\frac{1}{k_L^2-\xi^2}\right)B=0, \tag{6.126}$$

$$-\omega\xi\eta\gamma\left(\frac{1}{k_T^2-\xi^2}-\frac{1}{k_L^2-\xi^2}\right)\mu_0 A+\left(k_T^2-\frac{k_T^2\eta^2}{k_T^2-\xi^2}-\frac{k_L^2\xi^2}{k_L^2-\xi^2}\right)B=0. \tag{6.127}$$

With the above two equations, it follows that

$$\gamma^4 k_L^2+\gamma^2\left[(k_L^2+k_T^2)(\xi^2+\eta^2)-2k_L^2k_T^2\right]$$
$$+k_T^2(k_L^2-\xi^2-\eta^2)(k_T^2-\xi^2-\eta^2)=0. \tag{6.128}$$

The nonzero solutions are

$$\gamma_T=\sqrt{k_T^2-\lambda^2}, \tag{6.129}$$

$$\gamma_e=\sqrt{k_T^2-\frac{k_T^2}{k_L^2}\lambda^2}=\frac{k_T}{k_L}\gamma_L=\frac{k_T}{k_L}\sqrt{k_L^2-\lambda^2}, \tag{6.130}$$

where

$$\lambda^2 = \xi^2 + \eta^2. \tag{6.131}$$

The imaginary parts of γ_T and γ_e should be positive when λ is real. The coefficients A and B in Eqs. (6.126) and (6.127) should satisfy the equations

$$\frac{B_T}{A_T} = -\frac{\mu_0\omega\eta}{\xi\gamma_T}; \qquad \frac{B_e}{A_e} = -\frac{\mu_0\omega\xi\gamma_e}{k_T^2\eta}. \tag{6.132}$$

It yields

$$\tilde{B}_{2x} = \mu_0\left(A_T e^{-i\gamma_T z} + A_e e^{-i\gamma_e z}\right); \qquad z \le 0, \tag{6.133}$$

$$\tilde{E}_{2x} = -\frac{\mu_0\omega\eta}{\xi\gamma_T} A_T e^{-i\gamma_T z} + \frac{\nu_0\omega\xi\gamma_e}{k_T^2\eta} A_e e^{-i\gamma_e z}; \quad z \le 0. \tag{6.134}$$

Then, the remaining four components in Region 2 are obtained readily. They are

$$\tilde{B}_{2y} = \mu_0\left(\frac{\eta}{\xi} A_T e^{-i\gamma_T z} - \frac{\xi}{\eta} A_e e^{-i\gamma_e z}\right); \qquad z \le 0, \tag{6.135}$$

$$\tilde{B}_{2z} = \frac{\mu_0\lambda^2}{\xi\gamma_T} A_T e^{-i\gamma_T z}; \qquad z \le 0, \tag{6.136}$$

$$\tilde{E}_{2y} = \omega\mu_0\left(\frac{A_T}{\gamma_T} e^{-i\gamma_T z} + \frac{A_e\gamma_e}{k_T^2} e^{-i\gamma_e z}\right); \quad z \le 0, \tag{6.137}$$

$$\tilde{E}_{2z} = \frac{\omega\mu_0\lambda^2}{k_L^2\eta} A_e e^{-i\gamma_e z}; \qquad z \le 0. \tag{6.138}$$

With the boundary conditions $\tilde{B}_{1y} = \tilde{B}_{2y}$, $\tilde{B}_{1z} = \tilde{B}_{2z}$, $\tilde{E}_{1y} = \tilde{E}_{2y}$, and $k_1^2\tilde{E}_{1z} = k_L^2\tilde{E}_{2z}$ at $z = 0$, the coefficients C_1, C_2, A_T, and A_e can be obtained readily. We write

$$C_1 = \frac{I\,dl\,\xi\eta e^{i\gamma_1 d}}{\lambda^2 M_T N_e}\left(k_1^2\gamma_T\gamma_e - k_T^2\gamma_1^2\right); \tag{6.139}$$

$$C_2 = -\frac{I\,dl\,\omega\mu_0 e^{i\gamma_1 d}}{\lambda^2 M_T N_e}\left\{\lambda^2 N_e + \xi^2\left[\gamma_1\left(\gamma_T\gamma_e - k_T^2\right) - \lambda^2\gamma_e\right]\right\}, \tag{6.140}$$

$$A_T = -\frac{\xi\gamma_T}{\lambda^2(k_1^2 - \xi^2)}\left(\xi\gamma_1 C_1 + \omega\hat{\varepsilon}_1\eta C_2\right), \tag{6.141}$$

$$A_e = \frac{k_1^2\eta}{\omega\mu_0\lambda^2}\left(\frac{\omega\mu_0\eta C_1 - \xi\gamma_1 C_2}{k_1^2 - \xi^2} - \frac{\xi}{\omega\hat{\varepsilon}_1} e^{i\gamma_1 d}\right), \tag{6.142}$$

where

$$M_T = \gamma_1 + \gamma_T, \qquad N_e = k_T^2\gamma_1 + k_1^2\gamma_e = \frac{k_T}{k_L}\left(k_L k_T\gamma_1 + k_1^2\gamma_L\right). \tag{6.143}$$

With Eqs. (6.139)–(6.142), and the Fourier integrals like Eq. (6.106), the six components of the electromagnetic field in Region 1 in the range $0 \le z \le d$ can be expressed as follows:

$$E_{1x} = \frac{I\,dl}{(2\pi)^2}\int_{-\infty}^{\infty} d\xi \int_{-\infty}^{\infty} d\eta e^{i(\xi x+\eta y)} \times \left(\{\lambda^2 N_e + \xi^2[\gamma_1(\gamma_T\gamma_e - k_T^2) - \lambda^2\gamma_e]\} - \frac{e^{i\gamma_1 d}}{i\omega\hat{\varepsilon}_1\gamma_1}\sin\gamma_1 z \right), \quad (6.144)$$

$$E_{1y} = \frac{I\,dl}{(2\pi)^2}\int_{-\infty}^{+\infty} d\xi \int_{-\infty}^{+\infty} d\eta \xi \eta e^{i(\xi x+\eta y)} \times \left[\frac{\omega\mu_0 e^{i\gamma_1(z+d)}}{\lambda^2 M_T N_e}(k_T^2\gamma_1 + \lambda^2\gamma_e - \gamma_1\gamma_T\gamma_e) + \frac{e^{i\gamma_1 d}}{i\omega\hat{\varepsilon}_1\gamma_1}\sin\gamma_1 z \right], \quad (6.145)$$

$$E_{1z} = \frac{I\,dl}{(2\pi)^2}\int_{-\infty}^{\infty} d\xi \int_{-\infty}^{\infty} d\eta e^{i(\xi x+\eta y)} \xi \left[\frac{\omega\mu_0\gamma_e e^{i\gamma_1(z+d)}}{N_e} - \frac{e^{i\gamma_1 d}}{\omega\hat{\varepsilon}_1}\cos\gamma_1 z \right], \quad (6.146)$$

$$B_{1x} = \frac{\mu_0 I\,dl}{(2\pi)^2}\int_{-\infty}^{\infty} d\xi \int_{-\infty}^{\infty} d\eta e^{i(\xi x+\eta y)} \xi\eta \frac{e^{i\gamma_1(z+d)}}{\lambda^2 M_T N_e}(k_1^2\gamma_T\gamma_e - k_T^2\gamma_1^2), \quad (6.147)$$

$$B_{1y} = -\frac{\mu_0 I\,dl}{(2\pi)^2}\int_{-\infty}^{\infty} d\xi \int_{-\infty}^{\infty} d\eta e^{i(\xi x+\eta y)} \times \left\{ \frac{e^{i\gamma_1(z+d)}}{\lambda^2 M_T N_e}[\lambda^2\gamma_1 N_e + \xi^2(k_1^2\gamma_T\gamma_e - k_T^2\gamma_1^2)] - e^{i\gamma_1 d}\cos\gamma_1 z \right\}, \quad (6.148)$$

$$B_{1z} = \frac{\mu_0 I\,dl}{(2\pi)^2}\int_{-\infty}^{\infty} d\xi \int_{-\infty}^{\infty} d\eta e^{i(\xi x+\eta y)} \eta \left[\frac{e^{i\gamma_1(z+d)}}{M_T} - \frac{i e^{i\gamma_1 d}}{\gamma_1}\sin\gamma_1 z \right]. \quad (6.149)$$

With the relations

$$x = \rho\cos\phi; \qquad y = \rho\sin\phi; \qquad \xi = \lambda\cos\phi'; \qquad \eta = \lambda\sin\phi', \quad (6.150)$$

and the integral representation of the Bessel functions of the first kind, viz.,

$$J_n(\lambda\rho) = \frac{i^{-n}}{2\pi}\int_0^{2\pi} e^{i\lambda\rho\cos\theta} e^{in\theta}\, d\theta, \quad (6.151)$$

the final formulas of the six components in cylindrical coordinates ρ, ϕ, z in Region 1 can be expressed as follows:

$$E_{1\rho}(\rho,\phi,z) = -\frac{\mu_0\omega I\,dl}{4\pi k_1^2}\cos\phi \times \left(\int_0^{\infty} \left\{ k_1^2 J_0(\lambda\rho) - \frac{\lambda^2}{2}[J_0(\lambda\rho) - J_2(\lambda\rho)] \right\} \frac{e^{i\gamma_1|z-d|}}{\gamma_1}\lambda\, d\lambda \right.$$

$$+\int_0^\infty \left\{\frac{\gamma_1 Q_e}{2}\big[J_0(\lambda\rho) - J_2(\lambda\rho)\big] - \frac{k_1^2 P_T}{2\gamma_1}\big[J_0(\lambda\rho) + J_2(\lambda\rho)\big]\right\}$$

$$\times\, \mathrm{e}^{\mathrm{i}\gamma_1(z+d)}\lambda\,\mathrm{d}\lambda\Bigg), \tag{6.152}$$

$$E_{1\phi}(\rho,\phi,z) = \frac{\mu_0\omega I\,\mathrm{d}l}{4\pi k_1^2}\sin\phi$$

$$\times\left(\int_0^\infty \left\{k_1^2 J_0(\lambda\rho) - \frac{\lambda^2}{2}\big[J_0(\lambda\rho) + J_2(\lambda\rho)\big]\right\}\frac{\mathrm{e}^{\mathrm{i}\gamma_1|z-d|}}{\gamma_1}\lambda\,\mathrm{d}\lambda\right.$$

$$+\int_0^\infty \left\{\frac{\gamma_1 Q_e}{2}\big[J_0(\lambda\rho) + J_2(\lambda\rho)\big] - \frac{k_1^2 P_T}{2\gamma_1}\big[J_0(\lambda\rho) - J_2(\lambda\rho)\big]\right\}$$

$$\times\, \mathrm{e}^{\mathrm{i}\gamma_1(z+d)}\lambda\,\mathrm{d}\lambda\Bigg), \tag{6.153}$$

$$E_{1z}(\rho,\phi,z) = \frac{\mathrm{i}\omega\mu_0 I\,\mathrm{d}l}{4\pi k_1^2}\cos\phi\int_0^\infty \big[\pm\mathrm{e}^{\mathrm{i}\gamma_1|z-d|} + Q_e\mathrm{e}^{\mathrm{i}\gamma_1(z+d)}\big]J_1(\lambda\rho)\lambda^2\,\mathrm{d}\lambda, \tag{6.154}$$

$$B_{1\rho}(\rho,\phi,z) = -\frac{\mu_0 I\,\mathrm{d}l}{4\pi}\sin\phi$$

$$\times\left(\pm\int_0^\infty J_0(\lambda\rho)\mathrm{e}^{\mathrm{i}\gamma_1|z-d|}\lambda\,\mathrm{d}\lambda\right.$$

$$+\int_0^\infty \left\{\frac{Q_e}{2}\big[J_0(\lambda\rho) + J_2(\lambda\rho)\big] - \frac{P_T}{2}\big[J_0(\lambda\rho) - J_2(\lambda\rho)\big]\right\}$$

$$\times\, \mathrm{e}^{\mathrm{i}\gamma_1(z+d)}\lambda\,\mathrm{d}\lambda\Bigg), \tag{6.155}$$

$$B_{1\phi}(\rho,\phi,z) = -\frac{\mu_0 I\,\mathrm{d}l}{4\pi}\cos\phi$$

$$\times\left(\pm\int_0^\infty J_0(\lambda\rho)\mathrm{e}^{\mathrm{i}\gamma_1|z-d|}\lambda\,\mathrm{d}\lambda\right.$$

$$+\int_0^\infty \left\{\frac{Q_e}{2}\big[J_0(\lambda\rho) - J_2(\lambda\rho)\big] - \frac{P_T}{2}\big[J_0(\lambda\rho) + J_2(\lambda\rho)\big]\right\}$$

$$\times\, \mathrm{e}^{\mathrm{i}\gamma_1(z+d)}\lambda\,\mathrm{d}\lambda\Bigg), \tag{6.156}$$

$$B_{1z}(\rho,\phi,z) = \frac{\mathrm{i}\mu_0 I\,\mathrm{d}l}{4\pi}\sin\phi\int_0^\infty \big[\mathrm{e}^{\mathrm{i}\gamma_1|z-d|} - P_T\mathrm{e}^{\mathrm{i}\gamma_1(z+d)}\big]\gamma_1^{-1}J_1(\lambda\rho)\lambda^2\,\mathrm{d}\lambda, \tag{6.157}$$

where

$$Q_e = \frac{k_1^2\gamma_e - k_T^2\gamma_1}{k_1^2\gamma_e + k_T^2\gamma_1} = \frac{k_1^2\gamma_L - k_L k_T\gamma_1}{k_1^2\gamma_L + k_L k_T\gamma_1}; \qquad P_T = \frac{\gamma_T - \gamma_1}{\gamma_T + \gamma_1}. \tag{6.158}$$

In Eqs. (6.152)–(6.157), the lower sign applies to the range $0 < z < d$, while the upper sign indicates the range $z > d$. By now, the integrated formulas for the six components of the electromagnetic field have been obtained readily.

6.4.2 Approximated Formulas for the Field Components

In practical computations in engineering, it is difficult to evaluate the integrals in Eqs. (6.152)–(6.157). In the next step, we will attempt to evaluate these integrals and obtain useful physical insights. With the available work by Pan (1985), the approximated formulas are obtained for the six components of the electromagnetic field in Region 1. They can be expressed as follows:

$$\begin{aligned} E_{1\rho} = & -\frac{I\,\mathrm{d}l\omega\mu_0}{2\pi k_1^2}\cos\phi \\ & \times \left(\mathrm{e}^{\mathrm{i}k_1(z+d)}\left\{\mathrm{e}^{\mathrm{i}k_L\rho}\left[\frac{\mathrm{i}k_L k_T}{\rho} - \frac{k_T}{\rho^2} - \frac{k_L^2 k_T^2}{k_1}\sqrt{\frac{\pi}{k_L\rho}}\mathrm{e}^{-\mathrm{i}k_L\rho\frac{k_T^2}{2k_1^2}}\mathbf{F}(p_e)\right]\right.\right. \\ & \left.\left. -\frac{\mathrm{i}}{\rho^3}\mathrm{e}^{\mathrm{i}k_T\rho}\right\} - \mathrm{e}^{\mathrm{i}k_1 r_1}\left(\frac{k_1}{\rho^2} + \frac{\mathrm{i}}{\rho^3}\right)\right), \end{aligned} \tag{6.159}$$

$$\begin{aligned} E_{1\phi} = & \frac{I\,\mathrm{d}l\omega\mu_0}{\pi k_1^2}\sin\phi \\ & \times \left(\mathrm{e}^{\mathrm{i}k_1(z+d)}\left\{\mathrm{e}^{\mathrm{i}k_L\rho}\left[\frac{k_T}{2\rho^2} + \frac{\mathrm{i}k_L^2 k_T^2}{2k_1}\frac{\sqrt{\pi}}{(k_L\rho)^{\frac{3}{2}}}\mathrm{e}^{-\mathrm{i}k_L\rho\frac{k_T^2}{2k_1^2}}\mathbf{F}(p_e)\right]\right.\right. \\ & \left. + \mathrm{e}^{\mathrm{i}k_T\rho}\left(\frac{k_T}{2\rho^2} + \frac{\mathrm{i}}{\rho^3}\right)\right\} + \mathrm{e}^{\mathrm{i}k_1 r_2}\left(\frac{\mathrm{i}k_1^2}{2\rho} - \frac{k_1}{\rho^2} - \frac{\mathrm{i}}{\rho^3}\right) \\ & \left. -\frac{1}{4}\left(\mathrm{e}^{\mathrm{i}k_1 r_1} + \mathrm{e}^{\mathrm{i}k_1 r_2}\right)\left(\frac{\mathrm{i}k_1^2}{\rho} - \frac{k_1}{\rho^2} - \frac{\mathrm{i}}{\rho^3}\right)\right), \end{aligned} \tag{6.160}$$

$$\begin{aligned} E_{1z} = & \frac{\mathrm{i}I\,\mathrm{d}l\omega\mu_0}{2\pi k_1^2}\cos\phi \\ & \times \left(\frac{k_L k_T}{k_1}\left\{\mathrm{e}^{\mathrm{i}k_1(z+d)}\mathrm{e}^{\mathrm{i}k_L\rho}\left[\frac{k_L}{\rho} + \frac{\mathrm{i}}{\rho^2} + \frac{\mathrm{i}k_T k_L^2}{k_1}\sqrt{\frac{\pi}{k_L\rho}}\mathrm{e}^{-\mathrm{i}k_L\rho\frac{k_L^2}{2k_1^2}}\mathbf{F}(p_e)\right]\right.\right. \end{aligned}$$

$$- \frac{e^{ik_1 r_2}}{\rho^2} \Bigg\} - \frac{1}{2}\left(\frac{z-d}{\rho} e^{ik_1 r_1} + \frac{z+d}{\rho} e^{ik_1 r_2}\right)$$
$$\times \left(\frac{k_1^2}{\rho} + \frac{3ik_1}{\rho^2} - \frac{3}{\rho^3}\right)\Bigg)\Bigg), \quad (6.161)$$

$$B_{1\rho} = -\frac{I\,dl\,\mu_0}{2\pi k_1} \sin\phi$$
$$\times \Bigg(e^{ik_1(z+d)} \Bigg\{ e^{ik_L\rho} \left[\frac{k_T}{\rho^2} + \frac{ik_L k_T^2}{k_1 \rho} \sqrt{\frac{\pi}{k_L \rho}} e^{-ik_L\rho \frac{k_L^2}{2k_1^2}} \mathbf{F}(p_e) \right]$$
$$+ e^{ik_T\rho} \left(\frac{k_T}{\rho^2} + \frac{2i}{\rho^3} \right) \Bigg\} + e^{ik_1 r_2} \left(\frac{ik_1^2}{\rho} - \frac{2k_1}{\rho^2} - \frac{2i}{\rho^3} \right)$$
$$- \frac{1}{2}\left(\frac{z-d}{\rho} e^{ik_1 r_1} + \frac{z+d}{\rho} e^{ik_1 r_2}\right)\left(\frac{ik_1^2}{\rho} + \frac{2i}{\rho^3} - \frac{3}{k_1\rho^4}\right)\Bigg)\Bigg), \quad (6.162)$$

$$B_{1\phi} = -\frac{I\,dl\,\mu_0}{2\pi k_1} \cos\phi$$
$$\times \Bigg(e^{ik_1(z+d)} \Bigg\{ e^{ik_L\rho} \left[\frac{ik_L k_T}{\rho} - \frac{k_T}{\rho^2} - \frac{k_L^2 k_T^2}{k_1} \sqrt{\frac{\pi}{k_L \rho}} e^{-ik_L\rho \frac{k_L^2}{2k_1^2}} \mathbf{F}(p_e) \right]$$
$$- \frac{i}{\rho^3} e^{ik_T\rho} \Bigg\} + \frac{1}{2} e^{ik_1 r_2} \left(\frac{2}{\rho^3} + \frac{3i}{k_1\rho^4} \right)$$
$$+ \frac{1}{2}\left(\frac{z-d}{\rho} e^{ik_1 r_1} + \frac{z+d}{\rho} e^{ik_1 r_2}\right)\left(\frac{ik_1^2}{\rho} - \frac{k_1}{\rho^2}\right)\Bigg)\Bigg), \quad (6.163)$$

$$B_{1z} = \frac{-iI\,dl}{2\pi k_1^2} \sin\phi$$
$$\times \Bigg[e^{ik_1(z+d)} e^{ik_T\rho} \left(\frac{ik_T^2}{\rho^2} - \frac{3k_T}{\rho^3} - \frac{3i}{\rho^4} \right) - e^{ik_1 r_2} \left(\frac{ik_1^2}{\rho^2} - \frac{3k_1}{\rho^3} - \frac{3i}{\rho^4} \right)$$
$$+ \frac{1}{2}(e^{ik_1 r_1} - e^{ik_1 r_2}) \left(\frac{k_1^3}{\rho} + \frac{ik_1^2}{\rho^2} \right) \Bigg]. \quad (6.164)$$

In these formulas, $r_1 = \sqrt{\rho^2 + (z-d)^2}$, $r_2 = \sqrt{\rho^2 + (z+d)^2}$, $p_e = k_L\rho(\frac{k_T^2}{2k_1^2})$, and the Fresnel integral $\mathbf{F}(p_e)$ are defined by

$$\mathbf{F}(p_e) = \frac{1}{2}(1+i) - C_2(p_e) - iS_2(p_e), \quad (6.165)$$

where

$$C_2(p_e) + iS_2(p_e) = \int_0^{p_e} \frac{e^{it}}{(2\pi t)^{1/2}}\, dt. \quad (6.166)$$

In Eqs. (6.159)–(6.164), the four factors $e^{ik_1 r}$, $e^{ik_1 r_2}$, $e^{ik_1(z+d)}e^{ik_L\rho}$, and $e^{ik_1(z+d)} \times e^{ik_T\rho}$ are included. The preceding two terms including the factor $e^{ik_1 r}$ and $e^{ik_2 r}$ represent the directed wave and the ideal reflected wave, which the two terms can be neglected in the case of $\rho \gg z$ and $\rho \gg d$. The latter two terms represent the lateral wave, which travels from the transmitting source a distance d in Region 1 (sea water) to the boundary, then radially along the boundary in Region 2 (rock) a distance ρ, and finally a distance z in Region 2 to the receiving point. Because of the anisotropy of the conductivity of the rock, the lateral wave can be divided into electric-type and magnetic-type terms, which propagate outwardly with the wave numbers k_L and k_T, respectively. The wave number k_L of the electric-type (TM) wave is determined primarily by the conductivity σ_L in longitudinal direction, while the wave number k_T of the magnetic-type (TE) wave is determined by the conductivity σ_T in transverse direction. When the conditions of $\rho \gg z$ and $\rho \gg d$ are satisfied, Eqs. (6.159)–(6.164) are simplified as

$$E_{1\rho}^L = -\frac{\omega\mu_0 k_T I\,\mathrm{d}l}{2\pi k_1^2}\cos\phi e^{ik_1(z+d)} \times \left\{e^{ik_L\rho}\left[\frac{ik_L}{\rho} - \frac{1}{\rho^2} - \frac{k_L^2 k_T}{k_1}\sqrt{\frac{\pi}{k_L\rho}}e^{-ik_L\rho\frac{k_L^2}{2k_1^2}}\mathbf{F}(p_e)\right] - \frac{i}{k_T\rho^3}e^{ik_T\rho}\right\}, \tag{6.167}$$

$$E_{1\phi}^L = \frac{\omega\mu_0 k_T I\,\mathrm{d}l}{2\pi k_1^2}\sin\phi e^{ik_1(z+d)} \times \left\{e^{ik_L\rho}\left[\frac{1}{\rho^2} + \frac{ik_L k_T}{k_1\rho}\sqrt{\frac{\pi}{k_L\rho}}e^{-ik_L\rho\frac{k_L^2}{2k_1^2}}\mathbf{F}(p_e)\right] + e^{ik_T\rho}\left(\frac{1}{\rho^2} + \frac{2i}{k_T\rho^3}\right)\right\}, \tag{6.168}$$

$$E_{1z}^L = \frac{\omega\mu_0 k_L k_T I\,\mathrm{d}l}{2\pi k_1^3}\cos\phi e^{ik_1(z+d)}\phi e^{ik_L\rho}\left(\frac{ik_L}{\rho} - \frac{1}{\rho^2} - \frac{k_L^2 k_T}{k_1}\sqrt{\frac{\pi}{k_L\rho}}e^{-ik_L\rho\frac{k_L^2}{2k_1^2}}\mathbf{F}(p_e)\right), \tag{6.169}$$

$$B_{1\rho}^L = -\frac{k_1}{\omega}E_{1\phi}^L, \tag{6.170}$$

$$B_{1\phi}^L = \frac{k_1}{\omega}E_{1\rho}^L, \tag{6.171}$$

$$B_{1z}^L = \frac{\mu_0 k_T^2 I\,\mathrm{d}l}{2\pi k_1^2}\sin\phi e^{ik_1(z+d)}e^{ik_T\rho}\left(\frac{1}{\rho^2} + \frac{3i}{k_T\rho^3} - \frac{3}{k_T^2\rho^4}\right). \tag{6.172}$$

In the near and intermediate distances with $|k_L\rho \cdot \frac{k_T^2}{2k_1^2}| < 1$, the Fresnel-integral terms have a negligible effect on the nature of the field. With them, the components of the field can be expressed in the following simplified forms:

$$E_{1\rho}^{L} \approx -\frac{\omega\mu_0 I\,\mathrm{d}l}{2\pi k_1^2}\cos\phi\,\mathrm{e}^{\mathrm{i}k_1(z+d)}\left[\left(\frac{\mathrm{i}k_L k_T}{\rho}-\frac{k_T}{\rho^2}\right)\mathrm{e}^{\mathrm{i}k_L\rho}-\frac{\mathrm{i}}{\rho^3}\mathrm{e}^{\mathrm{i}k_T\rho}\right], \quad (6.173)$$

$$E_{1\phi}^{L} \approx \frac{\omega\mu_0 I\,\mathrm{d}l}{\pi k_1^2}\sin\phi\,\mathrm{e}^{\mathrm{i}k_1(z+d)}\left[\mathrm{e}^{\mathrm{i}k_L\rho}\frac{k_T}{2\rho^2}+\mathrm{e}^{\mathrm{i}k_T\rho}\left(\frac{k_T}{2\rho^2}+\frac{\mathrm{i}}{\rho^3}\right)\right], \quad (6.174)$$

$$E_{1z}^{L} \approx \frac{i\omega\mu_0 I\,\mathrm{d}l}{2\pi k_1^2}\cos\phi\,\mathrm{e}^{\mathrm{i}k_1(z+d)}\mathrm{e}^{\mathrm{i}k_L\rho}\frac{k_L k_T}{k_1}\left(\frac{k_L}{\rho}+\frac{\mathrm{i}}{\rho^2}\right), \quad (6.175)$$

$$B_{1\rho}^{L} = -\frac{\mu_0 I\,\mathrm{d}l}{2\pi k_1}\sin\phi\,\mathrm{e}^{\mathrm{i}k_1(z+d)}\left[\mathrm{e}^{\mathrm{i}k_L\rho}\frac{k_T}{\rho^2}+\mathrm{e}^{\mathrm{i}k_T\rho}\left(\frac{k_T}{\rho^2}+\frac{2\mathrm{i}}{\rho^3}\right)\right], \quad (6.176)$$

$$B_{1\phi}^{L} = -\frac{\mu_0 I\,\mathrm{d}l}{2\pi k_1}\cos\phi\,\mathrm{e}^{\mathrm{i}k_1(z+d)}\left[\left(\frac{\mathrm{i}k_L k_T}{\rho}-\frac{k_T}{\rho^2}\right)\mathrm{e}^{\mathrm{i}k_L\rho}-\frac{\mathrm{i}}{\rho^3}\mathrm{e}^{\mathrm{i}k_T\rho}\right], \quad (6.177)$$

$$B_{1z}^{L} = -\frac{\mathrm{i}\mu_0 I\,\mathrm{d}l}{2\pi k_1^2}\sin\phi\,\mathrm{e}^{\mathrm{i}k_1(z+d)}\mathrm{e}^{\mathrm{i}k_T\rho}\left(\frac{\mathrm{i}k_T^2}{\rho^2}-\frac{3k_T}{\rho^3}-\frac{3\mathrm{i}}{\rho^4}\right). \quad (6.178)$$

6.4.3 Computations and Discussions

In order to verify the accuracy of the approximated formulas, for typical conditions, we will attempt to carry out the computations by using the approximated formulas and the directional numerical calculations on the integrated formula. With $f = 1$ Hz, $I\,dl = 1$ A·m, $\sigma_1 = 4$ S/m, $\sigma_T = 0.004$ S/m, $\sigma_L = 0.002$ S/m, $d = 30$ m, and $z = 1$ m, the graphs of $|E_{1\rho}|$ as a function of the radial distance ρ at $\phi = 0$ is shown in Fig. 6.23. The dash lines are for the computations by the approximated formula, the solid lines are for the numerical calculations on the integrated formula, and the dotted lines are for the computations of the lateral-wave term. The corresponding curves for $|E_{1\phi}|$ at $\phi = \frac{\pi}{2}$ are shown in Fig. 6.24.

From the above computations, it is shown that the results by using the approximated formula, the directional numerical calculation method, and the lateral-wave term are very close together, and the differences between the three computed results are within 0.7 dB for $\rho < 600$ m. For $\rho > 100$ m, the differences are enlarged over 5 dB.

6.4.4 Comparisons with Measurements

An interesting experiment for detecting the subbed conductivity by the measurements of the ELF electric field on the ocean floor radiated by a horizontal antenna also on the ocean floor has been reported (Young and Cox 1981). A diagram of the experimental set and the measured data are shown in Fig. 6.25. In the experiment,

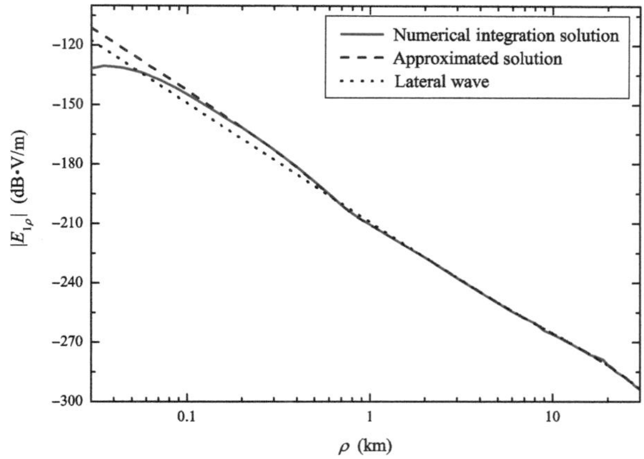

Fig. 6.23 The magnitudes of $E_{1\rho}(\rho, 0, z)$ due to an HED in sea water over one-dimensionally anisotropic rock

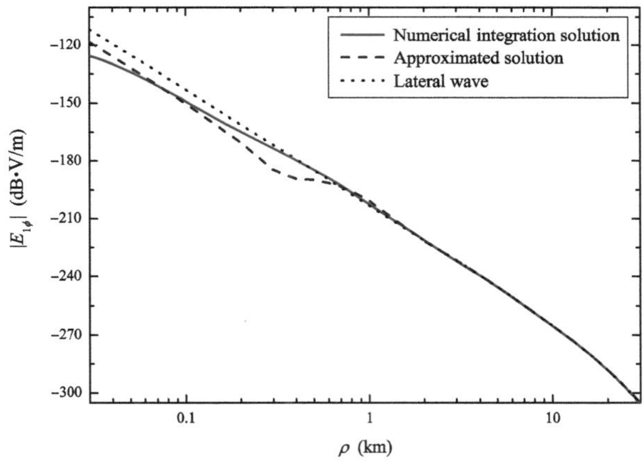

Fig. 6.24 The magnitudes of $E_{1\phi}(\rho, \frac{\pi}{2}, z)$ due to an HED in sea water over one-dimensionally anisotropic rock

the radial distance between the transmitting and receiving antennas is $\rho = 18.9$ km, and the operating frequency is in the range of 0.25–2.5 Hz. The actually measured components E_X and E_Y can be expressed in the terms of the components $E_{1\rho}(\rho, \phi, z)$ and $E_{1\phi}(\rho, \phi, z)$ of the field excited by the transmitter. We write

$$E_X = E_{1\rho}(\rho, \phi, z) \sin\psi + E_{1\phi}(\rho, \phi, z) \cos\psi, \tag{6.179}$$

$$E_Y = -E_{1\rho}(\rho, \phi, z) \cos\psi + E_{1\phi}(\rho, \phi, z) \sin\psi. \tag{6.180}$$

In the available reference (Young and Cox 1981), the angles ψ and ϕ are not given, but they can be estimated from the diagram in Fig. 6.25. Thus, we take $\psi = 5°$ and $\phi = 60°$. Since ψ is small, it is obtained for $E_X \sim E_{1\phi}(\rho, \phi, z)$ and $E_Y \sim -E_{1\rho}(\rho, \phi, z)$. The components E_X and E_Y as the functions of the operating frequency are computed and shown in Fig. 6.26. The solid and dot-dashed lines represent the one-dimensionally anisotropic rock, where the solid lines are for $\sigma_L < \sigma_T$

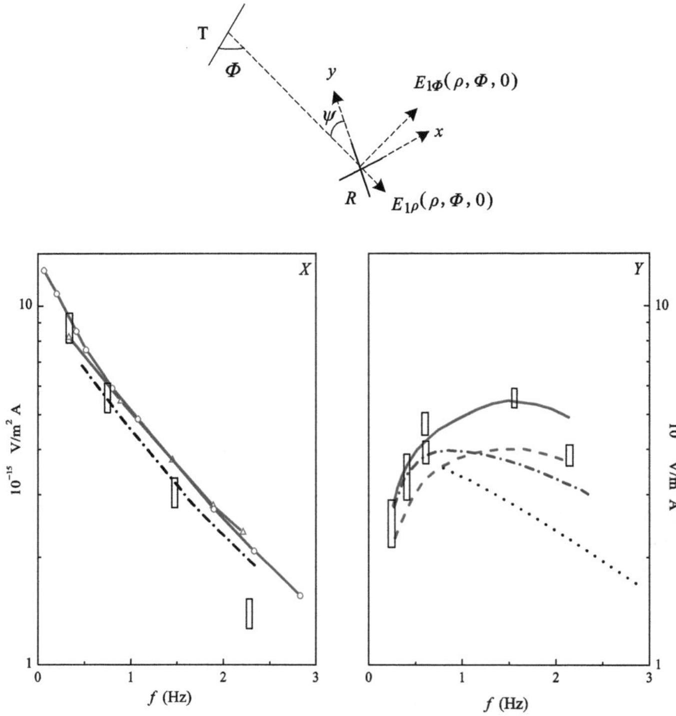

Fig. 6.25 Diagram of the experimental set and the measured data

and the dot-dashed lines for the physically unavailable but theoretical assumption $\sigma_L > \sigma_T$. The dashed and dotted lines represent isotropic rock, where the dashed lines are for the rock with lower conductivity and the dotted lines for the rock with high conductivity.

From Fig. 6.26, it is seen that for both E_X and E_Y only the solid lines are well in agreement with the measured data. For E_X, the solid and dot-dashed lines are nearly close together and both of them agree well with the measured data. It is found that E_X is not sensitive to the interchange of the two anisotropic models with $\sigma_L < \sigma_T$ and $\sigma_L > \sigma_T$. This is resulted by E_X being approximately equal to $E_{1\phi}$ in the case of $\psi \sim 0$, and meanwhile the magnitudes of the terms including $e^{ik_L\rho}$ are approximately equal to those of the terms including $e^{ik_T\rho}$ at small radial distance ($\rho = 18.9$ km).

Following the above computations and analyses, it may be reasonably concluded that the measured data are well represented as the one-dimensionally anisotropic model and the reasonable values are with $\sigma_L \sim 0.002$ to 0.0025 S/m and $\sigma_T \sim$ 0.004 to 0.005 S/m.

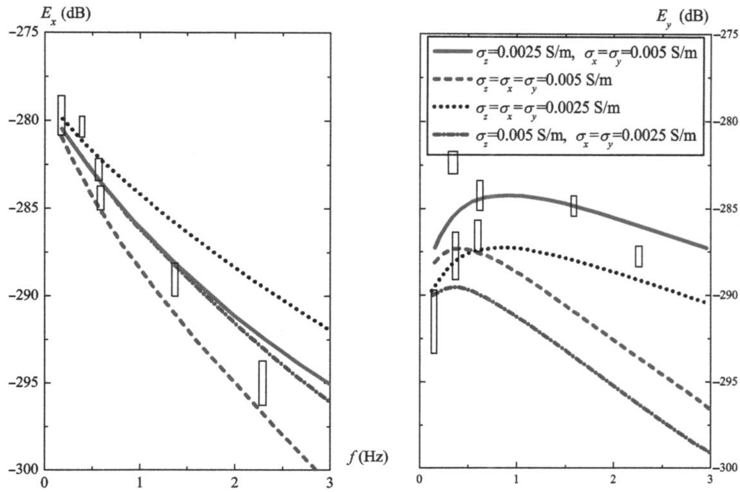

Fig. 6.26 The comparison of computations and ocean floor measurements by Young and Cox (1981)

6.5 The Electromagnetic Field Generated by HED for n-Layered Subbed

In the preceding sections, the analyses and computations are concerned with the electromagnetic field radiated by an HED on the boundary between the sea water and the sea floor. The actual oceanic crust is always composed of several layers with different conductivities. The multi-layered model is usually employed by geophysicists in the study of the ocean crust. Specifically, the measurements by Young and Cox (1981) described in Chaps. 8 and 9 in the book (King et al. 1992) are interpreted by using multi-layered model. Therefore, it is necessary to analyze the electromagnetic field excited by a horizontal transmitting antenna on the surface of multi-layered region.

In this section, the speeding numerical convergence algorithm will be used to compute the field components of the electromagnetic field radiated by an HED in the multi-layered region.

6.5.1 The Integrated Formulas for the Electromagnetic Field Generated by HED for n-Layered Subbed

The geometry under consideration and cylinder coordinate system are shown in Fig. 6.27, where an HED in the direction of $\phi = 0$ is located at (0, 0, d). Region 0 ($z > 0$) is the upper half-space occupied by sea water, the lower half-space is

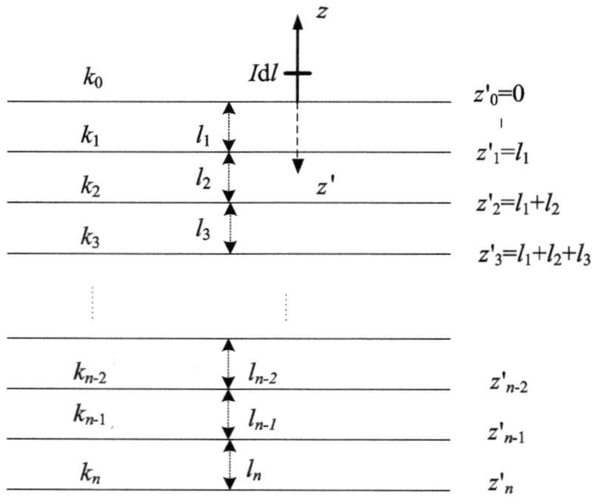

Fig. 6.27 Geometry of an HED in the sea water above the horizontally n-layered seabed

composed of a succession of n horizontally layers, each with arbitrary thickness l_j $(j = 1, 2, \ldots, n)$, and arbitrary wave number k_j.

In Chap. 11 of the book by King et al. (1992), the integrated formulas for the field component radiated by an HED in the sea water above the horizontally n-layered seabed were derived readily. They are

$$\begin{aligned} E_{0\rho} = &-\frac{\omega\mu_0 I\,\mathrm{d}l}{4\pi k_0^2}\cos\phi \\ &\times \Bigg(\int_0^\infty \left\{k_0^2 J_0(\lambda\rho) - \frac{\lambda^2}{2}\big[J_0(\lambda\rho) - J_2(\lambda\rho)\big]\right\}\gamma_0^{-1}\mathrm{e}^{\mathrm{i}\gamma_0|z-d|}\lambda\,\mathrm{d}\lambda \\ &+ \int_0^\infty \left\{\frac{\gamma_0 Q_n}{2}\big[J_0(\lambda\rho) - J_2(\lambda\rho)\big]\right. \\ &\left. - \frac{k_0^2 P_n}{2\gamma_0}\big[J_0(\lambda\rho) + J_2(\lambda\rho)\big]\right\}\mathrm{e}^{\mathrm{i}\gamma_0(z+d)}\lambda\,\mathrm{d}\lambda\Bigg), \end{aligned} \tag{6.181}$$

$$\begin{aligned} E_{0\phi} = &\frac{\omega\mu_0 I\,\mathrm{d}l}{4\pi k_0^2}\sin\phi \\ &\times \Bigg(\int_0^\infty \left\{k_0^2 J_0(\lambda\rho) - \frac{\lambda^2}{2}\big[J_0(\lambda\rho) + J_2(\lambda\rho)\big]\right\}\gamma_0^{-1}\mathrm{e}^{\mathrm{i}\gamma_0|z-d|}\lambda\,\mathrm{d}\lambda \\ &+ \int_0^\infty \left\{\frac{\gamma_0 Q_n}{2}\big[J_0(\lambda\rho) + J_2(\lambda\rho)\big]\right. \\ &\left. - \frac{k_0^2 P_n}{2\gamma_0}\big[J_0(\lambda\rho) - J_2(\lambda\rho)\big]\right\}\mathrm{e}^{\mathrm{i}\gamma_0(z+d)}\lambda\,\mathrm{d}\lambda\Bigg), \end{aligned} \tag{6.182}$$

$$E_{0z} = \frac{\mathrm{i}\omega\mu_0 I\,\mathrm{d}l}{4\pi k_0^2}\cos\phi\left[\int_0^\infty \left(\pm\mathrm{e}^{\mathrm{i}\gamma_0|z-d|} + Q_n\mathrm{e}^{\mathrm{i}\gamma_0(z+d)}\right)J_1(\lambda\rho)\lambda^2\,\mathrm{d}\lambda\right], \tag{6.183}$$

$$
\begin{aligned}
B_{0\rho} = & -\frac{\mu_0 I\,\mathrm{d}l}{4\pi}\sin\phi \\
& \times\left(\pm\int_0^\infty J_0(\lambda\rho)\mathrm{e}^{\mathrm{i}\gamma_0|z-d|}\lambda\,\mathrm{d}\lambda + \int_0^\infty\left\{\frac{Q_n}{2}\left[J_0(\lambda\rho)+J_2(\lambda\rho)\right]\right.\right. \\
& \left.\left. -\frac{P_n}{2}\left[J_0(\lambda\rho)-J_2(\lambda\rho)\right]\right\}\mathrm{e}^{\mathrm{i}\gamma_0(z+d)}\lambda\,\mathrm{d}\lambda\right),
\end{aligned}
\tag{6.184}
$$

$$
\begin{aligned}
B_{0\phi} = & -\frac{\mu_0 I\,\mathrm{d}l}{4\pi}\cos\phi \\
& \times\left(\pm\int_0^\infty J_0(\lambda\rho)\mathrm{e}^{\mathrm{i}\gamma_0|z-d|}\lambda\,\mathrm{d}\lambda + \int_0^\infty\left\{\frac{Q_n}{2}\left[J_0(\lambda\rho)-J_2(\lambda\rho)\right]\right.\right. \\
& \left.\left. -\frac{P_n}{2}\left[J_0(\lambda\rho)+J_2(\lambda\rho)\right]\right\}\mathrm{e}^{\mathrm{i}\gamma_0(z+d)}\lambda\,\mathrm{d}\lambda\right),
\end{aligned}
\tag{6.185}
$$

$$
B_{0z} = \frac{\mathrm{i}\mu_0 I\,\mathrm{d}l}{4\pi}\sin\phi\left[\int_0^\infty\left(\mathrm{e}^{\mathrm{i}\gamma_0|z-d|}-P_n\mathrm{e}^{\mathrm{i}\gamma_0(z+d)}\right)\gamma_0^{-1}J_1(\lambda\rho)\lambda^2\,\mathrm{d}\lambda\right],
\tag{6.186}
$$

where

$$
Q_n = -\frac{\gamma_0-\left(\frac{k_0^2}{\omega\mu_0}\right)Z_{s0}(0)}{\gamma_0+\left(\frac{k_0^2}{\omega\mu_0}\right)Z_{s0}(0)};\qquad P_n = -\frac{\gamma_0-\omega\mu_0 Y_{s0}(0)}{\gamma_0+\omega\mu_0 Y_{s0}(0)}.
\tag{6.187}
$$

In the above equations, $Z_{s0}(0)$ and $Y_{s0}(0)$ represent the surface impedance for the electric-type (TM) wave and the surface admittance for the magnetic-type (TE) wave, respectively. They are expressed in the following forms:

$$
\begin{aligned}
Z_{s0}(0) = & \frac{\omega\mu_0\gamma_0}{k_0^2}\tanh\left\{\tanh^{-1}\left[\left(\frac{\gamma_1 k_0^2}{\gamma_0 k_1^2}\right)\tanh\left[-\mathrm{i}\gamma_1 l_1\right.\right.\right. \\
& +\tanh^{-1}\left[\left(\frac{\gamma_2 k_1^2}{\gamma_1 k_2^2}\right)\tanh\left[-\mathrm{i}\gamma_2 l_2\right.\right. \\
& +\tanh^{-1}\left[\left(\frac{\gamma_3 k_2^2}{\gamma_2 k_3^2}\right)\tanh\left[-\mathrm{i}\gamma_3 l_3+\cdots\right.\right. \\
& +\tanh^{-1}\left[\left(\frac{\gamma_n k_{n-1}^2}{\gamma_{n-1} k_n^2}\right)\tanh\left[-\mathrm{i}\gamma_n l_n\right.\right. \\
& \left.\left.\left.\left.+\tanh^{-1}\left(\frac{\gamma_{n+1} k_n^2}{\gamma_n k_{n+1}^2}\right)\right]\right]\cdots\right]\right]\right]\right]\right]\right\},
\end{aligned}
\tag{6.188}
$$

$$
\begin{aligned}
Y_{s0}(0) = & \frac{\gamma_0}{\omega\mu_0}\tanh\left\{\tanh^{-1}\left[\left(\frac{\gamma_1}{\gamma_0}\right)\tanh\left[-\mathrm{i}\gamma_1 l_1\right.\right.\right. \\
& +\tanh^{-1}\left[\left(\frac{\gamma_2}{\gamma_1}\right)\tanh\left[-\mathrm{i}\gamma_2 l_2\right.\right.
\end{aligned}
$$

$$
\begin{aligned}
&+\tanh^{-1}\Bigg[\left(\frac{\gamma_3}{\gamma_2}\right)\tanh\Bigg[-\mathrm{i}\gamma_3 l_3+\cdots\\
&+\tanh^{-1}\Bigg[\left(\frac{\gamma_n}{\gamma_{n-1}}\right)\tanh\Bigg[-\mathrm{i}\gamma_n l_n\\
&+\tanh^{-1}\left(\frac{\gamma_{n+1}}{\gamma_n}\right)\Bigg]\Bigg]\cdots\Bigg]\Bigg]\Bigg]\Bigg]\Bigg]\Bigg]\Bigg\}.
\end{aligned}
\tag{6.189}
$$

When $l_1 = l_2 = \cdots = l_n = 0$, the space $z \le 0$ is uniform half-space with the parameters σ_{n+1} and k_{n+1}. For $n = 0$, they are reduced to

$$
Z_{s0}(0) = \frac{\omega\mu_0\gamma_{n+1}}{k_{n+1}^2} = \frac{\omega\mu_0\gamma_1}{k_1^2},
\tag{6.190}
$$

$$
Y_{s0}(0) = \frac{\gamma_{n+1}}{\omega\mu_0} = \frac{\gamma_1}{\omega\mu_0}.
\tag{6.191}
$$

Correspondingly, the factors Q_n and P_n are reduced to

$$
Q_n \underset{n\to 0}{\longrightarrow} \frac{k_0^2\gamma_{n+1} - k_{n+1}^2\gamma_0}{k_0^2\gamma_{n+1} + k_{n+1}^2\gamma_0} = \frac{k_0^2\gamma_1 - k_1^2\gamma_0}{k_0^2\gamma_1 + k_1^2\gamma_0},
\tag{6.192}
$$

$$
P_n \underset{n\to 0}{\longrightarrow} \frac{\gamma_{n+1} - \gamma_0}{\gamma_{n+1}\gamma_0} = \frac{\gamma_1 - \gamma_0}{\gamma_1\gamma_0}.
\tag{6.193}
$$

Obviously, when $n = 0$, all formulas are the same as those for the uniform half-space.

In another special case of $l_1 \to \infty$, namely, the thickness of the first layer is very large. Thus the remaining layers below have negligible effects on the nature of the field in the sea water, and this case should be like the uniform half-spaces. When $l_1 \to \infty$, it follows that

$$
\tanh(-\mathrm{i}\gamma_1 l_1 + \cdots) \underset{l_1\to\infty}{\longrightarrow} 1.
\tag{6.194}
$$

With the substitution of Eq. (6.94) into Eqs. (6.188) and (6.189), it yields

$$
Z_{s0}(0) \underset{l_1\to\infty}{\longrightarrow} \frac{\omega\mu_0\gamma_1}{k_1^2}; \qquad Y_{s0}(0) \underset{l_1\to\infty}{\longrightarrow} = \frac{\gamma_1}{\omega\mu_0}.
\tag{6.195}
$$

Obviously, the factors Q_n and P_n are reduced to Eqs. (6.192) and (6.193).

6.5.2 Speeding Numerical Convergence Algorithm

When $n \ge 3$, there have no available approximated formulas in engineering computations so far. By using the approximated formulas for the uniform half-space model

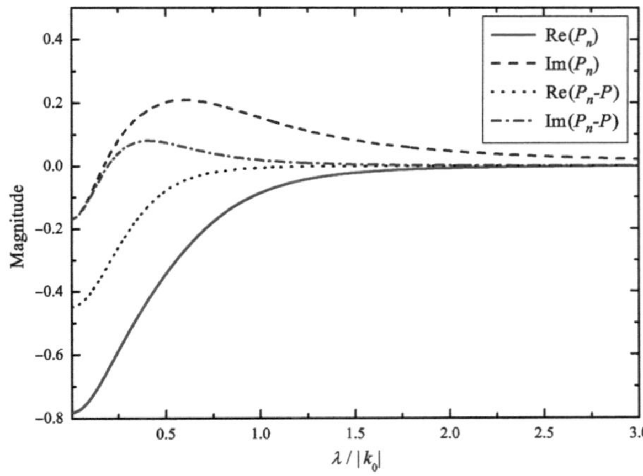

Fig. 6.28 The magnitudes of the factors P_n and $(P_n - P)$ as the functions of $\frac{\lambda}{|k_0|}$

and those for the three-layered model, which are available, the speed-up numerical convergence algorithm can be used to calculate numerically the integrated formulas for the n-layered model. In what follows, as an example for the electric field component $E_{0\rho}$, we will attempt to address the main ideas for the speeding numerical convergence algorithm.

We assume that $E^*_{0\rho}(\rho, \phi, z)$ is the electric field component in the $\hat{\rho}$ direction radiated by an HED in the sea water on or near the boundary between the sea water and the homogeneous ocean floor. From Sect. 6.2 in this chapter, the component $E^*_{0\rho}(\rho, \phi, z)$ is expressed as follows:

$$\begin{aligned} E^*_{0\rho}(\rho, \phi, z) = & -\frac{\omega\mu_0 I\,\mathrm{d}l}{4\pi k_0^2}\cos\phi \\ & \times \left(\int_0^\infty \left\{ k_0^2 J_0(\lambda\rho) - \frac{\lambda^2}{2}\left[J_0(\lambda\rho) - J_2(\lambda\rho)\right] \right\} \gamma_0^{-1} \mathrm{e}^{\mathrm{i}\gamma_0|z-d|} \lambda\,\mathrm{d}\lambda \right. \\ & + \int_0^\infty \left\{ \frac{\gamma_0 Q}{2}\left[J_0(\lambda\rho) - J_2(\lambda\rho)\right] \right. \\ & \left. \left. - \frac{k_0^2 P}{2\gamma_0}\left[J_0(\lambda\rho) + J_2(\lambda\rho)\right] \right\} \mathrm{e}^{\mathrm{i}\gamma_0(z+d)} \lambda\,\mathrm{d}\lambda \right). \end{aligned} \tag{6.196}$$

Here Q and P represent the reflection coefficient of the electric-type (TM) wave and that of the magnetic-type (TE) wave for the half-space model, respectively. It is noted that $E^*_{0\rho}(\rho, \phi, z)$ is expressed in the approximated analytical form, which can be computed accurately and easily (King et al. 1992).

With Eqs. (6.181) and (6.196), the integrated formulas for the component $E_{0\rho}$ of the electromagnetic field radiated by an HED in the sea water above the horizontally n-layered seabed can be rewritten in the following form:

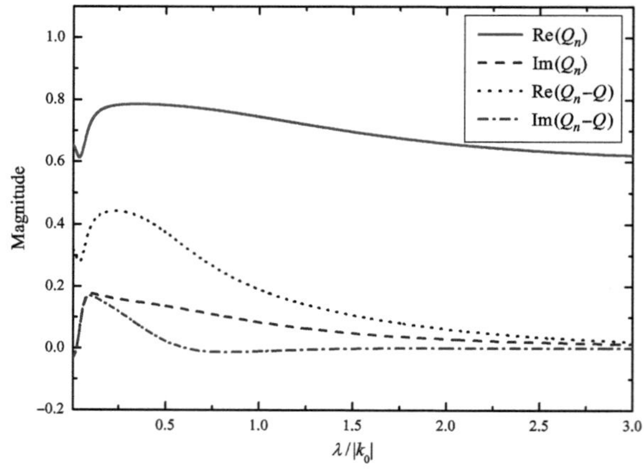

Fig. 6.29 The magnitudes of the factors Q_n and $(Q_n - Q)$ as the functions of $\frac{\lambda}{|k_0|}$

$$\begin{aligned}
E_{0\rho}(\rho,\phi,z) &= \left[E_{0\rho}(\rho,\phi,z) - E^*_{0\rho}(\rho,\phi,z)\right] + E^*_{0\rho}(\rho,\phi,z) \\
&= -\frac{\omega\mu_0 I\,\mathrm{d}l}{4\pi k_0^2}\cos\phi \\
&\quad\times \int_0^\infty \left\{ \frac{\gamma_0(Q_n - Q)}{2}\left[J_0(\lambda\rho) - J_2(\lambda\rho)\right] \right. \\
&\quad \left. - \frac{k_0^2(P_n - P)}{2\gamma_0}\left[J_0(\lambda\rho) + J_2(\lambda\rho)\right] \right\} \mathrm{e}^{\mathrm{i}\gamma_0(z+d)}\lambda\,\mathrm{d}\lambda \\
&\quad + E^*_{0\rho}(\rho,\phi,z).
\end{aligned} \tag{6.197}$$

It is seen that the factors $(Q_n - Q)$ and $(P_n - P)$ converge much more rapidly than the factors Q_n and P_n in the case of $\lambda \to \infty$ or λ being large enough. As a result, the field component $E_{0\rho}$ in Eq. (6.197) can be calculated efficiently with better accuracy. Similarly, the other five components can also be calculated efficiently with better accuracy by using the speed-up numerical convergence algorithm.

We take that the four-layered model and the operating frequency is $f = 1$ Hz. The upper space is occupied by sea water with $\sigma_0 = 4$ S/m, and both the transmitting antenna and the receiving antenna are located in Region 0 (sea water) with $d = 30$ m and $z = 1$ m. The lower space is composed of three layers with $\sigma_1 = 1$ S/m, $\sigma_2 = 0.01$ S/m, $\sigma_3 = 0.001$ S/m, $l_1 = 100$ m, and $l_2 = 1{,}000$ m. The half-space model is characterized by $\sigma_0 = 4$ S/m and $\sigma_1 = 1$ S/m. The magnitudes and phases of the factors Q_n, P_n, $(Q_n - Q)$, and $(P_n - P)$ as functions of $\frac{\lambda}{|k_0|}$ are shown in Figs. 6.28 and 6.29, respectively. It is seen that the factors $(Q_n - Q)$ and $(P_n - P)$ attenuate rapidly. As a result, the calculated efficiency is improved significantly with better accuracy.

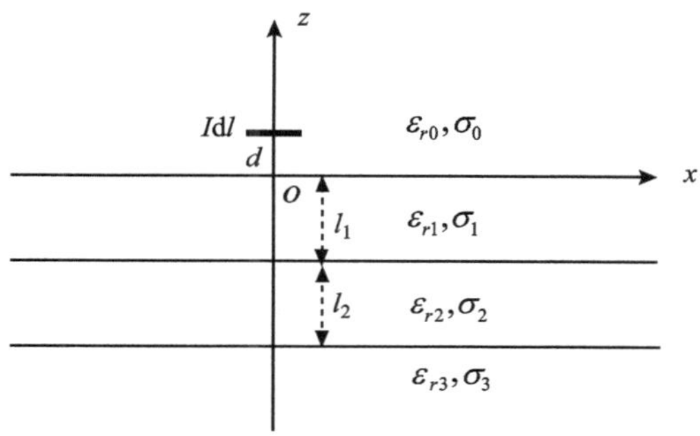

Fig. 6.30 Geometry of an HED in the four-layered region

6.5.3 Computations and Discussions on the Normalized Surface Impedance

The fundamental principle of the traditional passive electromagnetic sounding method (MT) can be summarized as follows: By receiving the electric component and magnetic component in ELF range meanwhile generated by the natural atmospheric noise and solar electromagnetic pulse on the Earth's surface, the surface impedance, which is defined as the ratio of the magnitude of the electric component and that of the magnetic component, as a function of the operating frequency can be concluded readily, and the conductivity distribution of the layered structure under the receiving point can be concluded.

Because the natural ELF radiation source is generally far away from the receiving point, the normalized surface impedance as a function of the operating frequency is only concerned with the conductivity structure under the receiving point, and has nothing to do with the location and intensity of the radiation source. When an artificial transmitting antenna is employed as the radiation source, the signal-to-noise ratio at the receiving point is improved significantly, while the measured time can be reduced greatly with better accuracy. On the other hand, the distance between the transmitting source and the receiving point is not too large. It is well known that the normalized surface impedance at the receiving point is not only concerned with the formation of the conductivity structure, but also is to do with the distance between the transmitting source and the receiving point. Thus, they should be considered in signal process and analyses.

In this section, it is estimated quantitatively how far the offset distance is between the transmitting source and the receiving point, the normalized surface impedance is not relevant to the offset distance.

Consider the four-layered region as shown in Fig. 6.30, where the HED is located at $(0, 0, d)$ in the $\hat{x}$. Region 0 ($z \geq 0$) is the upper space occupied by the sea water with the permeability μ_0, relative permittivity ε_{r0}, and conductivity σ_0; Region 1 ($-l_1 \leq z \leq 0$) is the sedimentary layer with the permeability μ_0, relative permittivity ε_{r1}, and conductivity σ_1; Region 2 ($-(l_1 + l_2) \leq z \leq -l_1$) is the high-impedance layer with the permeability μ_0, relative permittivity ε_{r2}, and conductivity σ_2; and Region 3 is the rest of the space ($z \leq z - (l_1 + l_2)$) occupied by the rock with the permeability μ_0, relative permittivity ε_{r3}, and conductivity σ_3.

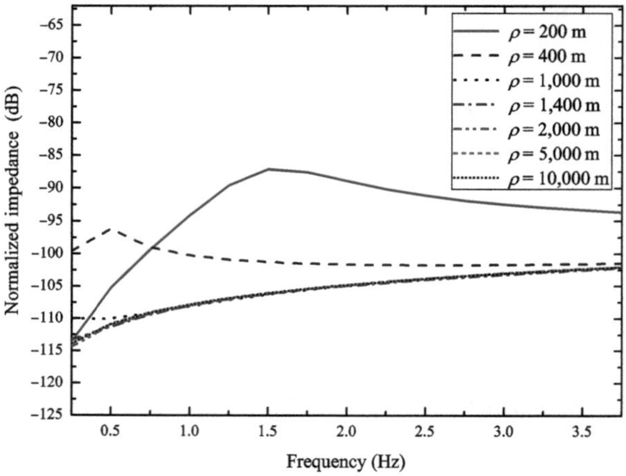

Fig. 6.31 The normalized surface impedance as a function of the operating frequency at different offset distances: $l_1 = 200$ m and $l_2 = 100$ m

Under the four-layered model, the integrated formulas can be given in Eqs. (6.181)–(6.186) by taking $n = 4$. With $\varepsilon_{r0} = 80$, $\varepsilon_{r1} = \varepsilon_{r2} = \varepsilon_{r3} = 10$, $\sigma_0 = 3.3$ S/m, $\sigma_0 = 1$ S/m, $\sigma_2 = 0.01$ S/m, $\sigma_3 = 1$ S/m, $d = 10$ m, and $z = 0$ m, at different depths and thicknesses of the high-impedance layer, the normalized surface impedance $\Delta = \eta \frac{E_\rho}{H_\phi}$ as the function of the operating frequency is computed and shown in Figs. 6.31, 6.32, 6.33, 6.34, 6.35, respectively.

From these curves in Figs. 6.31–6.34, it is seen that the normalized surface impedance as a function of the operating frequency is sensitive to the offset distance at small offset distance with $\rho < 1$ km. When the offset distance is larger than 1 km, the normalized impedance, for which the change is only within 1–2 dB, is not sensitive to the offset distance, while it changes significantly with the operating frequency. Considering ELF range, the efficiency of the magnetic field sensor is very low, and the sensitivity of the magnetic field sensor is much smaller than that of the electric field sensor. Therefore, in practical applications of the mCSEM method in geophysical prospecting and diagnostics, the electric component is only measured in the far distance, while both electric and magnetic components must be measured meanwhile in the near region.

6.6 The mCSEM Method and Its Applications

Since the early 1980s, the mCSEM method has been investigated widely because of its useful applications in geophysical prospecting and diagnostics. The fundamental principle of the mCSEM method is summarized as follows: Because the conductivity of sea water is much higher than that of the ocean floor, the attenuation of the electromagnetic wave in sea water is much larger than that in the ocean floor. If both ELF transmitter and receiver are placed on or near the surface of the ocean floor with a certain distance, it is well known that the lateral electromagnetic waves

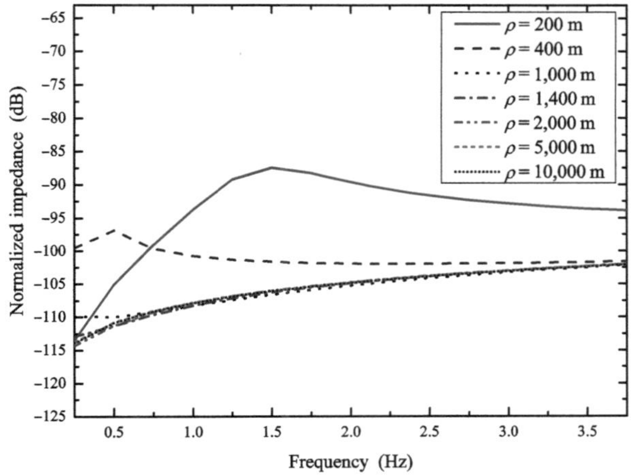

Fig. 6.32 The normalized surface impedance as a function of the operating frequency at different offset distances: $l_1 = 2{,}000$ m and $l_2 = 100$ m

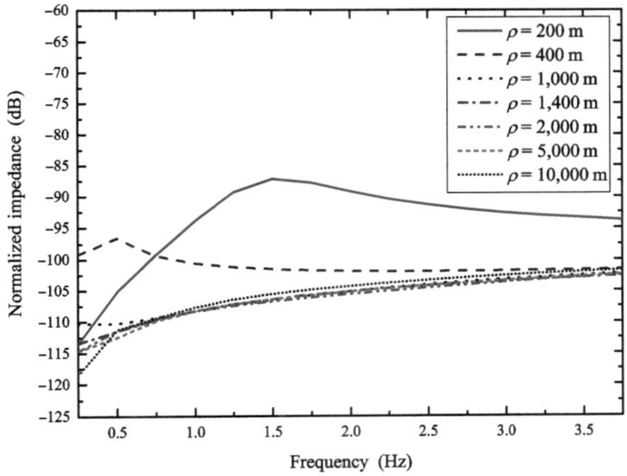

Fig. 6.33 The normalized surface impedance as a function of the operating frequency at different offset distances: $l_1 = 200$ m and $l_2 = 10$ m

can propagate along the boundary in the ocean floor. Based on the magnitudes and phases of ELF wave in the receiving point, the conductivity distribution of the sub-bottom structure can be concluded and inversed. Generally, the transmitting antenna is utilized a horizontal dual-electrode antenna, which is idealized as an HED.

In 2010, the analyses and computations were carried out for the changes of the measured anomaly of the electric field as the operating frequency, the offset distance, the target depth, the target thickness, and the conductivity (Lu et al. 2010). Then, the detecting ability of the mCSEM method can be evaluated. In this section, we will summarize the recent results by Pan's research group.

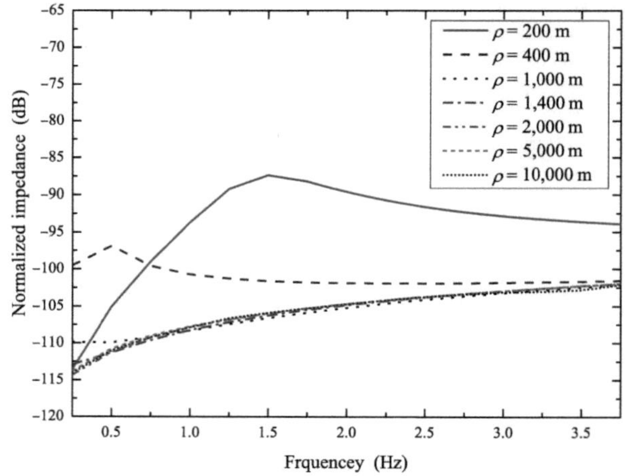

Fig. 6.34 The normalized surface impedance as a function of the operating frequency at different offset distances: $l_1 = 2{,}000$ m and $l_2 = 10$ m

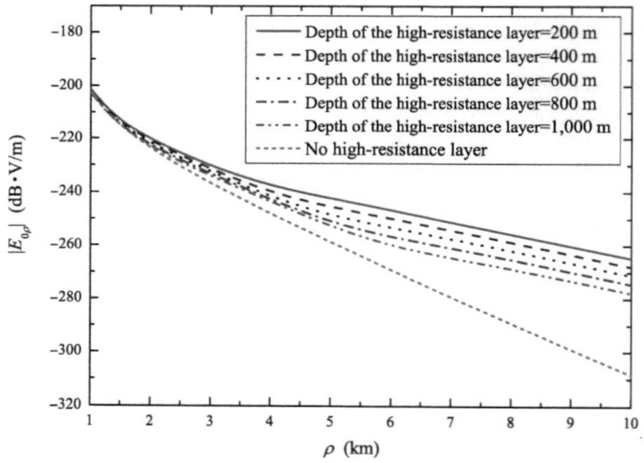

Fig. 6.35 The electric component $E_{0\rho}(\rho, 0, 0)$ as the function of the propagation distance ρ at different depths of the high-impedance layer: $f = 0.25$ Hz, $l_2 = 100$ m

6.6.1 The Electric Field Relative Anomaly Versus the Target's Depth

It is well known that the conductivity of the high-impedance layer is generally much lower than that of the surrounding rock. In the following computations, the four-layered model and the speed-up numerical convergence algorithm are employed. With $\varepsilon_{r0} = 80$, $\varepsilon_{r1} = \varepsilon_{r2} = \varepsilon_{r3} = 10$, $\sigma_0 = 3.3$ S/m, $\sigma_0 = 1$ S/m, $\sigma_2 = 0.01$ S/m, $\sigma_3 = 1$ S/m, $d = 10$ m, $z = 0$ m, and $l_2 = 100$ m, the electric component $E_{0\rho}(\rho, 0, 0)$ as a function of the propagation distance ρ is computed at $f = 0.25$ Hz and 0.5 Hz as shown in Figs. 6.35 and 6.36, respectively.

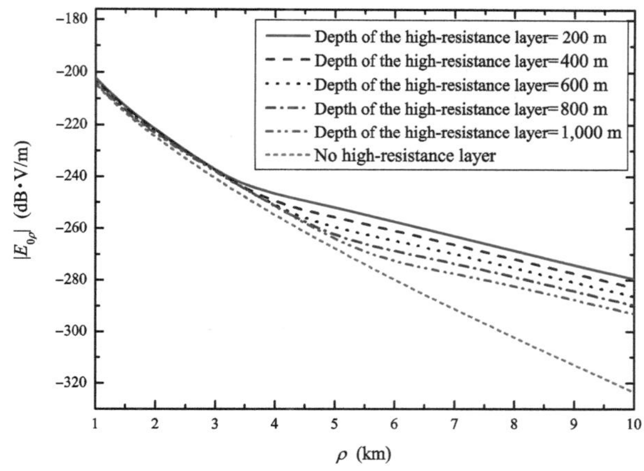

Fig. 6.36 The electric component $E_{0\rho}(\rho, 0, 0)$ as the function of the propagation distance ρ at different depths of the high-impedance layer: $f = 0.5$ Hz, $l_2 = 100$ m

In order to illustrate the effect of the measurements on the undersea electric field by the target layer, the relative anomaly of the electric field $\Delta E_{0\rho}$ is defined by

$$\Delta E_{0\rho} = 20\lg\left|\frac{E_{0\rho}(\rho, 0, 0)}{E_{0\rho}^{*}(\rho, 0, 0)}\right|. \tag{6.198}$$

It is noted that the electric component $E_{0\rho}^{*}(\rho, 0, 0)$ is for the case when the target layer does not exist or the ocean floor is homogeneous with the conductivity σ_1. Corresponding to Figs. 6.35 and 6.36, the relative anomaly of the electric field as a function of the offset distance at $f = 0.25$ Hz and 0.5 Hz are shown in Figs. 6.37 and 6.38, respectively.

From these computations, we conclude as follows:

- With the increase of the offset distance, the relative anomaly of the electric field is enlarged. Namely, with the same error of the measured equipment, the increase of the measured offset distance will follow that the "anomaly" is found more effectively.
- The deeper the target layer is, the smaller the relative anomaly of the electric field is under same conditions. Namely, in order to find the deep anomaly, the measured offset distance should be enlarged with the same error of the measured equipment. The principle is consistent with the analyses on the Poynting vector trajectory addressed in Sect. 6.2.3. It is indicated that the larger the distance between the transmitting and receiving points is, more likely the Poynting vector trajectory will penetrate into the deeper seabed, and the larger the maximum of the detecting depth.
- With the increases of the measured offset distance, the field amplitude decreases. Due to the limitations of the capacities of the transmitting and receiving systems, the measured offset distance will be limited by the equipment capacity, and the detecting depth will also be limited. With the mCSEM detection equipment used currently, the maximum of the detection distance is about 10 km, and the probing depth is about 3–4 km.

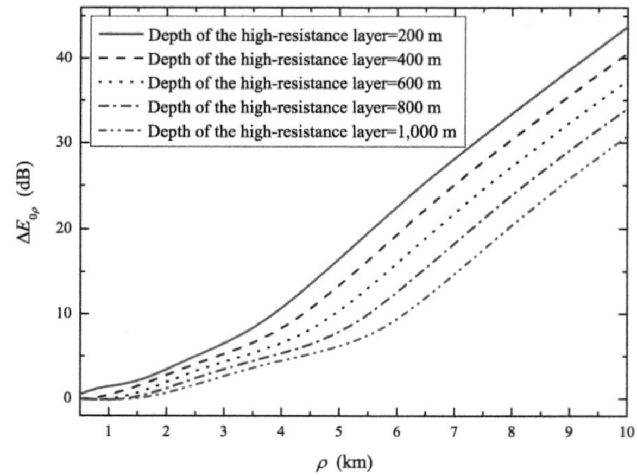

Fig. 6.37 The relative anomaly of the electric field $\Delta E_{0\rho}$ as a function of the propagation distance ρ at different depths of the high-impedance layer: $f = 0.25$ Hz, $l_2 = 100$ m

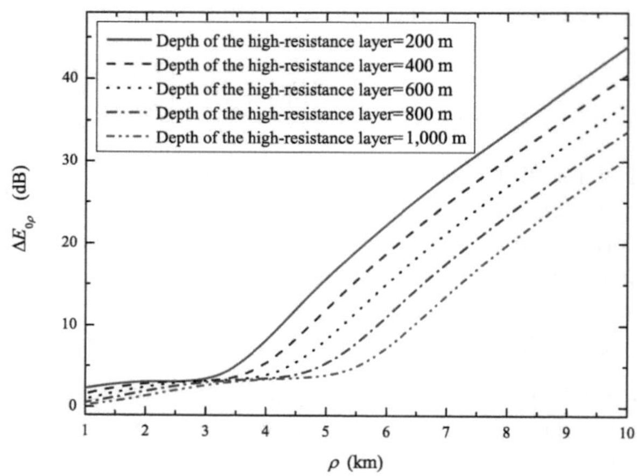

Fig. 6.38 The relative anomaly of the electric field $\Delta E_{0\rho}$ as a function of the propagation distance ρ at different depths of the high-impedance layer: $f = 0.5$ Hz, $l_2 = 100$ m

6.6.2 The Relative Anomaly of Electric Field Versus Target Thickness, Conductivity, and Operating Frequency

The curves in Figs. 6.39 and 6.40 are for the relative anomaly of the electric field versus the propagation distance ρ at different thicknesses of the high-impedance layer. All parameters are taken as those in Sect. 6.6.1. It is seen that at the same buried depth, the larger the thickness of the high-impedance layer, the more obvious is the relative anomaly of the electric field, and the more easily the target can be detected.

In Figs. 6.41 and 6.42, the curves are for the relative anomaly of the horizontal electric field as a function of the offset distance at different conductivities of the target layer (higher-impedance layer). It is seen that the greater the difference between

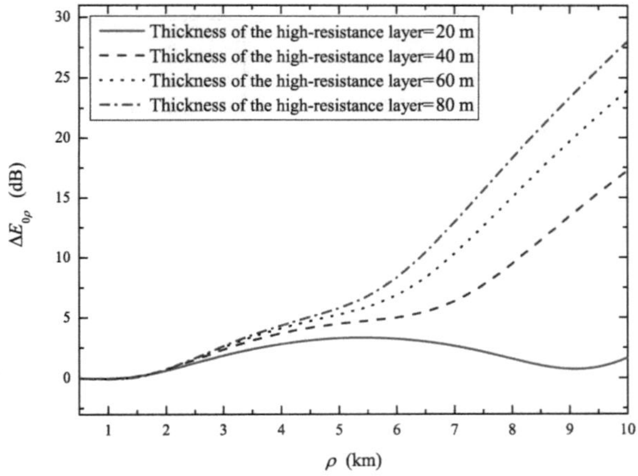

Fig. 6.39 The relative anomaly of the electric field $\Delta E_{0\rho}$ versus the propagation distance ρ at different thicknesses of the high-impedance layer: $f = 0.25$ Hz, $l_1 = 1{,}000$ m, $\sigma_2 = 0.01$ S/m

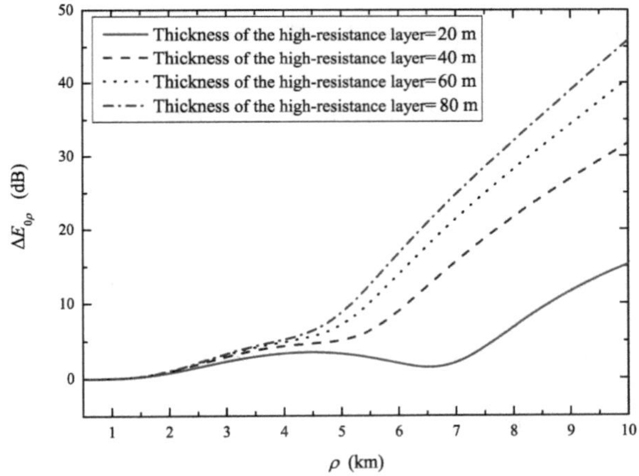

Fig. 6.40 The relative anomaly of the electric field $\Delta E_{0\rho}$ versus the propagation distance ρ at different thicknesses of the high-impedance layer: $f = 0.5$ Hz, $l_1 = 1{,}000$ m, $\sigma_2 = 0.01$ S/m

the conductivity of the more easily the target layer and that of the sedimentary layer, the greater the difference of the field strength is, and the more easily the target layer may be distinguished.

We assume that the depth and thickness of the target layer are taken as 1000 m and 100 m, respectively. The electric field $E_{0\rho}$ as a function of the offset distance in the sea water is computed at different operating frequencies and shown in Fig. 6.43. With the same parameters as in Fig. 6.43, the curves in Fig. 6.44 are for the relative anomaly of the electric field $\Delta E_{0\rho}$ as a function of the offset distance at different operating frequencies.

From Figs. 6.43 and 6.44, it is concluded as follows: With the decreasing operating frequency, the received field strength will increase, and the signal will be received more easily. The lower the operating frequency, the more significantly the relative anomaly of the electric field will reduce, and the detection resolution will decrease. Therefore, with the chosen operating frequency both effects should be considered.

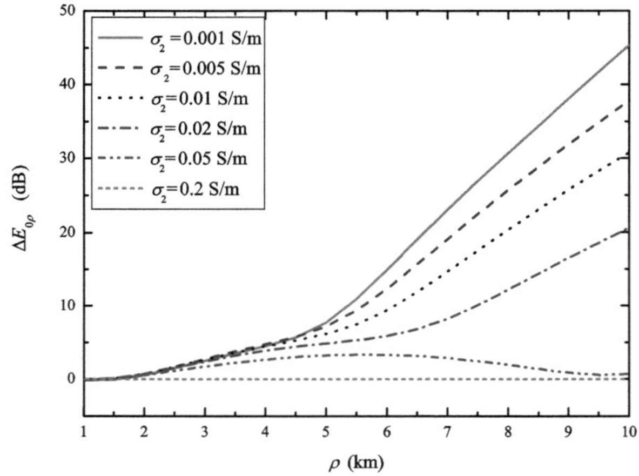

Fig. 6.41 The relative anomaly of the horizontal electric field $\Delta E_{0\rho}$ as a function of the offset distance at different conductivities of the high-impedance layer: $f = 0.25$ Hz, $l_1 = 1{,}000$ m, $l_2 = 100$ m

6.6.3 The Effect of Air Waves

When the mCSEM method is used to detect the subbottom structure by measuring the electromagnetic field components on the seabed, in order to detect the structure of the deep seabed, we always hope that the distance between the transmitting and receiving points is as large as possible. When the sea water is not too deep, while the distance between the transmitting and receiving points increases, except for the wave traveling on the ocean floor along the boundary between the sea water and ocean floor, there exists another wave, which travels upwardly to the air, and then along the sea surface, and finally vertically downward to the receiving antenna. When the sea water is very shallow, and the distance between the transmitting and receiving points is very large, the magnitude of the wave in the air may be larger than that of the wave on the ocean floor. Thus, it is necessary to analyze and discuss the effect of the air wave and the compensated method. In order to analyze the problem on the effect by the air, the five-layered model and Cartesian coordinate system is shown in Fig. 6.45, where the HED is located at (0, 0, d). It is noted that Region 0 with its thickness of l_0 is occupied by sea water, and the transmitting and receiving antennas are located in this region.

Under this five-layered model, with a series of complicated derivations, the integrated formulas are derived readily for the electromagnetic field in the sea water. They are written in the following forms:

$$\begin{aligned} E_{0\rho} = &-\frac{\omega\mu_0 I\,\mathrm{d}l\cos\phi}{4\pi\rho}\int_0^\infty \gamma_0^{-1}\left[\mathrm{e}^{\mathrm{i}\gamma_0|z-d|} + P^{\mathrm{TE}}\mathrm{e}^{\mathrm{i}\gamma_0 z} + Q^{\mathrm{TE}}\mathrm{e}^{\mathrm{i}\gamma_0(2l_0-z)}\right] \\ &\times J_1(\lambda\rho)\,\mathrm{d}\lambda \\ &-\frac{I\,\mathrm{d}l\cos\phi}{4\pi\omega\varepsilon_0}\int_0^\infty \gamma_0\left[\mathrm{e}^{\mathrm{i}\gamma_0|z-d|} - P^{\mathrm{TM}}\mathrm{e}^{\mathrm{i}\gamma_0 z} - Q^{\mathrm{TM}}\mathrm{e}^{\mathrm{i}\gamma_0(2l_0-z)}\right] \\ &\times\left[\lambda J_0(\lambda\rho) - \rho^{-1}J_1(\lambda\rho)\right]\mathrm{d}\lambda, \end{aligned} \tag{6.199}$$

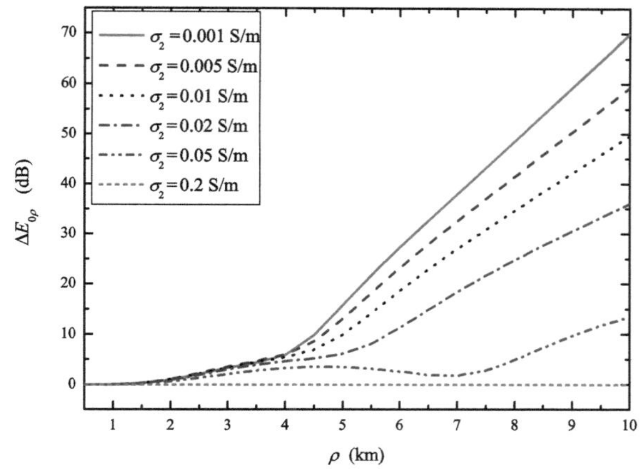

Fig. 6.42 The relative anomaly of the horizontal electric field $\Delta E_{0\rho}$ as a function of the offset distance at different conductivities of the high-impedance layer: $f = 0.5$ Hz, $l_1 = 1{,}000$ m, $l_2 = 100$ m

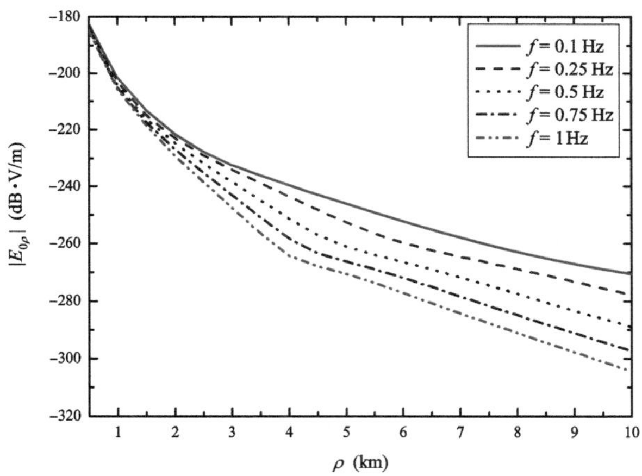

Fig. 6.43 The electric field $E_{0\rho}$ as a function of the offset distance at different operating frequency: $l_1 = 1{,}000$ m, $l_2 = 100$ m, $\sigma_2 = 0.01$ S/m

$$
\begin{aligned}
E_{0\phi} = {} & \frac{\omega\mu_0 I\,\mathrm{d}l \sin\phi}{4\pi} \int_0^\infty \gamma_0^{-1}\left[\mathrm{e}^{\mathrm{i}\gamma_0|z-d|} + P^{\mathrm{TE}}\mathrm{e}^{\mathrm{i}\gamma_0 z} + Q^{\mathrm{TE}}\mathrm{e}^{\mathrm{i}\gamma_0(2l_0-z)}\right] \\
& \times \left[\lambda J_0(\lambda\rho) - \rho^{-1} J_1(\lambda\rho)\right]\mathrm{d}\lambda \\
& + \frac{I\,\mathrm{d}l \sin\phi}{4\pi\omega\varepsilon_0\rho} \int_0^\infty \gamma_0\left[\mathrm{e}^{\mathrm{i}\gamma_0|z-d|} - P^{\mathrm{TM}}\mathrm{e}^{\mathrm{i}\gamma_0 z} - Q^{\mathrm{TM}}\mathrm{e}^{\mathrm{i}\gamma_0(2l_0-z)}\right] J_1(\lambda\rho)\,\mathrm{d}\lambda,
\end{aligned}
\tag{6.200}
$$

$$
\begin{aligned}
H_{0\rho} = {} & -\frac{I\,\mathrm{d}l \sin\phi}{4\pi} \int_{-\infty}^\infty \left[\pm\mathrm{e}^{\mathrm{i}\gamma_0|z-d|} + P^{\mathrm{TE}}\mathrm{e}^{\mathrm{i}\gamma_0 z} - Q^{\mathrm{TE}}\mathrm{e}^{\mathrm{i}\gamma_0(2l_0-z)}\right] \\
& \times \left[\lambda J_0(\lambda\rho) - \rho^{-1} J_1(\lambda\rho)\right]\mathrm{d}\lambda
\end{aligned}
$$

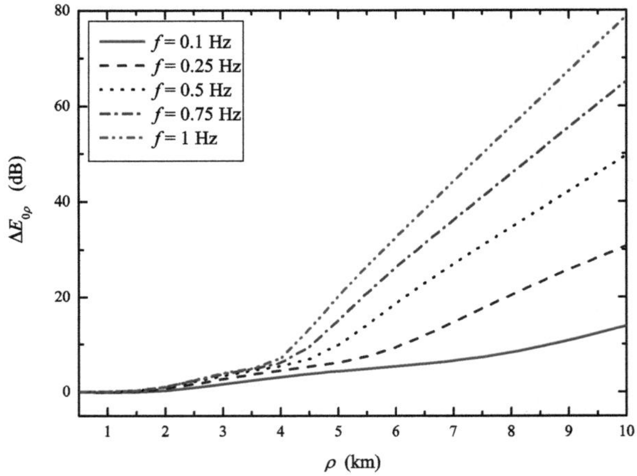

Fig. 6.44 The relative anomaly of the horizontal electric field $\Delta E_{0\rho}$ as a function of the offset distance at different operating frequency: $l_1 = 1{,}000$ m, $l_2 = 100$ m, $\sigma_2 = 0.01$ S/m

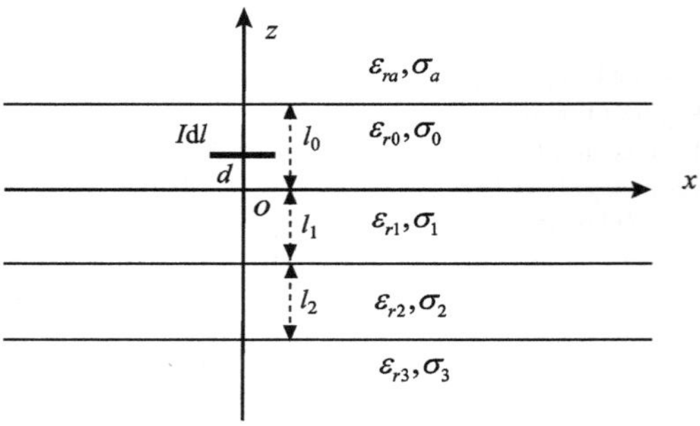

Fig. 6.45 Geometry of an HED in the five-layered region

$$-\frac{I\,\mathrm{d}l\sin\phi}{4\pi\rho}\int_{-\infty}^{\infty}\left[\pm e^{i\gamma_0|z-d|} - P^{\mathrm{TM}}e^{i\gamma_0 z} + Q^{\mathrm{TM}}e^{i\gamma_0(2l_0-z)}\right]J_1(\lambda\rho)\,\mathrm{d}\lambda, \tag{6.201}$$

$$\begin{aligned} H_{0\phi} = &-\frac{I\,\mathrm{d}l\cos\phi}{4\pi\rho}\int_{-\infty}^{\infty}\left[\pm e^{i\gamma_0|z-d|} + P^{\mathrm{TE}}e^{i\gamma_0 z} - Q^{\mathrm{TE}}e^{i\gamma_0(2l_0-z)}\right]J_1(\lambda\rho)\,\mathrm{d}\lambda \\ &-\frac{I\,\mathrm{d}l\cos\phi}{4\pi}\int_{-\infty}^{\infty}\left[\pm e^{i\gamma_0|z-d|} - P^{\mathrm{TM}}e^{i\gamma_0 z} + Q^{\mathrm{TM}}e^{i\gamma_0(2l_0-z)}\right] \\ &\times\left[\lambda J_0(\lambda\rho) - \rho^{-1}J_1(\lambda\rho)\right]\mathrm{d}\lambda, \end{aligned} \tag{6.202}$$

where the sign "+" represents the region of $z \geq d$, while the sign "−" represents the region of $0 < z < d$. In Eqs. (6.199)–(6.202), P^{TM}, Q^{TM}, P^{TE}, and Q^{TE} are expressed as follows:

$$P^{\mathrm{TM}} = \frac{(1 - R_{0+}^{\mathrm{TM}})R_{0-}^{\mathrm{TM}}}{1 - R_{0+}^{\mathrm{TM}}R_{0-}^{\mathrm{TM}}}e^{-i\gamma_0 d}, \tag{6.203}$$

$$Q^{\mathrm{TM}} = \frac{(1 - R_{0-}^{\mathrm{TM}}) R_{0+}^{\mathrm{TM}}}{1 - R_{0+}^{\mathrm{TM}} R_{0-}^{\mathrm{TM}}} \mathrm{e}^{\mathrm{i}\gamma_0 (d - 2l_0)}, \tag{6.204}$$

$$P^{\mathrm{TE}} = \frac{(1 + R_{0+}^{\mathrm{TE}}) R_{0-}^{\mathrm{TE}}}{1 - R_{0+}^{\mathrm{TE}} R_{0-}^{\mathrm{TE}}} \mathrm{e}^{-\mathrm{i}\gamma_0 d}, \tag{6.205}$$

$$Q^{\mathrm{TE}} = \frac{(1 + R_{0-}^{\mathrm{TE}}) R_{0+}^{\mathrm{TE}}}{1 - R_{0+}^{\mathrm{TE}} R_{0-}^{\mathrm{TE}}} \mathrm{e}^{\mathrm{i}\gamma_0 (d - 2l_0)}, \tag{6.206}$$

where R_{0+}^{TM} and R_{0-}^{TM} represent the reflection coefficients of the electric-type (TM) wave on the boundaries, while R_{0+}^{TE} and R_{0-}^{TE} represent those of the magnetic-type (TE) wave on the boundaries. It is noted that the sign "+" corresponds to the boundary between the air and the sea water, while the sign "−" corresponds to the boundary between the sea water and the sedimentary layer. They are

$$R_{0+}^{\mathrm{TM}} = \frac{\gamma_0 k_a^2 - \gamma_a k_0^2}{\gamma_0 k_a^2 + \gamma_a k_0^2} \cdot \mathrm{e}^{\mathrm{i}2\gamma_0 (l_0 - d)}, \tag{6.207}$$

$$R_{0-}^{\mathrm{TM}} = \frac{\mathrm{e}^{\mathrm{i}2\gamma_0 d} + \frac{M_1 \cdot M_2^{-1} \cdot \mathrm{e}^{\mathrm{i}2\gamma_2 l_2} + 1}{M_1 + M_2^{-1} \cdot \mathrm{e}^{\mathrm{i}2\gamma_2 l_2}} \cdot M_0 \cdot \mathrm{e}^{(\mathrm{i}2\gamma_0 d + \gamma_1 l_1)}}{M_0 + \frac{M_1 \cdot M_2^{-1} \cdot \mathrm{e}^{\mathrm{i}2\gamma_2 l_2} + 1}{M_1 + M_2^{-1} \cdot \mathrm{e}^{\mathrm{i}2\gamma_2 l_2}} \cdot \mathrm{e}^{\mathrm{i}2\gamma_1 l_1}}, \tag{6.208}$$

$$R_{0+}^{\mathrm{TE}} = \frac{\gamma_0 - \gamma_a}{\gamma_0 + \gamma_a} \cdot \mathrm{e}^{\mathrm{i}2\gamma_0 (l_0 - d)}, \tag{6.209}$$

$$R_{0-}^{\mathrm{TM}} = \frac{\mathrm{e}^{\mathrm{i}2\gamma_0 d} + \frac{M_1 \cdot M_2^{-1} \cdot \mathrm{e}^{\mathrm{i}2\gamma_2 l_2} + 1}{M_1 + M_2^{-1} \cdot \mathrm{e}^{\mathrm{i}2\gamma_2 l_2}} \cdot M_0 \cdot \mathrm{e}^{(\mathrm{i}2\gamma_0 d + \gamma_1 l_1)}}{M_0 + \frac{M_1 \cdot M_2^{-1} \cdot \mathrm{e}^{\mathrm{i}2\gamma_2 l_2} + 1}{M_1 + M_2^{-1} \cdot \mathrm{e}^{\mathrm{i}2\gamma_2 l_2}} \cdot \mathrm{e}^{\mathrm{i}2\gamma_1 l_1}}, \tag{6.210}$$

where

$$M_i = \frac{\gamma_i k_{i+1}^2 + \gamma_{i+1} k_i^2}{\gamma_i k_{i+1}^2 - \gamma_{i+1} k_i^2}; \qquad N_i = \frac{\gamma_i + \gamma_{i+1}}{\gamma_i - \gamma_{i+1}}. \tag{6.211}$$

In the above equations, γ_i $(i = 0, 1, 2)$ represents the wave number in the vertical direction of the ith layer, while γ_a represents the wave number in the vertical direction in the air.

In order to analyze the effect of the thickness of the sea water layer in detecting the ocean floor by using the mCSEM method, the following computations are carried out. We note that $E_{0\rho}(\rho, 0, z)$, which is computed by Eq. (6.199), represents the field component excited by unit HED in the sea water in the shallow-sea case with the depth being 300 m, while $E_{0\rho}^{g}(\rho, 0, z)$, which is computed by Eq. (6.181) with $n = 4$, represents the field component in the sea water for the four-layered model in the deep-sea case.

In Fig. 6.46, graphs are for the components $E_{0\rho}(\rho, 0, z)$ and $E_{0\rho}^{g}(\rho, 0, z)$ as functions of the offset distance ρ. In the computations, the parameters for all layers are

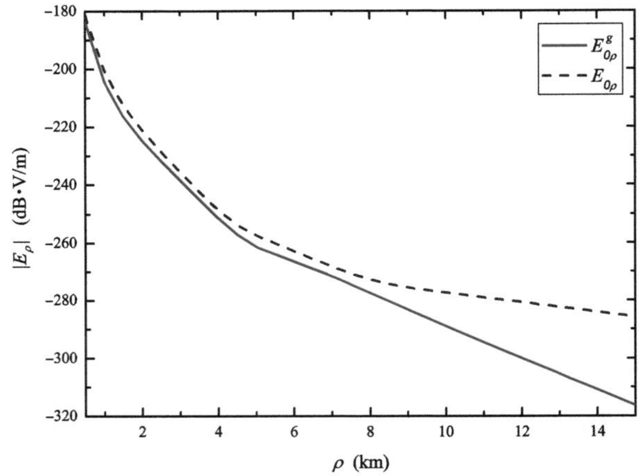

Fig. 6.46 In both the shallow-sea and the deep-sea cases, the electric field component versus the offset distance at $f = 0.5$ Hz

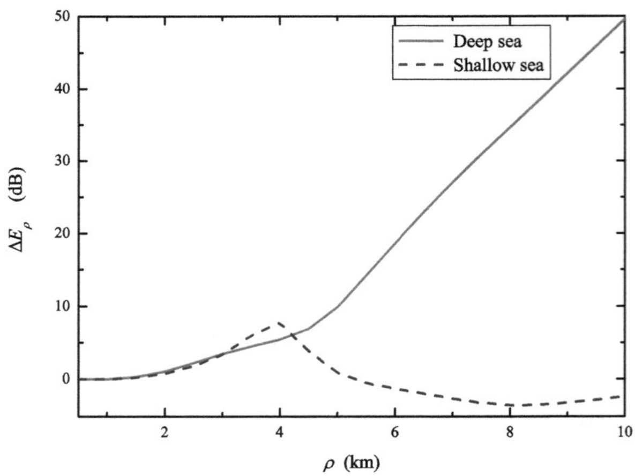

Fig. 6.47 In both the shallow-sea and the deep-sea cases, the relative anomaly of the electric field versus the offset distance at $f = 0.5$ Hz

as follows: $\varepsilon_{ra} = 1$, $\varepsilon_{r0} = 80$, $\varepsilon_{r1} = \varepsilon_{r2} = \varepsilon_{r3} = 10$, $\sigma_0 = 3.3$ S/m, $\sigma_1 = 1$ S/m, $\sigma_2 = 0.01$ S/m, and $\sigma_3 = 1$ S/m, the operating frequency is $f = 0.5$ Hz, and the height of the transmitting antenna and that of the receiving antenna are taken as $d = 10$ m and $z = 0$ m, respectively. It is seen that when the offset distance is less than 4 km, the response of the electric field for the shallow sea is basically consistent with that for the deep sea. When the offset distance is greater than 6 km, the electric component $E_{0\rho}$ in the shallow sea is enlarged significantly compared with that in the deep sea. Furthermore, with the increase in the offset distance, the difference is growing. It is shown that for the shallow-sea prospecting in the far zone, the impacts by the air wave will be more and more extensive.

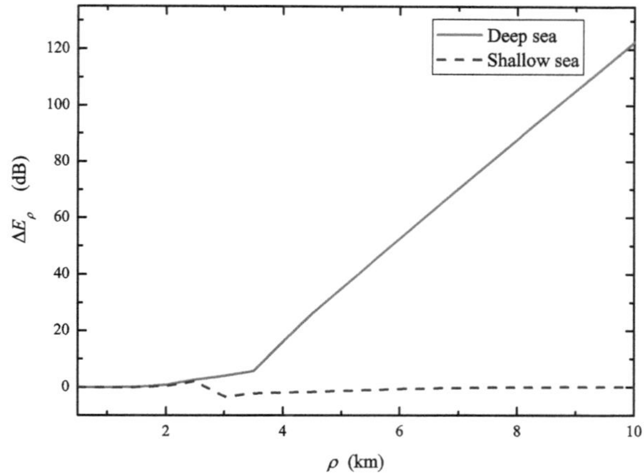

Fig. 6.48 In both the shallow-sea and the deep-sea cases, the relative anomaly of the electric field versus the offset distance is computed at $f = 2.0$ Hz

With the depth and thickness of the target being 1,000 m and 100 m, respectively, and the conductivity $\sigma_2 = 0.01$ S/m, the relative anomaly of the electric field versus the offset distance is computed at $f = 0.5$ Hz and 2 Hz and shown in Figs. 6.47 and 6.48, respectively. From these computations, it is seen that when the offset distance is less than 3 km, the relative anomaly of the electric field for the shallow sea is consistent with that for the deep sea. When the offset distance is larger than 3 km, the relative abnormality of the electric field for the shallow sea is reduced significantly, and it is unable to distinguish whether the high resistivity layer exists or not. This is resulted by the electric field, which is determined mainly by the air wave in the shallow-sea case. It is seen that in the shallow-sea case, it is necessary to carry out further works on the mCSEM method, otherwise both the probing depth and the resolving power will be limited.

6.7 Summary

In this chapter, under the half-space model, three- and multi-layered models, the ELF field in sea water radiated by an HED on the boundary between sea water and sea floor is investigated analytically and some computations are carried out carefully. As the applications of the electromagnetic field excited by an HED in layered region, the mCSEM method is addressed specifically.

References

Baños A Jr (1966) Dipole radiation in the presence of a conducting half-space. Pergamon, Oxford

Chave AD, Cox CS (1982) Controlled electromagnetic sources for measuring electrical conductivity beneath the oceans. I: Forward problem and model study. J Geophys Res 87:5327–5338

Collin RE (2004a) Some observations about the near zone electric field of a hertzian dipole above a lossy earth. IEEE Trans Antennas Propag 52(11):3133–3137

Collin RE (2004b) Hertzian dipole radiation over a lossy earth or sea: some early and late 20th century controversies. IEEE Antennas Propag Mag 46(2):64–79

Dunn JM (1984) Electromagnetic lateral waves in layered media. PhD Thesis, Harvard University, Cambridge, MA, USA

Dunn JM (1986) Lateral wave propagation in a three-layered medium. Radio Sci 21:787–796

Hoh JH, Li LW, Kooi PS, Yeo TS, Leong MS (1999) Dominant lateral waves in the canopy layer of a four-layered forest. Radio Sci 34(3):681–691

King RWP (1989) Lateral electromagnetic waves from a horizontal antenna for remote sensing in the ocean. IEEE Trans Antennas Propag 37:1250–1255

King RWP (1991) The electromagnetic field of a horizontal electric dipole in the presence of a three-layered region. J Appl Phys 69(12):7987–7995

King RWP (1993) The electromagnetic field of a horizontal electric dipole in the presence of a three-layered region: supplement. J Appl Phys 74(8):4548–4845

King RWP, Brown MF (1984) Lateral electromagnetic waves along plane boundaries: a summarizing approach. Proc IEEE 72:595–611

King RWP, Sandler BH, Shen LC (1980) A comprehensive study of subsurface propagation from horizontal electric dipoles. IEEE Trans Geosci Remote Sens GE-18:225–233

King RWP, Owens M, Wu TT (1986) Properties of lateral electromagnetic fields and their application. Radio Sci 21:12–23

King RWP, Owens M, Wu TT (1992) Lateral electromagnetic waves: theory and applications to communications, geophysical exploration, and remote sensing. Springer, New York

Li K (2009) Electromagnetic fields in stratified media. ZJU Press/Springer, Hangzhou/Berlin

Li K, Lu Y (2005) Electromagnetic field generated by a horizontal electric dipole near the surface of a planar perfect conductor coated with a uniaxial layer. IEEE Trans Antennas Propag 53(10):3191–3200

Li LW, Yeo TS, Kooi PS, Leong MS (1998) Radio wave propagation along mixed paths through a four-layered model of rain forest: an analytical approach. IEEE Trans Antennas Propag 46(7):1098–1111

Li LW, Lee CK, Yeo TS, Leong MS (2004) Wave mode and path characteristics in four-layered anisotropic forest environment. IEEE Trans Antennas Propag 52(9):2445–2455

Lu H, Peng HY, Pan WY (2010) The principles and properties on the wave propagation for the mCSEM method. Technical Report, Research Institute of Radiowave Propagation, Qingdao, China

Norton KA (1937) The propagation of radio waves over the surface of the earth and in the upper atmosphere. Proc IRE 25:1203–1236

Pan WY (1985) Surface wave propagation along the boundary between sea water and one-dimensionally anisotropic rock. J Appl Phys 58:3963–3974

Parkhomenko EI (1967) Electrical properties of rocks. Plenum, New York

Sommerfeld (1909) Propagation of waves in wireless telegraphy. Ann Phys 28:1135–1153

Van der Pol B (1935) Theory of the reflection of light from a point source by a finitely conducting flat mirror: with application to radiotelegraphy. Physics 2:843–853

Van der Pol B, Niessen KF (1930) The propagation of electromagnetic waves over a plane earth. Ann Phys 6:273–294

von Hörschelmann H (1911) Ueber die Wirkungsweise des geknickten Marconischen Senders in der drahtlosen Telegraphie. Jahrb Drahtlosen Telegr Teleph 5:14–34 and 188–211

Wait JR (1953) Propagation of radio waves over a stratified ground. Geophysics 18:416–422

Wait JR (1956) Low frequency radiation from a horizontal earth. Can J Phys 34:586–595

Wait JR (1961) The electromagnetic fields of a horizontal dipole in the presence of a conducting half-space. Can J Phys 39:1017–1027

Wait JR (1970) Electromagnetic waves in stratified media, 2nd edn. Pergamon, New York

Wu TT, King RWP (1987) Lateral electromagnetic pulses generated by a vertical electric dipole on the boundary between two dielectrics. J Appl Phys 62:4355–4543

Young PD, Cox CS (1981) Electromagnetic active source sounding near the East Pacific Rise. Geophys Res Lett 8(10):1043–1046

Zhang HQ, Pan WY, Li K, Shen KX (2005) Electromagnetic field for a horizontal electric dipole buried inside a dielectric layer coated high lossy half space. In: Progress in electromagnetics research, PIER, vol 50. EMW Publishing, Cambridge, pp 163–186

Chapter 7
Radiation and Propagation of SLF/ELF Electromagnetic Waves of Space Borne Transmitter

A space borne SLF/ELF transmitter for the communication with submarines can be a line antenna or loop antenna in the ionosphere at a heights of 300–450 km. In this chapter, the line antenna is idealized as an electric dipole, the ionosphere is idealized as homogeneous anisotropic plasma. The radiation of a SLF/ELF electric dipole in an infinite anisotropic plasma is treated analytically, while the electromagnetic field on the sea surface generated by a space borne SLF/ELF line antenna is analyzed. Some numerical results are also calculated.

7.1 Introduction

It is well known that SLF/ELF electromagnetic waves can be used for submarine communication. The available submarine communication systems require us to build huge ground-based transmitting stations with large investment and wartime sensitivity. Once damaged, they are difficult to repair. With the development of space technology and the decrease of satellite launching costs, it is possible to develop a space borne SLF/ELF transmitting system to use for submarine communication. Additionally, the ionosphere high-frequency heating can also be used to generate SLF/ELF waves. It is known that the programmes of the active SLF experiments with the use of a large loop antenna (with diameter 300 m) or line antenna onboard the spacecraft have been presented in Russia and the United States (Armand et al. 1988; Koshelev and Melnikov 1998; Bannister et al. 1993).

SLF/ELF waves in the ionosphere and the magnetosphere have been a subject of investigations already for more than 60 years. The theoretical and experimental works on SLF/ELF wave propagation have been treated by many investigators and the findings are well summarized by Wait (1962) and Galejs (1972). In the paper by Xia and Chen (2000), by using the matrix-exponent formulations, SLF/ELF wave radiation and propagation in anisotropic stratified ionosphere are analyzed and computed numerically. In the previous work by Li, the solution of the propagation of VLF/SLF/ELF electromagnetic waves penetrating the lower ionosphere is carried out (Li and Pan 1999; Li et al. 2011).

W. Pan, K. Li, *Propagation of SLF/ELF Electromagnetic Waves*,
Advanced Topics in Science and Technology in China,
DOI 10.1007/978-3-642-39050-0_7,

In this chapter, we will summarize our recent work on the radiation of an SLF/ELF electric dipole in an infinite anisotropic plasma and the electromagnetic field on the sea surface generated by the space borne SLF/ELF transmitter. Some new computations and conclusions are also presented.

7.2 Radiation of an SLF/ELF Electric Dipole in an Infinite Homogeneous Anisotropic Plasma

In this section, we will attempt to address the radiation of a SLF/ELF electric dipole in an infinite anisotropic plasma.

7.2.1 Radiation of a SLF/ELF Electric Dipole in an Infinite Homogeneous Anisotropic Plasma

In an infinite, homogeneous, anisotropic plasma, with the $\hat{z}$-axis of the coordinate system along the geomagnetic field, the electromagnetic properties of the plasma can be characterized by the tensor permittivity $\bar{\bar{\varepsilon}}$ and the scalar relative magnetic permeability $\mu_r = 1$. When the $\hat{z}$-axis is parallel to the direction of the geomagnetic field, the tensor permittivity $\bar{\bar{\varepsilon}}$ is expressed in the following form:

$$\bar{\bar{\varepsilon}} = \varepsilon_0 \begin{bmatrix} \varepsilon_1 & -\mathrm{i}\varepsilon_2 & 0 \\ \mathrm{i}\varepsilon_2 & \varepsilon_1 & 0 \\ 0 & 0 & \varepsilon_3 \end{bmatrix}, \tag{7.1}$$

where

$$\varepsilon_1 = 1 - \frac{X_e U_e}{U_e^2 - Y_e^2} - \frac{X_i U_i}{U_i^2 - Y_i^2}, \tag{7.2}$$

$$\varepsilon_2 = \frac{X_e Y_e}{U_e^2 - Y_e^2} + \frac{X_i Y_i}{U_i^2 - Y_i^2}, \tag{7.3}$$

$$\varepsilon_3 = 1 - \frac{X_e}{U_e} - \frac{X_i}{U_i}, \tag{7.4}$$

$$U_e = 1 + \mathrm{i}\frac{v_e}{\omega}; \qquad U_i = 1 + \mathrm{i}\frac{v_i}{\omega}, \tag{7.5}$$

$$X_e = \frac{\omega_{0e}^2}{\omega^2}; \qquad X_i = \frac{\omega_{0i}^2}{\omega^2}, \tag{7.6}$$

$$Y_e = \frac{\omega_H^{(e)}}{\omega}; \qquad Y_i = \frac{\omega_H^{(i)}}{\omega}. \tag{7.7}$$

In the above equations, $\omega_{0e}^2 = \frac{N_e e^2}{\varepsilon_0 m_e}$ and $\omega_{0i}^2 = \frac{N_i (z_i e)^2}{\varepsilon_0 m_i}$ are the angular plasma frequency of electron and of ion, respectively. The electron gyro-frequency $\omega_H^{(e)}$ and ions gyro-frequency $\omega_H^{(i)}$ are expressed as $\omega_H^{(e)} = |\frac{eB_0}{m_e}|$ and $\omega_H^{(i)} = -|\frac{ez_i B_0}{m_i}|$, respectively. It is noted that ω is the operating angular frequency, ν_e is the effective collision frequency of the electron, N_e is the electron density, m_e the electron mass, ν_i is the effective collision frequency of the ions, N_i is the ion density, B_0 is the geomagnetic field strength in Wb/m^2, and z_i, the order of the ions with positive charge, which is usually taken as 1 or 2, is determined by the ions' composition.

Maxwell's equations can be expressed in the following forms:

$$\nabla \times \boldsymbol{E} = \mathrm{i}\omega\mu_0 \boldsymbol{H}, \tag{7.8}$$

$$\nabla \times \boldsymbol{H} = \boldsymbol{J}\delta(\boldsymbol{r}) - \mathrm{i}\omega\varepsilon_0 \bar{\bar{\varepsilon}} \boldsymbol{E}, \tag{7.9}$$

where $\boldsymbol{J}$ is an unit electric dipole in the origin point. Introducing the 3D Fourier transforms, we have

$$\boldsymbol{E}(\boldsymbol{r}) = \frac{1}{(2\pi)} \iiint_{-\infty}^{\infty} \tilde{\boldsymbol{E}}(\boldsymbol{k}) \exp(-\mathrm{i}\boldsymbol{k} \cdot \boldsymbol{r})\, \mathrm{d}^3\boldsymbol{k}, \tag{7.10}$$

$$\boldsymbol{H}(\boldsymbol{r}) = \frac{1}{(2\pi)} \iiint_{-\infty}^{\infty} \tilde{\boldsymbol{H}}(\boldsymbol{k}) \exp(-\mathrm{i}\boldsymbol{k} \cdot \boldsymbol{r})\, \mathrm{d}^3\boldsymbol{k}, \tag{7.11}$$

$$\delta(\boldsymbol{r}) = \frac{1}{(2\pi)} \iiint_{-\infty}^{\infty} \exp(-\mathrm{i}\boldsymbol{k} \cdot \boldsymbol{r})\, \mathrm{d}^3\boldsymbol{k}. \tag{7.12}$$

Substituting Eqs. (7.10)–(7.12) into Eqs. (7.8) and (7.9), it follows that

$$\left.\begin{aligned}
&-\mathrm{i}k_y \tilde{E}_z + \mathrm{i}k_z \tilde{E}_y = \mathrm{i}\omega\mu_0 \tilde{H}_x \\
&-\mathrm{i}k_z \tilde{E}_x + \mathrm{i}k_x \tilde{E}_z = \mathrm{i}\omega\mu_0 \tilde{H}_y \\
&-\mathrm{i}k_x \tilde{E}_y + \mathrm{i}k_y \tilde{E}_x = \mathrm{i}\omega\mu_0 \tilde{H}_z \\
&-\mathrm{i}k_y \tilde{H}_z + \mathrm{i}k_z \tilde{H}_y = -\mathrm{i}\omega\varepsilon_0(\varepsilon_1 \tilde{E}_x - \mathrm{i}\varepsilon_2 \tilde{E}_y) + \tilde{J}_x \\
&-\mathrm{i}k_z \tilde{H}_x + \mathrm{i}k_x \tilde{H}_z = -\mathrm{i}\omega\varepsilon_0(\mathrm{i}\varepsilon_2 \tilde{E}_x + \varepsilon_1 \tilde{E}_y) + \tilde{J}_y \\
&-\mathrm{i}k_x \tilde{H}_y + \mathrm{i}k_y \tilde{H}_x = -\mathrm{i}\omega\varepsilon_0\varepsilon_3 \tilde{E}_z + \tilde{J}_z
\end{aligned}\right\}. \tag{7.13}$$

Letting $k_x = k \sin\theta_k \cos\phi_k$, $k_y = k \sin\theta_k \sin\phi_k$, $k_z = k \cos\theta_k$, we readily derive

$$\bar{\bar{\mathbf{A}}} \begin{bmatrix} \tilde{E}_x \\ \tilde{E}_y \\ \tilde{E}_z \end{bmatrix} = -\mathrm{i}\omega\mu_0 \begin{bmatrix} \tilde{J}_x \\ \tilde{J}_y \\ \tilde{J}_z \end{bmatrix}. \tag{7.14}$$

In the above equation, the 3×3 matrix is written in the form

$$\bar{\bar{\mathbf{A}}} = \begin{bmatrix} A_{11} & A_{12} & A_{13} \\ A_{21} & A_{22} & A_{23} \\ A_{31} & A_{32} & A_{33} \end{bmatrix}, \tag{7.15}$$

where

$$A_{11} = k_0^2\varepsilon_1 - k^2\left(\cos^2\theta_k + \sin^2\theta_k \sin^2\phi_k\right), \tag{7.16}$$

$$A_{22} = k_0^2\varepsilon_1 - k^2\left(\cos^2\theta_k + \sin^2\theta_k \cos^2\phi_k\right), \tag{7.17}$$

$$A_{33} = k_0^2\varepsilon_3 - k^2\sin^2\theta_k, \tag{7.18}$$

$$A_{12} = k^2\sin^2\theta_k \sin\phi_k \cos\phi_k + \mathrm{i}k_0^2\varepsilon_2, \tag{7.19}$$

$$A_{21} = k^2\sin^2\theta_k \sin\phi_k \cos\phi_k - \mathrm{i}k_0^2\varepsilon_2, \tag{7.20}$$

$$A_{13} = A_{31} = k^2\sin\theta_k \cos\theta_k \cos\phi_k, \tag{7.21}$$

$$A_{23} = A_{32} = k^2\sin\theta_k \cos\theta_k \sin\phi_k. \tag{7.22}$$

The solution of Eq. (7.14) can be expressed in the following form:

$$\tilde{\boldsymbol{E}}(\boldsymbol{k}) = -\mathrm{i}\omega\mu_0 \frac{\boldsymbol{F}(k, \theta_k, \phi_k)}{\det\bar{\bar{\mathbf{A}}}}, \tag{7.23}$$

where

$$\det\bar{\bar{\mathbf{A}}} = k_0^2\left(\varepsilon_1\sin^2\theta_k + \varepsilon_3\cos^2\theta_k\right)\left(k^2 - k_1^2\right)\left(k^2 - k_2^2\right), \tag{7.24}$$

$$k_{1,2}^2 = \frac{k_0^2}{2(\varepsilon_1\sin^2\theta_k + \varepsilon_3\cos^2\theta_k)} \cdot \Big[\varepsilon_1\varepsilon_3\left(1 + \cos^2\theta_k\right) + \left(\varepsilon_1^2 - \varepsilon_2^2\right)\sin^2\theta_k$$
$$\pm\sqrt{\left(\varepsilon_1^2 - \varepsilon_2^2 - \varepsilon_1\varepsilon_3\right)^2\sin^4\theta_k + 4\varepsilon_2^2\varepsilon_3^2\cos^2\theta_k}\,\Big]. \tag{7.25}$$

The vector $\boldsymbol{F}(k, \theta_k, \phi_k)$ is written as follows:

$$\boldsymbol{F}(k, \theta_k, \phi_k) = \hat{x}F_x + \hat{y}F_y + \hat{z}F_z. \tag{7.26}$$

In the case of the unit electric dipole being parallel to the direction of the geomagnetic field, $\boldsymbol{J} = \hat{z}J_z$, namely, $J_x = J_y = 0$ and $J_z = 1$, the three components F_x, F_y, and F_z can be expressed as follows:

$$F_x = \sin\theta_k\cos\theta_k\left[k^4\cos\phi_k - k^2k_0^2(\mathrm{i}\varepsilon_2\sin\phi_k + \varepsilon_1\cos\phi_k)\right], \tag{7.27}$$

$$F_y = \sin\theta_k\cos\theta_k\left[k^4\sin\phi_k - k^2k_0^2(\varepsilon_1\sin\phi_k - \mathrm{i}\varepsilon_2\cos\phi_k)\right], \tag{7.28}$$

$$F_z = k^4\cos^2\theta_k - k^2k_0^2\varepsilon_1\left(1 + \cos^2\theta_k\right) + k_0^4\left(\varepsilon_1^2 - \varepsilon_2^2\right). \tag{7.29}$$

In the case of the unit electric dipole being vertical to the direction of the geomagnetic field, and letting the electric dipole be in the $\hat{x}$ direction, we have $\boldsymbol{J} = \hat{x} J_x$, $J_x = 1$, and $J_y = J_z = 0$. With similar steps, the three elements F_x, F_y, and F_z can be expressed as follows:

$$F_x = k_0^4 \varepsilon_1 \varepsilon_3 - k_0^2 k^2 \left[\varepsilon_1 \sin^2\theta_k + \varepsilon_3 \left(\cos^2\theta_k + \sin^2\theta_k \cos^2\phi_k\right)\right] + k^4 \sin^2\theta_k \cos^2\phi_k, \tag{7.30}$$

$$F_y = -\mathrm{i} k_0^4 \varepsilon_2 \varepsilon_3 - k_0^2 k^2 \left(\varepsilon_3 \sin^2\theta_k \sin\phi_k \cos\phi_k - \mathrm{i}\varepsilon_2 \sin^2\theta_k\right) + k^4 \sin^2\theta_k \sin\phi_k \cos\phi_k, \tag{7.31}$$

$$F_z = -k_0^2 k^2 (\varepsilon_1 \cos\phi_k - \mathrm{i}\varepsilon_2 \sin\phi_k) \sin\theta_k \cos\theta_k + k^4 \sin\theta_k \cos\theta_k \cos\phi_k. \tag{7.32}$$

Thus, in the above two cases, the electromagnetic field can be expressed in the integrated forms:

$$\boldsymbol{E}(\boldsymbol{r}) = -\frac{\mathrm{i}\omega\mu_0}{(2\pi)^3} \int_0^{\pi} \int_0^{2\pi} \int_0^{\infty} \frac{\boldsymbol{F}(k, \theta_k, \phi_k)}{\det \bar{\bar{\mathbf{A}}}} \exp\left\{-\mathrm{i}kr\left[(\cos\theta\cos\theta_k + \sin\theta\sin\theta_k\cos(\phi - \phi_k)\right]\right\} k^2 \sin\theta_k \,\mathrm{d}\theta_k \,\mathrm{d}\phi_k \,\mathrm{d}k. \tag{7.33}$$

With the extended representation,

$$\exp\left\{-\mathrm{i}kr\left[(\cos\theta\cos\theta_k + \sin\theta\sin\theta_k\cos(\phi - \phi_k)\right]\right\} = \sqrt{\frac{\pi}{2kr}} \sum_{n=0}^{\infty} (2n+1)(-\mathrm{i})^n J_{n+\frac{1}{2}}(kr) \times P_n\left[\cos\theta\cos\theta_k + \sin\theta\sin\theta_k\cos(\phi - \phi_k)\right], \tag{7.34}$$

and substituting Eq. (7.34) into Eq. (7.33), we obtain readily

$$\boldsymbol{E}(\boldsymbol{r}) = \frac{\mathrm{i}\omega\mu_0}{(2\pi)^3} \sqrt{\frac{\pi}{2r}} \int_0^{\pi} \mathrm{d}\theta_k \int_0^{2\pi} \sum_{n=0}^{\infty} (2n+1)(-\mathrm{i})^n \times P_n\left[\cos\theta\cos\theta_k + \sin\theta\sin\theta_k\cos(\phi - \phi_k)\right] \mathrm{d}\phi_k \times \int_0^{\infty} \frac{k^{-\frac{1}{2}} J_{n+\frac{1}{2}}(kr) \boldsymbol{F}(k, \theta_k, \phi_k) k^2 \sin\theta_k}{k_0^2(\varepsilon_1 \sin^2\theta_k + \varepsilon_3 \cos^2\theta_k)(k^2 - k_1^2)(k^2 - k_2^2)} \mathrm{d}k. \tag{7.35}$$

We denote the latter integral in Eq. (7.35) as follows:

$$\boldsymbol{S}_n = \int_0^{\infty} \frac{k^{-\frac{1}{2}} J_{n+\frac{1}{2}}(kr) \boldsymbol{F}(k, \theta_k, \phi_k) k^2}{k_0^2 (k^2 - k_1^2)(k^2 - k_2^2)} \mathrm{d}k. \tag{7.36}$$

By examining the integral in Eqs. (7.35) and (7.36), it is seen that, when both θ_k and ϕ_k change to $\pi - \theta_k$ and $\pi + \phi_k$, the integrand $\boldsymbol{F}(k, \theta_k, \phi_k)$ in S_n is not changeable. In the same case, it is seen that the Legendre function P_n is an odd function of x when n is odd, while it is an even function of x when n is even. Obviously, when n is odd, the integration over the ranges for θ_k from 0 to π and ϕ_k from 0 to 2π will be zero. There only exists the contribution by the integration when n is even. Then, Eq. (7.35) is rewritten as follows:

$$\begin{aligned}\boldsymbol{E}(\boldsymbol{r}) = {} & \frac{\mathrm{i}\omega\mu_0}{(2\pi)^3}\sqrt{\frac{\pi}{2r}}\sum_{n=0}^{\infty}\int_0^{\pi}\mathrm{d}\theta_k\int_0^{2\pi}(4n+1)(-1)^n \\ & \times P_{2n}\big[\cos\theta\cos\theta_k+\sin\theta\sin\theta_k\cos(\phi-\phi_k)\big]\,\mathrm{d}\phi_k \\ & \times\int_0^{\infty}\frac{k^{-\frac{1}{2}}J_{2n+1/2}(kr)\boldsymbol{F}(k,\theta_k,\phi_k)k^2\sin\theta_k}{k_0^2(\varepsilon_1\sin^2\theta_k+\varepsilon_3\cos^2\theta_k)(k^2-k_1^2)(k^2-k_2^2)}\,\mathrm{d}k. \end{aligned} \tag{7.37}$$

Considering the relations

$$J^{(1)}_{2n+\frac{1}{2}}(kr)=\frac{1}{2}\big[H^{(1)}_{2n+\frac{1}{2}}(kr)+H^{(2)}_{2n+\frac{1}{2}}(kr)\big], \tag{7.38}$$

$$\mathrm{i}H^{(1)}_{2n+\frac{1}{2}}(-kr)=H^{(2)}_{2n+\frac{1}{2}}(kr), \tag{7.39}$$

and noting that both the function $\boldsymbol{F}$ and the integrand in Eq. (7.37) are even functions of the variable k, then it follows that

$$\begin{aligned}\boldsymbol{S}_{2n} &= \int_0^{\infty}\frac{k^{-\frac{1}{2}}J_{2n+\frac{1}{2}}(kr)\boldsymbol{F}(k,\theta_k,\phi_k)k^2}{k_0^2(k^2-k_1^2)(k^2-k_2^2)}\,\mathrm{d}k \\ &= \frac{1}{2}\int_{-\infty}^{\infty}\frac{k^{-\frac{1}{2}}H^{(1)}_{2n+\frac{1}{2}}(kr)\boldsymbol{F}(k,\theta_k,\phi_k)k^2}{k_0^2(k^2-k_1^2)(k^2-k_2^2)}\,\mathrm{d}k. \end{aligned} \tag{7.40}$$

With the residue theorem, we obtain readily

$$\begin{aligned}\boldsymbol{S}_{2n} &= -\frac{\pi\mathrm{i}}{2k_0^2(k_1^2-k_2^2)}\big[k_2^{\frac{1}{2}}\boldsymbol{F}(k_2,\theta_k,\phi_k)H^{(1)}_{2n+\frac{1}{2}}(k_2r) \\ &\quad - k_1^{\frac{1}{2}}\boldsymbol{F}(k_1,\theta_k,\phi_k)H^{(1)}_{2n+\frac{1}{2}}(k_1r)\big] \\ &= \boldsymbol{S}_n(r,\theta_k,\phi_k). \end{aligned} \tag{7.41}$$

Then, the electric field can be expressed in the following form:

$$\begin{aligned}\boldsymbol{E}(\boldsymbol{r}) = {} & \frac{\mathrm{i}\omega\mu_0}{(2\pi)^3}\sqrt{\frac{\pi}{2r}}\sum_{n=1}^{\infty}\int_0^{\pi}\sin\theta_k\,\mathrm{d}\theta_k\int_0^{2\pi}\frac{(4n+1)(-1)^n\boldsymbol{S}_{2n}(r,\theta_k,\phi_k)}{\varepsilon_1\sin^2\theta_k+\varepsilon_3\cos^2\theta_k} \\ & \times P_{2n}\big[\cos\theta\cos\theta_k+\sin\theta\sin\theta_k\cos(\phi-\phi_k)\big]\,\mathrm{d}\phi_k. \end{aligned} \tag{7.42}$$

If Y_{n_1} is the n_1-order scalar spherical harmonic function of θ_k and ϕ_k, i.e.

$$Y_{n_1}(\theta_k,\phi_k)=a_0P_{n_1}(\cos\theta_k)+\sum_{m=1}^{n_1}(a_m\cos m\phi_k+b_m\sin m\phi_k)P_{n_1}^m(\cos\theta_k), \quad (7.43)$$

and considering the orthogonality of spherical harmonic functions, it follows that

$$\int_0^{\pi}\sin\theta_k\,\mathrm{d}\theta_k\int_0^{2\pi}Y_{n_1}(\theta_k,\phi_k)P_n\big[\cos\theta\cos\theta_k+\sin\theta\sin\theta_k\cos(\phi-\phi_k)\big]\,\mathrm{d}\phi_k$$
$$=\begin{cases}0; & n\neq n_1,\\ \frac{4\pi}{2n+1}Y_n(\theta,\phi); & n=n_1.\end{cases} \quad (7.44)$$

Thus, the integrand function can be expanded in terms of the scalar spherical harmonic functions. We write

$$\frac{S_{2n}(r,\theta_k,\phi_k)}{\varepsilon_1\sin^2\theta_k+\varepsilon_3\cos^2\theta_k}=\sum_{n_1=0}^{\infty}\Bigg\{\boldsymbol{a}_{2n,n_1,0}(r)P_{n_1}(\cos\theta_k)$$
$$+\sum_{m=1}^{n_1}\big[\boldsymbol{a}_{2n,n_1,m}(r)\cos m\phi_k$$
$$+\boldsymbol{b}_{2n,n_1,m}(r)\sin m\phi\big]P_{n_1}^m(\cos\theta_k)\Bigg\}, \quad (7.45)$$

where

$$\boldsymbol{a}_{2n,n_1,0}(\boldsymbol{r})=\frac{2n_1+1}{4\pi}$$
$$\times\int_0^{2\pi}\mathrm{d}\phi_k\int_0^{\pi}\frac{S_{2n}(r,\theta_k,\phi_k)}{\varepsilon_1\sin^2\theta_k+\varepsilon_3\cos^2\theta_k}\sin\theta_kP_{n_1}(\cos\theta_k)\,\mathrm{d}\theta_k, \quad (7.46)$$

$$\boldsymbol{a}_{2n,n_1,m}(\boldsymbol{r})=\frac{2n_1+1}{2\pi}\frac{(n_1-m)!}{(n_1+m)!}$$
$$\times\int_0^{2\pi}\mathrm{d}\phi_k\int_0^{\pi}\frac{S_{2n}(r,\theta_k,\phi_k)P_{n_1}^m(\cos\theta_k)}{\varepsilon_1\sin^2\theta_k+\varepsilon_3\cos^2\theta_k}\sin\theta_k\cos m\phi_k\,\mathrm{d}\theta_k, \quad (7.47)$$

$$\boldsymbol{b}_{2n,n_1,m}(\boldsymbol{r})=\frac{2n_1+1}{2\pi}\frac{(n_1-m)!}{(n_1+m)!}$$
$$\times\int_0^{2\pi}\mathrm{d}\phi_k\int_0^{\pi}\frac{S_{2n}(r,\theta_k,\phi_k)P_{n_1}^m(\cos\theta_k)}{\varepsilon_1\sin^2\theta_k+\varepsilon_3\cos^2\theta_k}\sin\theta_k\sin m\phi_k\,\mathrm{d}\theta_k. \quad (7.48)$$

It is noted that in Eqs. (7.46)–(7.48), n_1 should be $2n$. If is not equal to $2n$, the above three coefficients should be zero.

Substituting Eq. (7.45) into Eq. (7.42), with Eq. (7.44), it yields

$$\boldsymbol{E}(\boldsymbol{r}) = -\frac{\mathrm{i}\omega\mu_0}{(2\pi)^3}\sqrt{\frac{\pi}{2r}}\sum_{n=0}^{\infty}(4\pi)(-1)^n\left\{\boldsymbol{a}_{2n,2n,0}(\boldsymbol{r})P_{2n}(\cos\theta)\right.$$
$$\left.+\sum_{m=1}^{n}\left[\boldsymbol{a}_{2n,2n,m}(\boldsymbol{r})\cos m\phi + \boldsymbol{b}_{2n,2n,m}(\boldsymbol{r})\sin m\phi\right]P_{2n}^{m}(\cos\theta)\right\}. \tag{7.49}$$

By using the following relations, the representations in spherical and cylindrical coordinates can be obtained, respectively:

$$\begin{bmatrix} E_\rho \\ E_\phi \\ E_z \end{bmatrix} = \begin{bmatrix} \cos\phi & \sin\phi & 0 \\ -\sin\phi & \cos\phi & 0 \\ 0 & 0 & 1 \end{bmatrix}\begin{bmatrix} E_x \\ E_y \\ E_z \end{bmatrix}, \tag{7.50}$$

$$\begin{bmatrix} E_r \\ E_\theta \\ E_\phi \end{bmatrix} = \begin{bmatrix} \sin\theta\cos\phi & \sin\theta\sin\phi & \cos\theta \\ \cos\theta\cos\phi & \cos\theta\sin\phi & -\sin\theta \\ -\sin\phi & \cos\phi & 0 \end{bmatrix}\begin{bmatrix} E_x \\ E_y \\ E_z \end{bmatrix}. \tag{7.51}$$

It is noted that because the function $\boldsymbol{S}_{2n}$ and its denominator in Eqs. (7.46)–(7.48) are only concerned with $\cos^2\theta_k$, the expanded coefficients are non-zero only when n_1 refers to the even term. In the case of the electric dipole being parallel to the direction of the geomagnetic field, F_x and F_y are proportional to the factors $\sin\phi_k$ and $\cos\phi_k$, while F_z is not involved in ϕ_k. Therefore, for F_x and F_y, there is only one non-zero term with $m = 1$, while for F_z, there is only one non-zero term with $m = 0$. In the case of the unit electric dipole being vertical to the direction of the geomagnetic field, the functions F_x and F_y include only the terms with the factors $\sin 2\phi_k$ and $\cos 2\phi_k$ and the terms which are not concerned with ϕ_k, while the function F_z is proportional to the factor $\sin\phi_k$ and $\cos\phi_k$. Therefore, for F_x and F_y, there are two non-zero terms with $m = 0$ and $m = 2$, while for F_z, there is only one non-zero term with $m = 1$.

7.2.2 The Approximation of the Far Field

At large distance between the observation point and dipole source, namely $k_1 r \gg 1$ and $k_2 r \gg 1$, Eq. (7.41) is approximated by

$$\boldsymbol{S}_{2n} = \frac{\pi}{k_0^2(k_1^2 - k_2^2)}\sqrt{\frac{1}{2\pi r}}(-1)^n\left[\boldsymbol{F}(k_1,\theta_k,\phi_k)\mathrm{e}^{\mathrm{i}k_1 r} - \boldsymbol{F}(k_2,\theta_k,\phi_k)\mathrm{e}^{\mathrm{i}k_2 r}\right]. \tag{7.52}$$

With the substitution of Eq. (7.52) into Eq. (7.42), it follows that

$$\boldsymbol{E}(\boldsymbol{r}) = \frac{\mathrm{i}\omega\mu_0}{(2\pi)^3}\frac{\pi}{k_0^2}\frac{1}{2r}\int_0^{2\pi}\mathrm{d}\phi_k\int_0^{\pi}\sum_{n=1}^{\infty}\frac{\boldsymbol{S}_{2n}^{*}(r,\theta_k,\phi_k)\sin\theta_k}{\varepsilon_1\sin^2\theta_k + \varepsilon_3\cos^2\theta_k}\frac{(4n+1)}{(k_1^2 - k_2^2)}$$
$$\times P_{2n}\left[\cos\theta\cos\theta_k + \sin\theta\sin\theta_k\cos(\phi - \phi_k)\right]\mathrm{d}\theta_k, \tag{7.53}$$

where

$$\boldsymbol{S}^*_{2n}(r,\theta_k,\phi_k)=\boldsymbol{F}(k_1,\theta_k,\phi_k)\mathrm{e}^{\mathrm{i}k_1 r}-\boldsymbol{F}(k_2,\theta_k,\phi_k)\mathrm{e}^{\mathrm{i}k_2 r}. \tag{7.54}$$

The integrand in Eq. (7.53) can be expanded in scalar spherical harmonic functions (Abamowitz and Stegun 1972; Gradshteyn and Ryzhik 1980). Let

$$\begin{aligned}\frac{\boldsymbol{S}_{2n_1}(r,\theta_k,\phi_k)}{(k_1^2-k_2^2)(\varepsilon_1\sin^2\theta_k+\varepsilon_3\cos^2\theta_k)}&=\sum_{n_1=0}^{\infty}\Bigg\{\boldsymbol{a}_{2n_1}(r)P_{2n_1}(\cos\theta_k)+\sum_{m=1}^{2n_1}\big[\boldsymbol{a}_{2n_1,m}(r)\cos m\phi_k\\&\quad+\boldsymbol{b}_{2n_1,m}(r)\sin m\phi\big]P_{2n_1}^m(\cos\theta_k)\Bigg\},\end{aligned} \tag{7.55}$$

then it is seen that the coefficients must be similar to those in Eqs. (7.46)–(7.48). Considering the orthogonality of spherical harmonic functions, we obtain readily

$$\begin{aligned}\boldsymbol{E}(\boldsymbol{r})=\left[\frac{\mathrm{i}\omega\mu_0}{(2\pi)^3}\frac{\pi}{k_0^2}\frac{1}{2r}\right]\frac{1}{4\pi}\sum_{n_1=1}^{\infty}\Bigg\{&\boldsymbol{a}_{2n_1}(r)P_{2n_1}(\cos\theta)\\&+\sum_{m=1}^{2n_1}\big[\boldsymbol{a}_{2n_1,m}(r)\cos m\phi_k+\boldsymbol{b}_{2n_1,m}(r)\sin m\phi\big]P_{n_1}^m(\cos\theta)\Bigg\}.\end{aligned} \tag{7.56}$$

Comparing Eqs. (7.55) and (7.56) yields

$$\boldsymbol{E}(\boldsymbol{r})=\frac{\mathrm{i}\omega\mu_0}{(2\pi)^3}\frac{1}{8k_0^2 r}\frac{\boldsymbol{S}(r,\theta,\phi)}{(\varepsilon_1\sin^2\theta+\varepsilon_3\cos^2\theta)(k_1^2-k_2^2)|_{\theta_k=\theta}}, \tag{7.57}$$

where

$$\boldsymbol{S}(r,\theta,\phi)=\boldsymbol{F}(k_1,\theta,\phi)\mathrm{e}^{\mathrm{i}k_1 r}-\boldsymbol{F}(k_2,\theta,\phi)\mathrm{e}^{\mathrm{i}k_2 r}. \tag{7.58}$$

So far, the approximated formulas of the electric field in the far-field region radiated by unit electric dipole in an infinite anisotropic plasma are derived readily.

If the unit electric dipole is parallel to the direction of the geomagnetic field, namely, $\boldsymbol{J}=\hat{z}$, then we have

$$\begin{aligned}\boldsymbol{F}\big(k_j(\theta),\theta\big)&=\hat{r}\big[k_j^4(\theta)-2k_j^2(\theta)k_0^2\varepsilon_1+k_0^4\big(\varepsilon_1^2-\varepsilon_2^2\big)\big]\cos\theta\\&\quad+\hat{\theta}\big[k_j^2(\theta)k_0^2\varepsilon_1-k_0^4\big(\varepsilon_1^2-\varepsilon_2^2\big)\big]\sin\theta\\&\quad+\hat{\phi}\mathrm{i}k_0^2k_j^2(\theta)\varepsilon_2\sin\theta\cos\theta;\quad j=1,2.\end{aligned} \tag{7.59}$$

If the unit electric dipole is vertical to the direction of the geomagnetic field, and the electric dipole is in the $\hat{x}$ direction, namely, $\boldsymbol{J} = \hat{x}$, then we have

$$\boldsymbol{F}\big(k_j(\theta), \theta, \phi\big) = \hat{r} F_r + \hat{\theta} F_\theta + \hat{\phi} F_\phi; \quad j = 1, 2, \tag{7.60}$$

where

$$F_r = \big[k_j^4 \cos\phi - k_0^2 k_j^2 (\varepsilon_1 \cos\phi + i\varepsilon_2 \sin\phi + \varepsilon_3 \cos\phi) + k_0^4 (\varepsilon_1 \cos\phi + i\varepsilon_2 \sin\phi)\varepsilon_3\big] \sin\theta, \tag{7.61}$$

$$F_\theta = \big[-k_0^2 k_j^2 \varepsilon_3 \cos\phi + k_0^4 (\varepsilon_1 \cos\phi + i\varepsilon_2 \sin\phi)\varepsilon_3\big] \cos\theta, \tag{7.62}$$

$$F_\phi = k_0^2 k_j^2 \big[(\varepsilon_1 \sin\phi - i\varepsilon_2 \cos\phi) \sin^2\theta + \varepsilon_3 \cos^2\theta \sin\phi\big] - k_0^4 (\varepsilon_1 \sin\phi - i\varepsilon_2 \cos\phi)\varepsilon_3. \tag{7.63}$$

In Eq. (7.57), k_1 and k_2 refer to the wave numbers of the ordinary and extraordinary waves, respectively. The wave numbers k_1 and k_2 are determined by Eq. (7.25) and the Sommerfeld radiation condition, namely, $\mathrm{Im}(k_j) > 0$ and $\mathrm{Re}(k_j) > 0$.

When the three components of the electric field are obtained, the three components of the magnetic field can be derived readily by using the following relations:

$$H_r = \frac{1}{i\omega\mu_0} \frac{1}{r \sin\theta} \left[\frac{\partial}{\partial\theta}(\sin\theta E_\phi) - \frac{\partial E_\theta}{\partial\phi}\right], \tag{7.64}$$

$$H_\theta = \frac{1}{i\omega\mu_0} \frac{1}{r} \left[\frac{1}{\sin\theta} \frac{\partial E_r}{\partial\phi} - \frac{\partial}{\partial r}(r E_\phi)\right], \tag{7.65}$$

$$H_\phi = \frac{1}{i\omega\mu_0} \frac{1}{r} \left[\frac{\partial}{\partial r}(r E_\theta) - \frac{\partial E_r}{\partial\theta}\right]. \tag{7.66}$$

7.2.3 Analyses, Discussions, and Computations

From Eqs. (7.57) and (7.58), it is seen that the electromagnetic wave, which is radiated by an electric dipole in an infinite anisotropic plasma, can be expressed in the sum of the two waves with the wave numbers k_1 and k_2. The wave numbers k_1 and k_2 are determined by the operating frequency, the ionospheric parameters ε_1, ε_2, and ε_3, and the angle θ between the propagating direction and that of the geomagnetic field. We write

$$\frac{k_{1,2}^2}{k_0^2} = \frac{1}{2(\varepsilon_1 \sin^2\theta + \varepsilon_3 \cos^2\theta)} \Big[\varepsilon_1\varepsilon_3\big(1 + \cos^2\theta\big) + \big(\varepsilon_1^2 - \varepsilon_2^2\big) \sin^2\theta \pm \sqrt{\big(\varepsilon_1^2 - \varepsilon_2^2 - \varepsilon_1\varepsilon_3\big)^2 \sin^4\theta + 4\varepsilon_2^2\varepsilon_3^2 \cos^2\theta}\Big]. \tag{7.67}$$

With the substitutions of Eqs. (7.2)–(7.7) into Eq. (7.67), the complex refractive index n can be expressed as follows:

$$n_{\substack{o\\e}}^2 = \frac{k_{1,2}^2}{k_0^2} = 1 - \frac{X}{U - \frac{Y_T^2}{2(U-X)} \pm \sqrt{\frac{Y_T^4}{4(U-X)^2} + Y_L^2}}, \tag{7.68}$$

where $Y_L = Y\cos\theta$ represents the component parallel to the direction of the geomagnetic field (longitudinal direction), while $Y_T = Y\sin\theta$ represents that vertical to the direction of the geomagnetic field. Equation (7.68) is the well-known *Appleton–Hartee Equation* in the ionospheric wave propagation.

In SLF/ELF ranges, it is known that $X \gg Y^2 \gg 1$. From Eq. (7.68), it is seen that n_o^2 is close to a large negative real number, and n_o has a positive imaginary part. That is to say, the wave of e^{ik_1r} is an evanescent wave. When $Y_L > \frac{Y_T^2}{2(U-X)}$, n_e^2 is close to a large positive real number, and n_e is a positive real number. That is to say, the wave of e^{ik_2r} is a propagating wave.

When the propagating direction is vertical to the direction of the geomagnetic field, namely, $\theta = \frac{\pi}{2}$, both n_o^2 and n_e^2 are close to a negative real number. Namely, in this case, both the o and e waves are evanescent. The critical angle θ_A exists. When $\theta < \frac{\pi}{2} - \theta_A$ or $\theta > \frac{\pi}{2} + \theta_A$, the extraordinary wave is a propagating wave, while the ordinary wave is evanescent. When $\frac{\pi}{2} - \theta_A < \theta < \frac{\pi}{2} + \theta_A$, both the extraordinary and the ordinary waves are evanescent. From Eq. (7.68), the critical angle θ_A is approximated by

$$\theta_A \approx \sin^{-1}\sqrt{\frac{1 + \frac{Y^2}{X}}{Y^2}}. \tag{7.69}$$

When the observation point is far away from the dipole source, there only exists the extraordinary wave. Then, the electric field can be simplified as follows:

$$\begin{aligned} \boldsymbol{E}(\boldsymbol{r}) &= \frac{i\omega\mu_0 k_0 I\,dl}{64\pi^3} \frac{\boldsymbol{F}(k_2,\theta,\phi)}{(\varepsilon_1\sin^2\theta + \varepsilon_3\cos^2\theta)k_0^2(k_2^2 - k_1^2)|_{\theta_k=\theta}} \frac{e^{ik_2r}}{k_0 r} \\ &= \frac{ik_0^2\eta I\,dl}{64\pi^3} \cdot \boldsymbol{G}(k_2,\theta,\phi) \cdot \frac{e^{ik_2r}}{k_0 r}, \end{aligned} \tag{7.70}$$

where

$$\boldsymbol{G}(k_2,\theta,\phi) = -\frac{\boldsymbol{F}(k_2,\theta,\phi)}{k_0^4\sqrt{(\varepsilon_1^2 - \varepsilon_2^2 - \varepsilon_1\varepsilon_3)^2\sin^4\theta + 4\varepsilon_2\varepsilon_3\cos^2\theta}}. \tag{7.71}$$

In the far-field region, we have $nk_0r \gg 1$, and a spherical wave excited by a dipole source can be approximated as a plane wave. In this case, the spatial changes in the wave front, which is perpendicular to the propagation direction, can be neglected. Then, we have $\frac{E_\phi}{E_\theta} \approx -i$, $H_r \approx 0$, $H_\theta = -\frac{n_e}{\eta}E_\phi$, and $H_\phi = \frac{n_e}{\eta}E_\theta$.

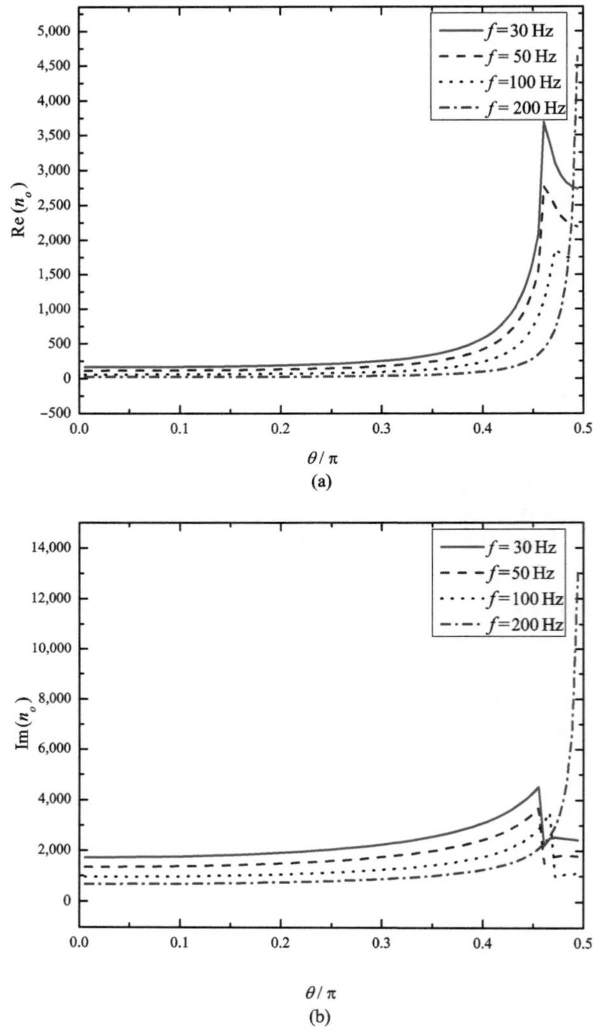

Fig. 7.1 The real and imaginary parts of n_o versus the angle θ in SLF range

The radiated field of an electric dipole in an infinite anisotropic ionosphere can be easily calculated. In this section, the parameters are taken as $B_0 = 0.5 \times 10^{-4}$ T, $\omega_{0e} = 6.67 \times 10^7$ arc/s, $\omega_{0i} = 3.575 \times 10^5$ arc/s, $\omega_H^{(e)} = 8.8 \times 10^6$ arc/s, $\omega_H^{(i)} = 2.532 \times 10^2$ arc/s, and $\nu_e = \nu_i = 10^3$ s^{-1}, which are the same as in the available reference (Bannister et al. 1993). The refractive indices n_o and n_e are computed at $f = 30$ Hz, 50 Hz, 100 Hz, and 200 Hz and shown in Figs. 7.1 and 7.2, respectively. It is found that in the direction near $\theta = 90°$, the changes of the refractive index are more complicated.

From the computations on $\text{Im}(n_o)$, it is seen that the imaginary part of n_o is very large. Namely, the ordinary wave is an evanescent wave. From the computations on $\text{Im}(n_e)$, it is seen that the imaginary part of n_e is the minimum in the direction of

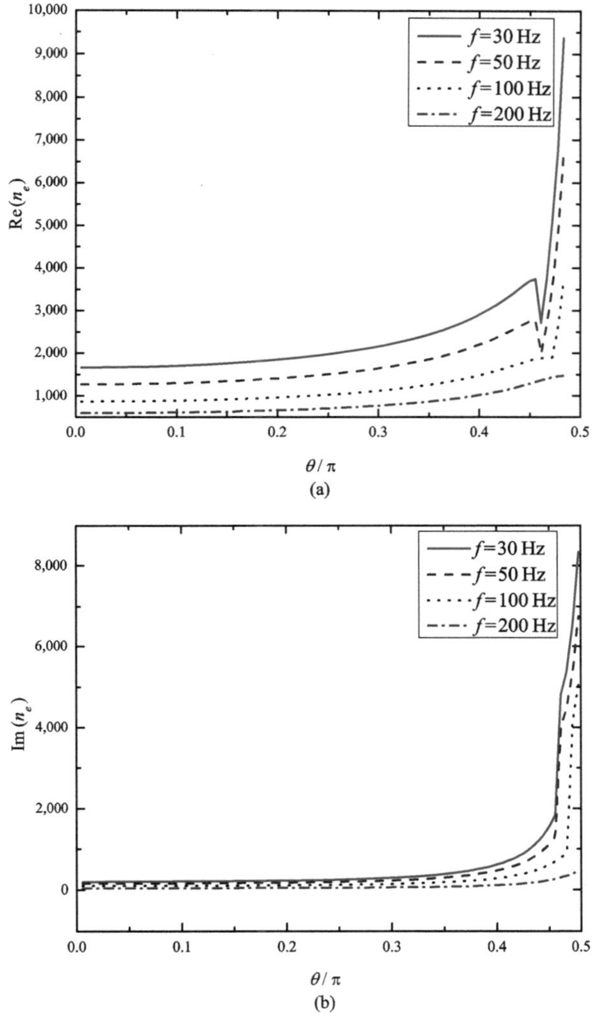

Fig. 7.2 The real and imaginary parts of n_e versus the angle θ in SLF range

$\theta = 0°$, i.e., the propagation loss along the magnetic field line is minimum. When θ is close to 90°, $\mathrm{Im}(n_e)$ increases rapidly. Furthermore, $\mathrm{Im}(n_e)$ is sensitive to the value of $\frac{\nu}{\omega}$. When the collision is neglected, $\mathrm{Im}(n_e)$ is very small. When the collision is considered, $\mathrm{Im}(n_e)$ increases as the operating frequency decreases.

In ELF range, because the ion collision frequency and ion gyro-frequency can be comparable to the operating frequency, the effect by the ions should not be neglected. With the same parameters as in Figs. 7.1 and 7.2, the real and imaginary parts of n_o and n_e versus the angle θ are computed for $f = 10$ Hz, and 30 Hz and shown in Figs. 7.3 and 7.4. It is seen that the effects on the real part of n_o and the imaginary part of n_e are very large, while the effects on the real part of n_e and the imaginary part of n_o are very small. In ELF range, the ion effect must be considered.

Fig. 7.3 The real and imaginary parts of n_o versus the angle θ at ELF range

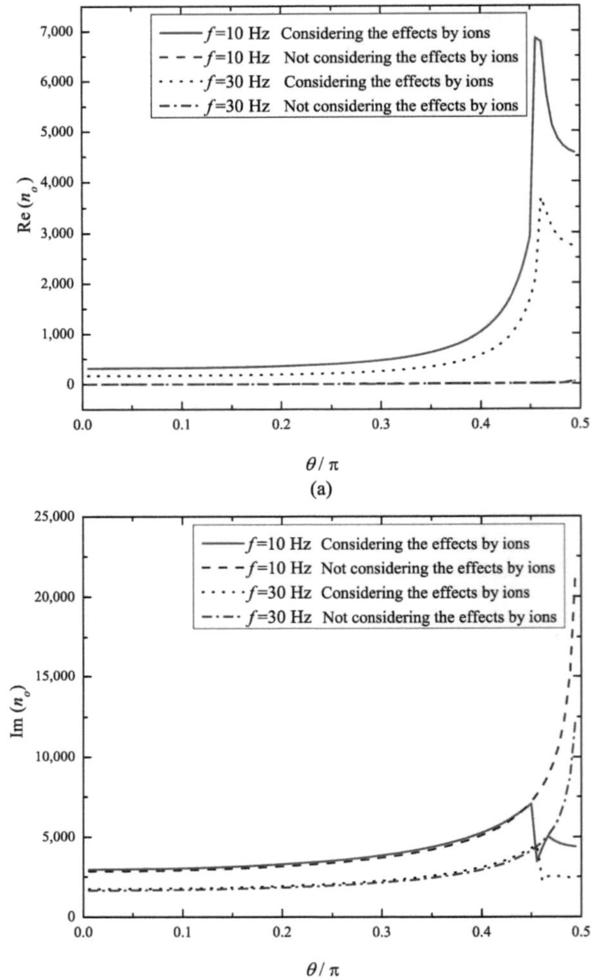

With the same parameters as in Fig. 7.1–7.4, as shown in Fig. 7.5, the function of the electric field component $|E_r|$ of the angle θ is computed at $r = 10$ km and 20 km, respectively. It is noted that the component $|E_r|$ is radiated by a unit electric dipole which is parallel to the direction of the geomagnetic field. From Fig. 7.5, it is seen that in the SLF range, the magnitude of the field component shows little changes when the angle θ is not close to 0°, 180°, or 90°. When θ is close to 90°, the field components decrease rapidly. The larger the distance, and the higher the operating frequency is, the more extensive the component decays is.

In the above computations, a unit dipole source is employed. Namely, it is taken as $I\,dl = 1$ A·m. If the moment of an electric dipole is assumed to be 100 A × 1 km $= 10^5$ A·m, the propagating distance is $r = 100$ km, the angle θ is taken as any angular degree except for 0°, 90°, and 180°, and the rest of the parameters are

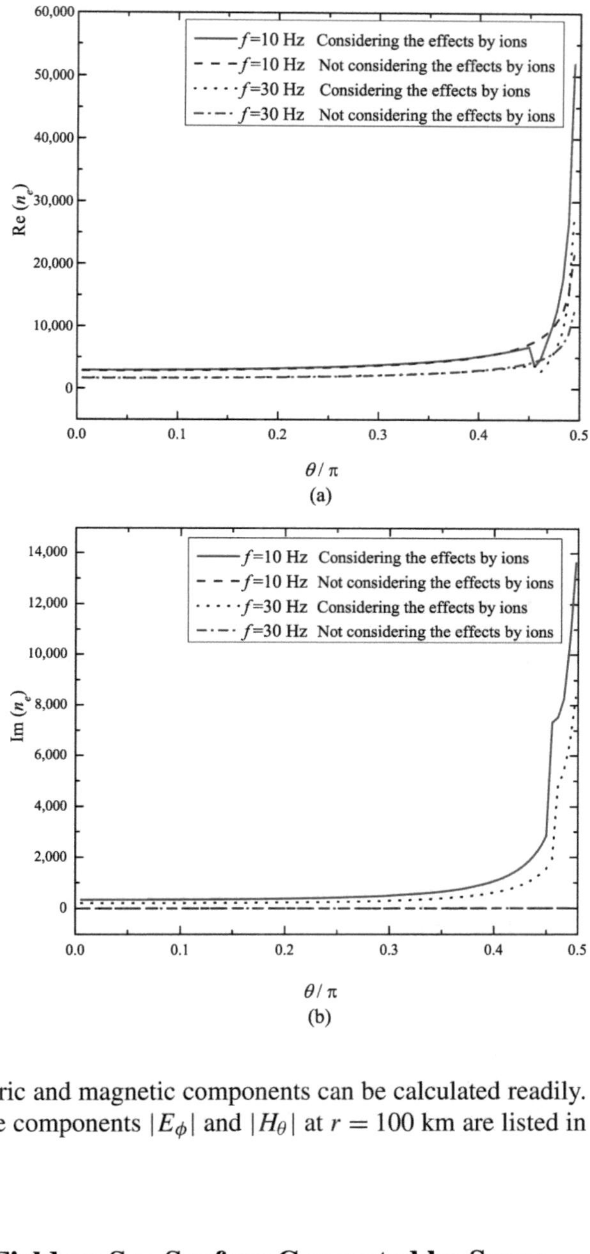

Fig. 7.4 The real and imaginary parts of n_e versus the angle θ at ELF range

assumed as above, the electric and magnetic components can be calculated readily. The evaluated ranges for the components $|E_\phi|$ and $|H_\theta|$ at $r = 100$ km are listed in Table 7.1.

7.3 Electromagnetic Field on Sea Surface Generated by Space Borne SLF Transmitter

In this section, the ionosphere is regarded as a homogeneous anisotropic plasma with sharp boundary, we will treat analytically the electromagnetic field on the sea surface generated by the space borne SLF transmitting antenna.

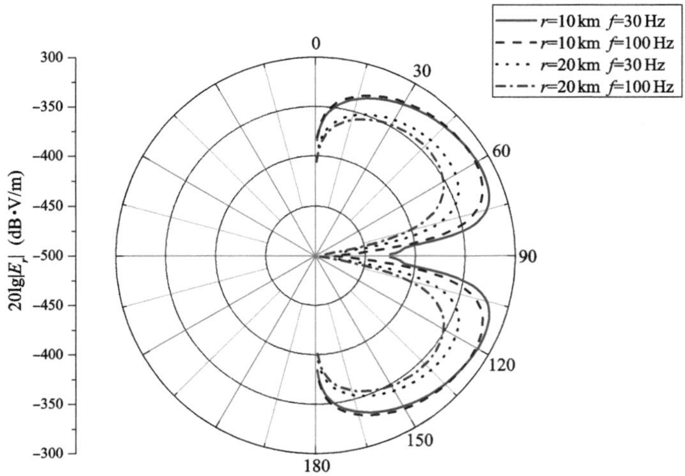

Fig. 7.5 The magnitude of $|E_r|$ versus the angle θ at $f = 30$ Hz: $r = 10$ km and 20 km

Table 7.1 The evaluated ranges for $|E_\phi|$ and $|H_\theta|$ at different frequencies

Frequency	Component	
	$\lvert E_\phi \rvert$ dB·V/m	$\lvert H_\theta \rvert$ dB·A/m
$f = 30$ Hz	$[-368, -346]$	$[-361, -333]$
$f = 100$ Hz	$[-354, -325]$	$[-349, -321]$
$f = 300$ Hz	$[-340, -313]$	$[-332, -311]$

7.3.1 The Representations of the Field Components in the Ionosphere

The geometry and notation of consideration are shown in Fig. 7.6. The geomagnetic field $\boldsymbol{B}_0$, which has the angle θ_b with the z-direction, is in the x–z plane. The ionosphere is regarded as a homogeneous anisotropic plasma with sharp boundary, which is characterized by the tensor permittivity $\bar{\bar{\varepsilon}}$ (Budden 1961).

Letting l, m, and n be the direction cosines of the geomagnetic field $\boldsymbol{B}_0$ in the $\hat{x}$, $\hat{y}$, and $\hat{z}$ directions, we have

$$l_b = \sin\theta_b; \qquad m_b = 0; \qquad n_b = \cos\theta_b. \tag{7.72}$$

In VLF range and the higher end of SLF range, the effect of the ions can be neglected. Then, the tensor permittivity $\bar{\bar{\varepsilon}}$ of the ionosphere can be written in the 3×3 matrix form. It is

$$\hat{\varepsilon} = \varepsilon_0[\mathbf{I} + \mathbf{M}], \tag{7.73}$$

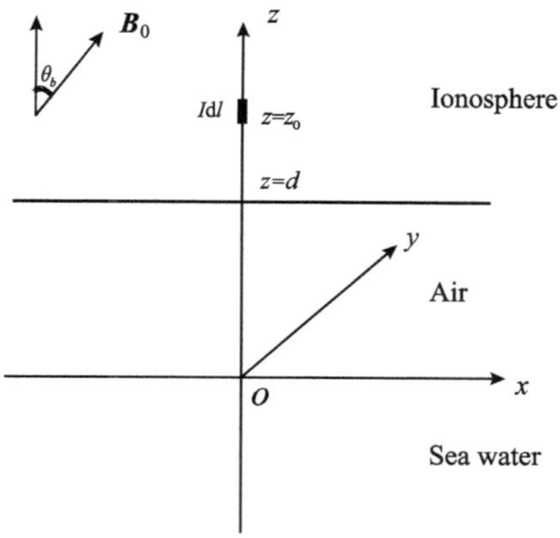

Fig. 7.6 The physical model for the space borne SLF transmitting antenna

where ε_0 is the free-space permittivity, **I** is 3 × 3 unit matrix, and **M** is the susceptibility of the ionosphere. We write

$$\mathbf{M} = \frac{-X}{U(U^2 - Y^2)} \begin{bmatrix} U^2 - l_b^2 Y^2 & \mathrm{i} n_b Y U & -l_b n_b Y^2 \\ -\mathrm{i} n_b Y U & U^2 & \mathrm{i} l_b Y U \\ -l_b n_b Y^2 & -\mathrm{i} l_b Y U & U^2 - n_b^2 Y^2 \end{bmatrix}, \tag{7.74}$$

where

$$U = 1 + \mathrm{i}\frac{\nu}{\omega}, \tag{7.75}$$

$$X = \frac{\omega_0^2}{\omega^2}; \qquad \omega_0^2 = \frac{Ne^2}{\varepsilon_0 m}, \tag{7.76}$$

$$Y = \frac{\omega_H}{\omega}; \qquad \omega_H = \left| \frac{eB_0}{m} \right|. \tag{7.77}$$

In the above formulas, ω is the operating angular frequency; **M** is the susceptibility matrix for electrons, respectively; ω_0 is the angular plasma frequency for electrons; and N is the electron density; ω_H is the electron gyro-frequency; l_b and n_b are the directional cosine of the geomagnetic field in the x and z direction, respectively.

In the ionosphere, the electromagnetic field should satisfy Maxwell's equations. We write

$$\nabla \times \boldsymbol{E}_j = \mathrm{i}\omega\mu_0 \boldsymbol{H}_j, \tag{7.78}$$

$$\nabla \times \boldsymbol{H}_j = -\mathrm{i}\omega\hat{\varepsilon}_j \boldsymbol{E} + \hat{z} I \,\mathrm{d}l \delta(x)\delta(y)\delta(z - z_0), \tag{7.79}$$

where the subscript $j = i, a$. It is noted that i and a indicate the ionosphere and the air, respectively.

We introduce the Fourier transformations of the electromagnetic field,

$$\boldsymbol{e}_j(\boldsymbol{k}, z) = \frac{1}{4\pi^2}\int_{-\infty}^{\infty}\int_{-\infty}^{\infty} \boldsymbol{E}_j(\boldsymbol{r})\mathrm{e}^{-\mathrm{i}\boldsymbol{k}\cdot\boldsymbol{\rho}}\,\mathrm{d}x\,\mathrm{d}y, \tag{7.80}$$

$$\boldsymbol{h}_j(\boldsymbol{k}, z) = \frac{\eta}{4\pi^2}\int_{-\infty}^{\infty}\int_{-\infty}^{\infty} \boldsymbol{H}_j(\boldsymbol{r})\mathrm{e}^{-\mathrm{i}\boldsymbol{k}\cdot\boldsymbol{\rho}}\,\mathrm{d}x\,\mathrm{d}y. \tag{7.81}$$

The corresponding inverse transformations are written in the forms

$$\boldsymbol{E}_j(\boldsymbol{r}) = \int_{-\infty}^{\infty}\int_{-\infty}^{\infty} \boldsymbol{e}_j(\boldsymbol{k}, z)\mathrm{e}^{\mathrm{i}\boldsymbol{k}\cdot\boldsymbol{\rho}}\,\mathrm{d}k_x\,\mathrm{d}k_y, \tag{7.82}$$

$$\boldsymbol{H}_j(\boldsymbol{r}) = \frac{1}{\eta}\int_{-\infty}^{\infty}\int_{-\infty}^{\infty} \boldsymbol{h}_j(\boldsymbol{k}, z)\mathrm{e}^{\mathrm{i}\boldsymbol{k}\cdot\boldsymbol{\rho}}\,\mathrm{d}k_x\,\mathrm{d}k_y, \tag{7.83}$$

where η is the wave impedance of free space, $\boldsymbol{k} = \hat{x}k_x + \hat{y}k_y$, and $\boldsymbol{\rho} = \hat{x}x + \hat{y}y$.

Substituting Eqs. (7.82) and (7.83) into the Maxwell equations in the ionosphere, and eliminating the vertical components e_z and h_z, the following matrix equation is obtained readily:

$$\frac{\mathrm{d}\boldsymbol{W}}{\mathrm{d}z} = -\mathrm{i}k_0 \cdot \mathbf{T} \cdot \boldsymbol{W} + \boldsymbol{f}_e\delta(z - z_0), \tag{7.84}$$

where k_0 is the wave number in free space. The column vectors $\boldsymbol{W}$ and $\boldsymbol{f}_e$ are represented as follows:

$$\boldsymbol{W} = \begin{bmatrix} e_x^i & e_y^i & h_x^i & h_y^i \end{bmatrix}^{\mathrm{T}}, \tag{7.85}$$

$$\boldsymbol{f}_e = \frac{I\,\mathrm{d}l \cdot \eta}{4\pi^2(1 + M_{zz})}\begin{bmatrix} \frac{k_x}{k_0} & \frac{k_y}{k_0} & -M_{yz} & M_{xz} \end{bmatrix}^T. \tag{7.86}$$

The 4×4 matrix $\mathbf{T}$ is written as

$$\mathbf{T} = \begin{bmatrix} \frac{k_x M_{zx}}{k_0(1+M_{zz})} & \frac{k_x M_{zy}}{k_0(1+M_{zz})} & -\frac{k_x k_y}{k_0^2(1+M_{zz})} & \frac{k_x^2}{k_0^2(1+M_{zz})} - 1 \\ \frac{k_y M_{zx}}{k_0(1+M_{zz})} & \frac{k_y M_{zy}}{k_0(1+M_{zz})} & 1 - \frac{k_y^2}{k_0^2(1+M_{zz})} & \frac{k_x k_y}{k_0^2(1+M_{zz})} \\ T_{31} & T_{32} & \frac{k_y M_{yz}}{k_0(1+M_{zz})} & -\frac{k_x M_{yz}}{k_0(1+M_{zz})} \\ T_{41} & T_{42} & -\frac{k_y M_{xz}}{k_0(1+M_{zz})} & \frac{k_x M_{xz}}{k_0(1+M_{zz})} \end{bmatrix}, \tag{7.87}$$

where

$$T_{31} = M_{yx} - \frac{M_{yz}M_{zx}}{1 + M_{zz}} + \frac{k_x k_y}{k_0^2}; \qquad T_{32} = 1 + M_{yy} - \frac{M_{yz}M_{zy}}{1 + M_{zz}} - \frac{k_x^2}{k_0^2}, \tag{7.88}$$

$$T_{41} = -1 - M_{xx} + \frac{M_{xz}M_{zx}}{1 + M_{zz}} + \frac{k_y^2}{k_0^2}; \qquad T_{42} = -M_{xy} + \frac{M_{xz}M_{zy}}{1 + M_{zz}} - \frac{k_x k_y}{k_0^2}. \tag{7.89}$$

In this section, the ionosphere is idealized as a homogeneous anisotropic plasma with uniformly sharp boundary, it is seen that all elements in the matrix **T**, which do not vary with the changes in spatial position, are determined by the parameters including electron density, collision frequency, and the operating frequency. For the matrix equation (7.87), it is seen that there are four eigenvalues and the corresponding four eigenvectors. Generally, considering that the ionosphere is a loss medium with $\nu \neq 0$, the four eigenvalues are complex numbers, where the two in the lower half plane with $\mathrm{Im}\,\lambda < 0$ are denoted λ_1 and λ_2, and the other two in the upper lower half plane with $\mathrm{Im}\,\lambda > 0$ are denoted λ_3 and λ_4. Obviously, the eigenvalues λ_1 and λ_2 correspond to the downward characteristic waves, while λ_3 and λ_4 correspond to the upward characteristic waves. We assume that $\boldsymbol{W}^{(j)}$ represents the eigenvector, which corresponds to the eigenvalue λ_j. Then, the transformations of the electromagnetic field components, which is generated by an electric dipole, can be expressed as follows:

$$\begin{aligned}\widetilde{\boldsymbol{W}} &= \left[C_1 \mathrm{e}^{\mathrm{i}k_0\lambda_1(z-z_0)}\boldsymbol{W}^{(1)} + C_2 \mathrm{e}^{\mathrm{i}k_0\lambda_2(z-z_0)}\boldsymbol{W}^{(2)}\right]\cdot U(z_0 - z) \\ &\quad + \left[C_3 \mathrm{e}^{\mathrm{i}k_0\lambda_3(z-z_0)}\boldsymbol{W}^{(3)} + C_4 \mathrm{e}^{\mathrm{i}k_0\lambda_4(z-z_0)}\boldsymbol{W}^{(4)}\right]\cdot U(z - z_0). \end{aligned} \tag{7.90}$$

It is noted that the Heaviside step function $U(x)$ in Eq. (7.90) is defined as follows:

$$U(x) = \begin{cases} 1; & x \geq 0, \\ 0; & x < 0. \end{cases} \tag{7.91}$$

With the substitution of Eq. (7.90) into Eq. (7.84), we obtain readily

$$-C_1\boldsymbol{W}^{(1)} - C_2\boldsymbol{W}^{(2)} + C_3\boldsymbol{W}^{(3)} + C_4\boldsymbol{W}^{(4)} = \boldsymbol{f}_e. \tag{7.92}$$

When all eigenvalues and eigenvectors of the matrix **T** are obtained, the excitation coefficients C_j $(j = 1, 2, 3, 4)$ can be solved by the following equation:

$$\begin{bmatrix} C_1 & C_2 & C_3 & C_4 \end{bmatrix} = \mathbf{G}^{-1}\cdot \boldsymbol{f}_e, \tag{7.93}$$

where the matrix **G** is expressed in the form of

$$\mathbf{G} = \begin{bmatrix} -W_1^{(1)} & -W_1^{(2)} & W_1^{(3)} & W_1^{(4)} \\ -W_2^{(1)} & -W_2^{(2)} & W_2^{(3)} & W_2^{(4)} \\ -W_3^{(1)} & -W_3^{(2)} & W_3^{(3)} & W_3^{(4)} \\ -W_4^{(1)} & -W_4^{(2)} & W_4^{(3)} & W_4^{(4)} \end{bmatrix}. \tag{7.94}$$

Considering the path that the waves propagate along from the ionosphere to the air until the sea surface, the waves will encounter the reflections by the air–ionosphere interface. Thus, the transformations of the electromagnetic field components can be expressed as follows:

$$\boldsymbol{W} = R_1\boldsymbol{W}^{(3)}\mathrm{e}^{\mathrm{i}k_0\lambda_3(z-d)} + R_2\boldsymbol{W}^{(4)}\mathrm{e}^{\mathrm{i}k_0\lambda_4(z-d)} + \widetilde{\boldsymbol{W}}. \tag{7.95}$$

When $d < z < z_0$, we write

$$\begin{aligned} \boldsymbol{W} &= R_1 \boldsymbol{W}^{(3)} \mathrm{e}^{\mathrm{i}k_0\lambda_3(z-d)} + R_2 \boldsymbol{W}^{(4)} \mathrm{e}^{\mathrm{i}k_0\lambda_4(z-d)} \\ &\quad + C_1 \boldsymbol{W}^{(1)} \mathrm{e}^{\mathrm{i}k_0\lambda_1(z-z_0)} + C_2 \boldsymbol{W}^{(2)} \mathrm{e}^{\mathrm{i}k_0\lambda_2(z-z_0)}. \end{aligned} \tag{7.96}$$

7.3.2 The Representations of the Field Components in the Air

Similarly, with the substitutions of Eqs. (7.80) and (7.81) into the Maxwell equations in the air, the components e_y^0, e_z^0, h_y^0, and h_z^0 can be expressed in terms of the components e_x^0 and h_x^0. The components e_x^0 and h_x^0 should satisfy the following differential equations. We write

$$\frac{\partial^2}{\partial^2 z} e_x^0 + \gamma^2 e_x^0 = 0; \qquad \frac{\partial^2}{\partial^2 z} h_x^0 + \gamma^2 h_x^0 = 0, \tag{7.97}$$

where $\gamma = k_0^2 - k_x^2 - k_y^2$. Obviously, the general solution of Eq. (7.97) can be represented in the form of $\mathrm{e}^{\pm i\gamma z}$.

On the sea surface, $z = 0$, the field components should satisfy the impedance condition, which is expressed in the form of

$$-\left.\frac{e_x^0}{h_y^0}\right|_{z=0} = \left.\frac{e_y^0}{h_x^0}\right|_{z=0} = \Delta_{\text{sea}}, \tag{7.98}$$

where Δ_{sea} is the normalized surface impedance, $\Delta_{\text{sea}} = \sqrt{\mathrm{i}\omega\varepsilon_0/\sigma_{\text{sea}}}$. In SLF/ELF ranges, Δ_{sea} can be approximately taken as 0. Thus, the Fourier transformations e_x^0 and h_x^0 are expressed as follows:

$$e_x^0 = T_1 \sin(\gamma z); \qquad h_x^0 = T_2 \cos(\gamma z). \tag{7.99}$$

At the air–ionosphere interface, $z = d$, the tangential components in the air are written readily. They are

$$e_x^0\big|_{z=d} = T_1 \sin(\gamma d), \tag{7.100}$$

$$e_y^0\big|_{z=d} = \frac{\sin(\gamma d)}{k_0^2 - k_x^2}(-k_x k_y T_1 - \mathrm{i}k_0 \gamma T_2), \tag{7.101}$$

$$h_x^0\big|_{z=d} = T_1 \cos(\gamma d), \tag{7.102}$$

$$h_y^0\big|_{z=d} = \frac{\cos(\gamma d)}{k_0^2 - k_x^2}(-\mathrm{i}k_0 \gamma T_1 - k_x k_y T_2). \tag{7.103}$$

With boundary conditions at the air–ionosphere interface, $z = d$, the following four equations are obtained readily. We write

$$R_1 W_1^{(3)} + R_2 W_1^{(4)} + C_1 W_1^{(1)} \mathrm{e}^{\mathrm{i}k_0\lambda_1(d-z_0)} + C_2 W_1^{(2)} \mathrm{e}^{\mathrm{i}k_0\lambda_2(d-z_0)}$$
$$= T_1 \sin(\gamma d), \tag{7.104}$$
$$R_1 W_2^{(3)} - R_2 W_2^{(4)} + C_1 W_2^{(1)} \mathrm{e}^{\mathrm{i}k_0\lambda_1(d-z_0)} + C_2 W_2^{(2)} \mathrm{e}^{\mathrm{i}k_0\lambda_2(d-z_0)}$$
$$= \frac{\sin(\gamma d)}{k_0^2 - k_x^2}(-k_x k_y T_1 - \mathrm{i}k_0\gamma T_2), \tag{7.105}$$
$$R_1 W_3^{(3)} + R_2 W_3^{(4)} + C_1 W_3^{(1)} \mathrm{e}^{\mathrm{i}k_0\lambda_1(d-z_0)} + C_2 W_3^{(2)} \mathrm{e}^{\mathrm{i}k_0\lambda_2(d-z_0)}$$
$$= T_2 \cos(\gamma d), \tag{7.106}$$
$$R_1 W_4^{(3)} + R_2 W_4^{(4)} + C_1 W_4^{(1)} \mathrm{e}^{\mathrm{i}k_0\lambda_1(d-z_0)} + C_2 W_4^{(2)} \mathrm{e}^{\mathrm{i}k_0\lambda_2(d-z_0)}$$
$$= \frac{\cos(\gamma d)}{k_0^2 - k_x^2}(-k_x k_y T_2 - \mathrm{i}k_0\gamma T_1). \tag{7.107}$$

In Eqs. (7.104)–(7.107), there are four unknown coefficients T_1, T_2, R_1, and R_2. When the excitation coefficients C_1 and C_2 are obtained, T_1 and T_2 are determined readily. The Fourier transformations of the field components e_x^0 and h_x^0 can be obtained easily by Eq. (7.99). The transformations of the remaining four components e_y^0, e_z^0, h_y^0, and h_z^0 are also obtained readily.

7.4 The Quasi-Longitudinal Approximation

In what follows, we will attempt to treat the case of the quasi-longitudinal approximation. When the horizontal wave numbers k_x and k_y are given, the eigenvalues and eigenvectors of the matrix $\mathbf{T}$ may be easily obtained by using the numerical method. For SLF/ELF waves traveling downward from the ionosphere into the air, due to the relative refractive index in the ionosphere being much larger than 1, SLF/ELF waves propagate mainly along the vertical direction. Thus we may only consider the case which is near the normal of the boundary, namely, $k_x \to 0$ and $k_y \to 0$. Additionally, because of $m_e \ll m_i$, generally, the condition of $Y_i \ll Y$ is satisfied. Thus, when the angle θ_b does not approach $\frac{\pi}{2}$, by ignoring some small quantities, the matrix $\mathbf{T}$ is simplified as follows:

$$\mathbf{T} = \begin{bmatrix} \frac{l_b k_x}{n_b k_0} & 0 & 0 & -1 \\ \frac{l_b k_y}{n_b k_0} & 0 & 1 & 0 \\ -\frac{\mathrm{i}X}{n_b y} & \frac{X}{n_b^2 y^2} & 0 & 0 \\ -\frac{X}{n_b^2 y^2} & -\frac{\mathrm{i}X}{n_b y} & -\frac{l_b k_y}{n_b k_0} & \frac{l_b k_x}{n_b k_0} \end{bmatrix}. \tag{7.108}$$

The characteristic equation of the matrix **T** is written in the form

$$\lambda^4 - \frac{2l_b k_x}{n_b k_0}\lambda^3 - \left(\frac{2X}{n_b^2 y^2} - \frac{l_b^2 k_x^2}{n_b^2 k_0^2}\right)\lambda^2 + \frac{2l_b k_x}{n_b k_0}\frac{X}{n_b^2 y^2}\lambda$$
$$-\frac{X^2}{n_b^2 y^2} + \frac{X^2}{n_b^4 y^4} - \frac{X}{n_b^2 y^2}\frac{l_b^2(k_x^2+k_y^2)}{n_b^2 k_0^2} = 0. \tag{7.109}$$

By solving the above characteristic equation, the eigenvalues of the matrix **T** can be obtained readily. They are

$$\lambda_1 = \left[\left(\frac{l_b^4}{4n_b^4} + \frac{X^2}{n_b^2 Y^2}\right)^{\frac{1}{2}} + \left(1 + \frac{l_b^2}{2n_b^2} + \frac{UX}{n_b^2 Y^2}\right)\right]^{\frac{1}{2}}, \tag{7.110}$$

$$\lambda_2 = -\mathrm{i}\left[\left(\frac{l_b^4}{4n_b^4} + \frac{X^2}{n_b^2 Y^2}\right)^{\frac{1}{2}} - \left(1 + \frac{l_b^2}{2n_b^2} + \frac{UX}{n_b^2 Y^2}\right)\right]^{\frac{1}{2}}, \tag{7.111}$$

$$\lambda_3 = -\lambda_1, \tag{7.112}$$

$$\lambda_4 = -\lambda_2. \tag{7.113}$$

The corresponding eigenvectors are

$$\mathbf{W}^{(1)} = \begin{bmatrix} -P & \mathrm{i}P\,\mathrm{sign}(n_b) & \mathrm{i}\,\mathrm{sign}(n_b) & 1 \end{bmatrix}^{\mathrm{T}}, \tag{7.114}$$

$$\mathbf{W}^{(2)} = \begin{bmatrix} -\mathrm{i}P & P\,\mathrm{sign}(n_b) & -\mathrm{i}\,\mathrm{sign}(n_b) & 1 \end{bmatrix}^{\mathrm{T}}, \tag{7.115}$$

$$\mathbf{W}^{(3)} = \begin{bmatrix} P & -\mathrm{i}P\,\mathrm{sign}(n_b) & \mathrm{i}\,\mathrm{sign}(n_b) & 1 \end{bmatrix}^{\mathrm{T}}, \tag{7.116}$$

$$\mathbf{W}^{(4)} = \begin{bmatrix} \mathrm{i}P & -P\,\mathrm{sign}(n_b) & -\mathrm{i}\,\mathrm{sign}(n_b) & 1 \end{bmatrix}^{\mathrm{T}}, \tag{7.117}$$

where $P = \sqrt{\frac{|n|Y}{X}}$, $\mathrm{sign}(n_b)$ is the sign function.

By now, the eigenvalues and eigenvectors are derived analytically for the quasi-longitudinal approximation of the electromagnetic field in the lower ionosphere.

From the above analyses, it is seen that the extraordinary wave, which corresponds with the eigenvalue λ_1, may travel efficiently downward into the air. The ordinary wave, which corresponds with the eigenvalue λ_2, is an attenuated wave. It is well known that the attenuation is caused by the electric collision, which can be described by the imaginary part of the eigenvalue λ_1.

7.4.1 SLF Field on the Sea Surface

Substituting the eigenvectors into Eq. (7.93), the excitation coefficients C_1 and C_2 can be determined readily:

$$C_1 \approx -\frac{I\,\mathrm{d}l\eta[M_{xz}+\mathrm{i}\,\mathrm{sign}(n_b)M_{yz}]}{16\pi^2(1+M_{zz})} \approx -\frac{I\,\mathrm{d}l\eta l_b[n_b+\frac{\mathrm{sign}(n_b)}{y}]}{16\pi^2 n_b^2}, \tag{7.118}$$

$$C_2 \approx \frac{I\,\mathrm{d}l\eta l_b[n_b-\frac{\mathrm{sign}(n_b)}{y}]}{16\pi^2 n_b^2}. \tag{7.119}$$

With the boundary conditions, the coefficients T_1 and T_2 are obtained readily:

$$T_1 = -\frac{2PC_1}{\sin(\gamma d)}\mathrm{e}^{\mathrm{i}k_0\lambda_1(d-z_0)}, \tag{7.120}$$

$$T_2 = -\frac{2PC_1}{\sin(\gamma d)}\left[\frac{\mathrm{sign}(n_b)(k_0^2-k_x^2)+\mathrm{i}k_xk_y}{k_0\gamma}\right]\mathrm{e}^{\mathrm{i}k_0\lambda_1(d-z_0)}. \tag{7.121}$$

With Eqs. (7.120) and (7.121), and by using the inverse Fourier transform, the transforms of the magnetic field components on the sea surface generated by a VED in the ionosphere can be expressed in the following forms:

$$h_x^{\mathrm{air}}\big|_{z=0} = -\frac{2PC_1}{\sin(\gamma d)}\left[\frac{\mathrm{sign}(n_b)(k_0^2-k_x^2)-\mathrm{i}k_xk_y}{k_0\gamma}\right]\mathrm{e}^{\mathrm{i}k_0\lambda_1(d-z_0)}, \tag{7.122}$$

$$h_y^{\mathrm{air}}\big|_{z=0} = -\frac{2PC_1}{\sin(\gamma d)}\left[\frac{\mathrm{i}(k_0^2-k_x^2)-k_xk_y\,\mathrm{sign}(n_b)}{k_0\gamma}\right]\mathrm{e}^{\mathrm{i}k_0\lambda_1(d-z_0)}. \tag{7.123}$$

In cylindrical coordinates, we have

$$\begin{aligned} h_\rho^{\mathrm{air}}\big|_{z=0} = {}& -\frac{2PC_1\,\mathrm{sign}(n_b)}{k_0\gamma\sin(\gamma d)}\big[k_0^2\mathrm{e}^{\mathrm{i}\,\mathrm{sign}(n_b)\phi}-k_0^2\mathrm{e}^{\mathrm{i}\,\mathrm{sign}(n_b)\phi'} \\ & \times\cos(\phi-\phi')\big]\mathrm{e}^{\mathrm{i}k_0\lambda_1(d-z_0)}, \end{aligned} \tag{7.124}$$

$$\begin{aligned} h_\phi^{\mathrm{air}}\big|_{z=0} = {}& -\frac{2PC_1\,\mathrm{sign}(n_b)}{k_0\gamma\sin(\gamma d)}\big[\mathrm{i}k_0^2\mathrm{e}^{\mathrm{i}\,\mathrm{sign}(n_b)\phi}+\mathrm{sign}(n_b)k_0^2\mathrm{e}^{\mathrm{i}\,\mathrm{sign}(n_b)\phi'} \\ & \times\sin(\phi-\phi')\big]\mathrm{e}^{\mathrm{i}k_0\lambda_1(d-z_0)}. \end{aligned} \tag{7.125}$$

With the substitution of Eqs. (7.124) and (7.125) into Eqs. (7.82) and (7.83), it follows that

$$\begin{aligned} H_\rho\big|_{z=0} = {}& \frac{I\,\mathrm{d}l[M_{xz}+\mathrm{i}\,\mathrm{sign}(n_b)M_{yz}]P}{8k_0\pi(1+M_{zz})}\mathrm{e}^{\mathrm{i}\,\mathrm{sign}(n_b)\phi}\mathrm{e}^{\mathrm{i}k_0\lambda_1(d-z_0)}\,\mathrm{sign}(n_b) \\ & \times\int_{-\infty}^{\infty}\frac{\gamma^2H_0^{(1)}(k\rho)+\frac{k}{\rho}H_1^{(1)}(k\rho)}{\gamma\sin\gamma d}k\,\mathrm{d}k, \end{aligned} \tag{7.126}$$

$$
H_\phi|_{z=0} = \frac{\mathrm{i}I\,\mathrm{d}l[M_{xz} + \mathrm{i}\,\mathrm{sign}(n_b)M_{yz}]P}{8k_0\pi(1+M_{zz})}\mathrm{e}^{\mathrm{i}\,\mathrm{sign}(n_b)\phi}\mathrm{e}^{\mathrm{i}k_0\lambda_1(d-z_0)} \times \int_{-\infty}^{\infty} \frac{k_0^2 H_0^{(1)}(k\rho) - \frac{k}{\rho}H_1^{(1)}(k\rho)}{\gamma\sin\gamma d} k\,\mathrm{d}k. \tag{7.127}
$$

In the above two formulas, there exist poles at $\gamma d = \sqrt{k_0^2 - k_j^2} = j\pi$ ($j = 1, 2, 3, \ldots$) for the integrand function. We denote

$$
\bar{k}_j = \begin{cases} \sqrt{k_0^2 - (\frac{j\pi}{d})^2}; & k_0 \geq \frac{j\pi}{d}, \\ \mathrm{i}\sqrt{(\frac{j\pi}{d})^2 - k_0^2}; & k_0 < \frac{j\pi}{d}. \end{cases} \tag{7.128}
$$

Considering that $\gamma = 0$ is a removable pole for the integrand function, Eqs. (7.126) and (7.127) can be written as follows:

$$
H_\rho(\rho,\phi)|_{z=0} = \frac{\mathrm{i}I\,\mathrm{d}l[M_{xz} + \mathrm{i}\,\mathrm{sign}(n_b)M_{yz}]}{4k_0(1+M_{zz})d}P\mathrm{e}^{\mathrm{i}\,\mathrm{sign}(n_b)\phi}\mathrm{e}^{\mathrm{i}k_0\lambda_1(d-z_0)} \times \sum_{j=0}^{\infty}(-1)^j\left[\left(\frac{j\pi}{d}\right)^2 H_0^{(1)}(\bar{k}_j\rho) + \frac{\bar{k}_j}{\rho}H_1^{(1)}(\bar{k}_j\rho)\right], \tag{7.129}
$$

$$
H_\phi(\rho,\phi)|_{z=0} = -\frac{I\,\mathrm{d}l[M_{xz} + \mathrm{i}\,\mathrm{sign}(n_b)M_{yz}]}{4k_0(1+M_{zz})d}P\mathrm{e}^{\mathrm{i}\,\mathrm{sign}(n_b)\phi}\mathrm{e}^{\mathrm{i}k_0\lambda_1(d-z_0)} \times \sum_{j=0}^{\infty}(-1)^j\left[k_0^2 H_0^{(1)}(\bar{k}_j\rho) - \frac{\bar{k}_j}{\rho}H_1^{(1)}(\bar{k}_j\rho)\right]. \tag{7.130}
$$

When the observation point is located on the sea surface near the bottom of the transmitting antenna, the series in Eqs. (7.129) and (7.130) converge very slowly. In this case, the expanded expression of the factor $\frac{1}{\sin\gamma d}$ can be obtained readily. We write

$$
\frac{1}{\sin\gamma d} = -2\mathrm{i}\mathrm{e}^{\mathrm{i}\gamma d}\sum_{j=0}^{\infty}\mathrm{e}^{\mathrm{i}(2jd)\gamma} = -2\mathrm{i}\sum_{j=0}^{\infty}\mathrm{e}^{\mathrm{i}\gamma(2j+1)d}. \tag{7.131}
$$

Then, the magnetic field component on the sea surface is expressed in the following forms:

$$
H_\rho(\rho,\phi) = \frac{\mathrm{i}I\,\mathrm{d}l[M_{xz} + \mathrm{i}\,\mathrm{sign}(n_b)M_{yz}]}{2k_0\pi(1+M_{zz})}P\mathrm{e}^{\mathrm{i}\,\mathrm{sign}(n_b)\phi}\mathrm{e}^{\mathrm{i}k_0\lambda_1(d-z_0)}\,\mathrm{sign}(n_b) \times \sum_{j=0}^{\infty}\mathrm{e}^{\mathrm{i}k_0R_j}\left[\frac{2}{R_j^3} - \frac{2\mathrm{i}k_0}{R_j^2} - \frac{(2j+1)^2d^2}{R_j^2}\left(\frac{3}{R_j^3} - \frac{3\mathrm{i}k_0}{R_j^2} - \frac{k_0^2}{R_j}\right)\right], \tag{7.132}
$$

$$H_\phi(\rho,\phi) = -\frac{I\,\mathrm{d}l[M_{xz} + \mathrm{i}\,\mathrm{sign}(n_b)M_{yz}]}{2k_0\pi(1+M_{zz})} P\mathrm{e}^{\mathrm{i}\,\mathrm{sign}(n_b)\phi}\mathrm{e}^{\mathrm{i}k_0\lambda_1(d-z_0)}$$
$$\times \sum_{j=0}^{\infty} e^{\mathrm{i}k_0 R_j}\left(-\frac{1}{R_j^3} + \frac{\mathrm{i}k_0}{R_j^2} + \frac{k_0^2}{R_j}\right), \tag{7.133}$$

where $r = [(2j+1)^2 d^2 + \rho^2]^{\frac{1}{2}}$.

By now, in the case of the quasi-longitudinal approximation, the approximated formulas have been obtained for the magnetic field components on the sea surface.

7.4.2 SLF Field on the Sea Surface for Several Special Cases

When the geomagnetic field is in the vertical direction, $\theta_0 = 0°$ or $180°$. For the susceptibility matrix, we have $M_{xy} = M_{yx} = 0$. It is found that the downward wave excited by a VED is very small. In this case, the excitation coefficient of the downward wave is represented as follows:

$$C_1 \approx \frac{I\,\mathrm{d}l\eta[k_x + \mathrm{i}\,\mathrm{sign}(n_b)k_y]}{16\pi^2(1+M_{zz})Pk_0}. \tag{7.134}$$

With the substitution of Eq. (7.134) into Eqs. (7.124) and (7.125), with a similar procedure as in Sect. 7.4.1, the approximated formulas are derived readily for the magnetic field component on the sea surface. We write

$$H_\rho(\rho,\phi)|_{z=0} = -\frac{I\,\mathrm{d}l\mathrm{e}^{\mathrm{i}k_0\lambda_1(d-z_0)}}{2\pi(1+M_{zz})k_0^2}\cdot\left\{\sum_{j=0}^{\infty}\frac{\rho\mathrm{e}^{\mathrm{i}k_0R_j}}{R_j^3}\left[\frac{3}{R_j^2} - \frac{3\mathrm{i}k_0}{R_j} - k_0^2\right.\right.$$
$$\left. + \frac{(2j+1)^2d^2}{R_j^2}\left(-\frac{15}{R_j^2} + \frac{15\mathrm{i}k_0}{R_j} + 6k_0^2 - \mathrm{i}k_0^3R_j\right)\right]$$
$$\left. + k_0^2[1+\mathrm{sign}(n_b)]\sum_{j=0}^{\infty}\frac{\rho\mathrm{e}^{\mathrm{i}k_0R_j}}{R_j}\left(-\frac{1}{R_j^2} + \frac{\mathrm{i}k_0}{R_j}\right)\right\}, \tag{7.135}$$

$$H_\phi(\rho,\phi)|_{z=0} = \frac{\mathrm{i}I\,\mathrm{d}l\mathrm{e}^{\mathrm{i}k_0\lambda_1(d-z_0)}}{2\pi(1+M_{zz})}\sum_{j=0}^{\infty}\frac{\rho}{R_j}\left(-\frac{1}{R_j^2} + \frac{\mathrm{i}k_0}{R_j}\right)\mathrm{e}^{\mathrm{i}k_0R_j}. \tag{7.136}$$

When the geomagnetic field is in the horizontal direction, i.e., $n_b = \cos\theta_0 = 0$, then we have $M_x y = M_x z = M_y x = M_z x = 0$. In SLF range, the eigenvalues of the matrix **T** can be derived easily. We write

$$\lambda^2 = -X + \mathrm{i}Y\frac{\sqrt{Xk_xk_y}}{k_0} + \frac{Y^2}{2}\left(1 - \frac{k_y^2}{k_0^2}\right). \tag{7.137}$$

Because of $X \gg Y^2$, all the eigenvalues will have large imaginary parts, and the wave cannot travel downward into the air. When the dipole is on the equator of the geomagnetic field, the wave cannot go down directly.

When the electric dipole is located in the ionosphere with an arbitrary direction, Eq. (7.79) can be rewritten as follows:

$$\nabla \times \boldsymbol{H}_j = -\mathrm{i}\omega\hat{\varepsilon}_j \boldsymbol{E} + [l_e\hat{x} + m_e\hat{y} + n_e\hat{z}] I\, \mathrm{d}l\delta(x)\delta(y)\delta(z - z_0), \tag{7.138}$$

where l_e, m_e, and n_e are the directional cosines of the electric dipole. Thus, Eq. (7.84) is written in the form

$$\frac{\mathrm{d}\boldsymbol{W}}{\mathrm{d}z} = -\mathrm{i}k_0\mathbf{T} \cdot \boldsymbol{W} + \boldsymbol{f}_e \cdot \delta(z - z_0), \tag{7.139}$$

where

$$\boldsymbol{f}_e = \frac{I\,\mathrm{d}l\eta}{4\pi^2}\left[\frac{n_e k_x}{k_0(1+M_{zz})} \quad \frac{n_e k_y}{k_0(1+M_{zz})} \quad m_e - \frac{n_e M_{yz}}{1+M_{zz}} \quad l_e + \frac{n_e M_{xz}}{1+M_{zz}}\right]^T. \tag{7.140}$$

Then, the excitation coefficient C_1' is approximated as follows:

$$C_1' \approx \frac{I\,\mathrm{d}l\eta}{16\pi^2}\left[l_e + \frac{n_e M_{xz}}{1 + M_{zz}} + \mathrm{i}\,\mathrm{sign}(n_b)\left(m_e - \frac{n_e M_{yz}}{1 + M_{zz}}\right)\right]. \tag{7.141}$$

In the above derivation, replacing the excitation coefficient C_1 by the excitation coefficient C_1', the components of the electromagnetic field on the sea surface may be obtained readily.

7.5 Computations and Discussions

Following the above derivations and analyses, some numerical calculations will be carried out.

In the middle-latitude regions ($10° < \theta < 80°$ or $100° < \theta < 170°$), the geomagnetic field has a vertical component and $n_b^2 Y^2 \gg 1$, and the quasi-longitudinal approximation can be used. In this case, the downward waves include two eigenwaves. The first one, with small attenuation, is defined as the extraordinary wave, while the second one, which is an evanescent wave, is defined as the ordinary wave. In the computations, the parameters are taken as follows: $B_0 = 0.5 \times 10^{-4}$ T, $N = 1.4 \times 10^{12}$ m^{-3}, and $\nu = 10^3$ s^{-1}. Then, we have $\omega_0 = 6.67 \times 10^7$ arc/s and $\omega_H = 8.8 \times 10^6$ arc/s. The magnitudes of the components H_ρ and H_ϕ on the sea surface radiated by a unit electric dipole at the height of 350 km are computed at different angles θ and shown in Figs. 7.7 and 7.8, respectively. Similar calculations are carried out at several different frequencies and shown in Figs. 7.9 and 7.10, respectively.

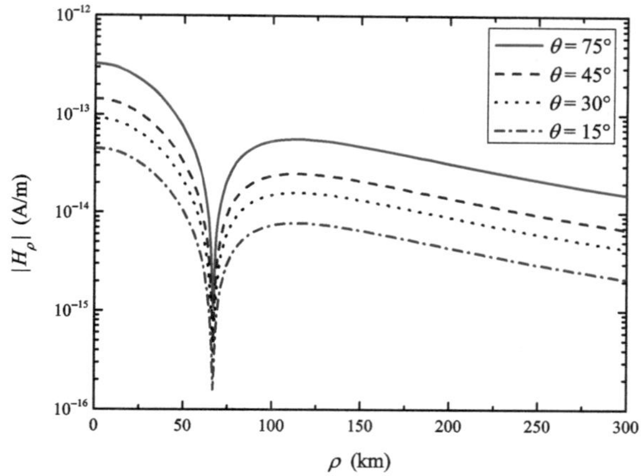

Fig. 7.7 The magnitudes of $|H_\rho|$ in A/m on the sea surface radiated by a space borne VED at different angle θ: $f = 100$ Hz, $I\,dl = 1$ A·m, $d = 90$ km, and $z_0 = 350$ km

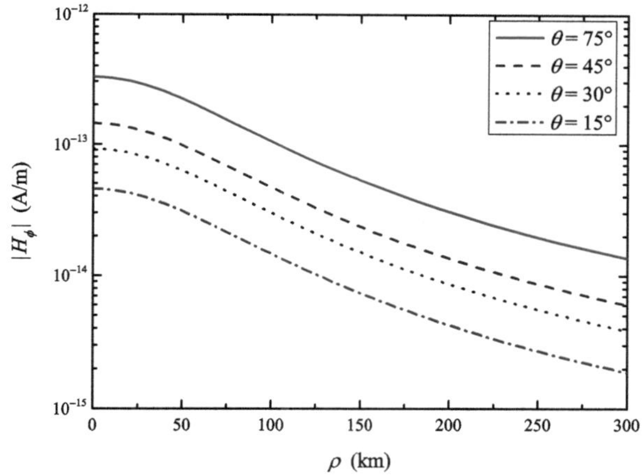

Fig. 7.8 The magnitudes of $|H_\phi|$ in A/m on the sea surface radiated by a space borne VED at different angle θ: $f = 100$ Hz, $I\,dl = 1$ A·m, $d = 90$ km, and $z_0 = 350$ km

From Figs. 7.7 and 7.8, it is seen that when the angle θ varies from 30° to 75°, the attenuation of the downward extraordinary wave will decrease. Correspondingly, the magnetic field component strength on the sea surface will be enhanced. From Figs. 7.9 and 7.10, it is seen that when the operating frequency increases, the attenuation of the downward extraordinary wave will increase. Correspondingly, the magnetic field component strength on the sea surface will decrease. The exciting efficiency of the downward wave is affected by the change of the ionosphere susceptibility parameters with the angular θ. In the region of 50–100 km on the Earth's surface, there exists a minimum value for the magnitude of the component H_ρ. It is explained that this is caused by the interference of the reflections of the electromagnetic wave in the space between the ground and the lower boundary of the ionosphere.

In the neighborhood of the equator of the geomagnetic field, both of the downgoing ordinary and extraordinary waves have a large attenuation rate, thus the SLF wave cannot go down into the air directly.

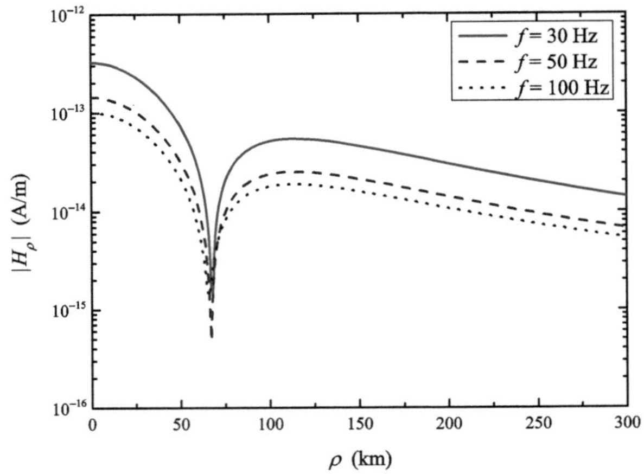

Fig. 7.9 The magnitudes of $|H_\rho|$ in A/m on the sea surface radiated by a space borne VED at different frequencies: $I\,dl = 1$ A·m, $d = 90$ km, $z_0 = 350$ km, and $\theta = 45°$

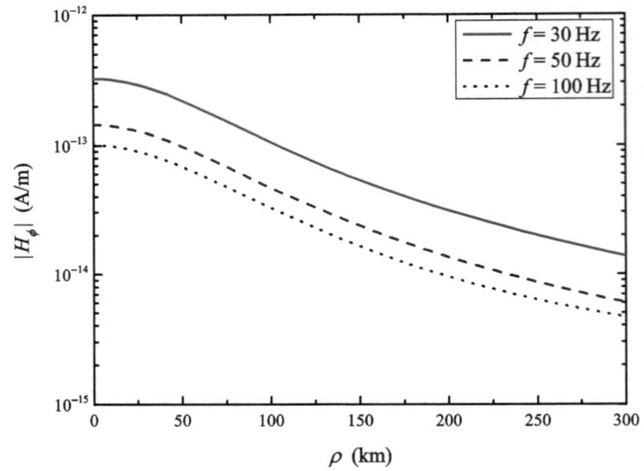

Fig. 7.10 The magnitudes of $|H_\phi|$ in A/m on the sea surface radiated by a space borne VED at different frequencies: $I\,dl = 1$ A·m, $d = 90$ km, $z_0 = 350$ km, and $\theta = 45°$

When the direction of the geomagnetic field is perpendicular to the ground, the excitation efficiency of the downward wave by a VED is reduced extensively. In order to obtain a significant field strength on the sea surface, the transmitting antenna must be placed with a tilt angle to the vertical direction.

References

Abamowitz M, Stegun IA (1972) Handbook of mathematical functions. Dover, New York

Armand NA et al (1988) Experimental researches in the ionosphere of the Earth of the radiation of loop antenna in a range VLF waves, installed onboard the orbital complex "Progress-28"–"Souz TM-2". Radiotech Electron 33:2225–2233

Bannister PB et al (1993) Orbiting transmitter and antenna for space borne communications at ELF/VLF to submerged submarine. In: AGARD conference proceeding, vol 529, pp 33.1–33.14

Budden KG (1961) Radio waves in the ionosphere. Cambridge University Press, Cambridge

Galejs J (1972) Terrestrial propagation of long electromagnetic waves. Pergamon, New York

Gradshteyn IS, Ryzhik IM (1980) Table of integrals, series, and products. Academic Press, New York

Ipatov EB, Lukin DS, Palkin PA (1990) Maslov canonical operator in problems of numerical simulation of diffraction and propagation of waves in inhomogeneous media. Russ J Numer Anal Math Model 5(6):465–488

Kondrat'ev IG, Kudrin AV, Zaboronkova TM (1994) Application of self-consistent VLF plasma antennas in the near Earth space. Turk J Phys 18(11):1248–1253

Koshelev VA, Melnikov VM (1998) Large space structures formed by centrifugal forces. CRC Press, Baca Raton

Li K, Pan WY (1999) Propagation of VLF electromagnetic waves penetrating the low ionosphere. Indian J Radio Space Phys 33:87–94

Li K, Sun XY, Zhai HT (2011) Propagation of ELF electromagnetic waves in the lower ionosphere. IEEE Trans Antennas Propag 59(2):661–666

Markov GL, Mironov VA, Sergeev AM, Sokolova IA (1981) Multibeam self-channeling of plasma waves. Sov Phys JETP 53(6):1183–1186

Morfitt DG, Shellman CH (1976) MODESRCH, an improved computer program for obtaining ELF/VLF/LF mode constants in an Earth–Ionosphere Waveguide. Naval Electronics Laboratory Center Interim Report 77T, NTIS Accession No ADA032573, National Technical Information Service Springfield, VA 22161, USA

Wait JR (1962) Electromagnetic waves in stratified media. Pergamon, London

Xia MY, Chen ZY (2000) Matrix-exponent formulations for wave radiation and propagation in anisotropic stratified ionosphere. IEEE Trans Antennas Propag 48(2):339–341

Zaboronkova IM, Kondratev IG, Kurdin AV (1991) Whistlers in magnetoactive plasma I. Radiofizika 34(9):990–1000. (In Russian)

Chapter 8
Atmospheric Noises in SLF/ELF Ranges

In this chapter, we will attempt to treat the atmospheric noise monitoring, data analysis, and forecasting distribution of atmospheric noise in SLF/ELF ranges.

8.1 Introduction

The performance evaluation of the radio communication system and the received minimum acceptable signal level are dependent on the study of the statistical characteristics of both signal and noise. Especially, the signal-to-noise ratio (SNR) is an important parameter for system design. The radio noise can be classified into internal and external noises. The internal noise is generated by the thermal noise of the receiving system. The external noise includes the background noise and disturbances which originate from the universe, Earth's atmosphere, ground, and man-made industrial system. Generally, the external noise is of non-Gaussian type and has pulse characteristics.

In SLF/ELF ranges, the external noise is mainly from lightning and man-made industrial disturbances, and the atmosphere noise is dominant on the sea surface (Feldman 1972; Enge and Sawate 1988; Meloni et al. 1992; Chrissan and Fraser-Smith 1996a, 1996b). The atmospheric noise with a statistical distribution has significant and complex changes in time domain, geographical domain, and frequency domain (Fraser-Smith and Bowen 1992; Fraser-Smith and Turtle 1993; Chrissan 1998; Fu et al. 2010). So far, not only are there no organizations to do work on SLF/ELF atmospheric noise analysis, but also no independent systematic measured data in China. From the 1970s, with the demands of SLF/ELF submarine communication, a lot of works on the experimental measurements have been carried out in the USA (Feldman 1972; Evans and Griffiths 1974; Ginsberg 1974), and most of measured results have not been published.

In this chapter, based on available materials, including public literature and non-public AD reports, and our recent work, synthesized summaries are carried out. By using the observation data in HF and VLF ranges, the global atmospheric noise distributions in SLF/ELF ranges can be speculated on readily.

W. Pan, K. Li, *Propagation of SLF/ELF Electromagnetic Waves*, Advanced Topics in Science and Technology in China,
DOI 10.1007/978-3-642-39050-0_8,

8.2 The Distribution of SLF/ELF Noise Sources and Its Statistical Properties

In SLF ranges, the dominant noise source is from lightning discharges. Within the Earth's surface area, which is approximately 500 million square kilometers, no matter what the time is, there is always lightning happening somewhere (Price and Rind 1992). On the whole of Earth's surface, there are about 30 thunderstorms per minute. According to statistical data, more than 100 times per second a lightning occurs in the world. It is seen that the lightning is very frequent. However, the geographical distribution of lightning is very uneven. In some places there are dozens, even over 100 of thunderstorm days a year, while it is extremely rare in some other places. The equatorial regions and tropics are the most active thunderstorm areas in the world, where the thunderstorm appears mostly and each thunderstorm lasts the longest time.

In general, the number of thunderstorms decreases from the equator to the Arctic and Antarctic circles in one year. There are about 75 to 100 thunderstorm days in one year in the regions near the equator, and 20 to 50 thunderstorm days in middle-altitude regions, and only a few thunderstorm days near the Arctic and Antarctic circles. North of 82° north-altitude and south of the 65° south-latitude, there occur thunderstorms. In summary, the distribution of thunderstorm in the northern hemisphere is much wider than that in the southern hemisphere.

There are several world-wide thunderstorm centers, which are located in the equatorial and tropical regions. For example, there are about 140 thunderstorm days in Southeast Asia each year, about 180 thunderstorm days in Central Africa, and about 200 thunderstorm days in central Brazil. In Chino fabric, a city on the island of Java in Indonesia, there are 332 days of thunderstorm in average in one year.

In the tropical region, there are thunderstorms throughout the whole year. Generally, a thunderstorm usually occurs on afternoon, and also often occurs at night. In mid-latitude regions of the northern hemisphere, the thunderstorms concentrate mostly in the period from June to July, while the thunderstorms are rare in winter. But there are thunderstorms in winter in Iceland, Norway, and the UK. On a global scale, the thunderstorms usually occur in 16 : 00–18 : 00 (local time), while seldom in 6 : 00–10 : 00 (local time). In continental interior regions a thunderstorm usually occurs in afternoon, while often at night in the coastal region or on an island.

It is well known that there exist complex convective activities in the atmosphere. A thundercloud with a separation of positive and negative charges is the result, while a voltage difference between cloud and cloud or between cloud and ground is caused. When positive and negative charges in the cloud have accumulated enough, and the electric field strength between cloud and cloud or between cloud and ground exceeds the breakdown strength, a discharge current between cloud and cloud or cloud and ground, named a lightning, is generated.

For a lightning, the peak of the discharge current is approximately in the range from thousands of ampères to several hundreds of thousands of ampères, and sometimes over one million ampères. The main discharge duration is about 50–100 microseconds. Since it is a short-term transient pulse, a lightning current has a very

Table 8.1 The divisions of season in one year and intervals in one day in CCIR 322-3

Months in one year	Season		Intervals in one day
	Northern hemisphere	Southern hemisphere	
12, 1, 2	Winter	Summer	00 : 00–04 : 00, 04 : 00–
3, 4, 5	Spring	Autumn	08 : 00, 08 : 00–12 : 00,
6, 7, 8	Summer	Winter	12 : 00–06 : 00, 16 : 00–
9, 10, 11	Autumn	Spring	20 : 00, 20 : 00–24 : 00

wide spectrum. From the excellent monograph (Galejs 1972), it is known that the main spectrum of the lightning flow is in VLF/SLF ranges.

The lightning current between the ground and the cloud is a radiation source with a very broad spectrum. The high-frequency components decay rapidly, while the propagation losses of SLF/ELF components are very small. As addressed in the monograph by Galejs (1972), it is known from the statistical results that the lightning region is the largest in the interval of 12 : 00–20 : 00 UT. In practice, thunderstorms occur at any time in the whole world, and the exact location, time, and intensity of a thunderstorm are unpredictable. Essentially, the atmosphere noise in the receiving point is contributed totally by the lightning radiated field. Therefore, the atmosphere noise in SLF range has complex variances in time domain, geographic domain, and frequency domain, while there is some statistical regularity. The investigation on the statistical distribution of the SLF atmosphere noises is important and useful for the overall design of SLF engineering.

8.3 Atmospheric Noise Data in HF and VLF Ranges

In HF/MF ranges, the long-term measurements in the 27 sites world-wide are carried out for the atmosphere noise data, which organized by URSI. The obtained data are summarized in CCIR Rept-322-2 (1983), CCIR Rept-322-3 (1988), and ITU-RP327-7 (2001), which are important and useful in radio engineering designs in HF/MF ranges. In the two reports, in order to address the distributions of atmosphere noises in different seasons and different time intervals, one year and one day are divided into four seasons and six time intervals. The specific divisions are shown in Table 8.1. It is noted that the distribution in the spectral density of the atmospheric noises may be available in the above two reports. But they are only suitable for the range of 10 kHz–30 MHz. For the SLF/ELF regions, they cannot be obtained from the predicted data.

Based on the accumulations by the communication and navigation engineering in the United Station, the famous monograph *VLF Radio Engineering* was published (Watt 1967), and the global distribution of atmospheric noise in VLF range was included. The inquiry methods for the atmospheric noise in VLF range in Watt's book

Table 8.2 The measured data of the atmospheric noise at $f = 10$ kHz for given location

Year or season	Measured data (10 kHz)		
	Values	Time interval	Location
Year	−91.8 (Max)	Summer 20 : 00–24 : 00	California
	−110.2 (Min)	Summer 04 : 00–08 : 00	The Antarctic
Winter	−94 (Max)	00 : 00–04 : 00	New Zealand
	−108 (Min)	12 : 00–16 : 00	The Antarctic
Spring	−98 (Max)	20 : 00–24 : 00	California
	−110 (Min)	12 : 00–16 : 00	The Antarctic
Summer	−91.8 (Max)	20 : 00–24 : 00	California
	−110.2 (Min)	04 : 00–08 : 00	The Antarctic
Autumn	−98.2 (Max)	16 : 00–20 : 00	New Zealand
	−109.5 (Min)	08 : 00–12 : 00	The Antarctic

Table 8.3 The measured data of the atmospheric noise at $f = 80$ Hz for given location

Year or season	Measured data (80 Hz)		
	Values	Time interval	Location
Year	−128.04 (Max)	Summer 20 : 00–24 : 00	California
	−142.90 (Min)	Autumn 04 : 00–08 : 00	The Antarctic
Winter	−133.97 (Max)	08 : 00–12 : 00	New Zealand
	−142.30 (Min)	00 : 00–04 : 00	The Antarctic
Spring	−134.38 (Max)	20 : 00–24 : 00	California
	−141.90 (Min)	00 : 00–04 : 00	The Antarctic
Summer	−128.04 (Max)	20 : 00–24 : 00	California
	−142.51 (Min)	00 : 00–04 : 00	The Antarctic
Autumn	−132.41 (Max)	20 : 00–24 : 00	California
	−142.90 (Min)	00 : 00–04 : 00	The Antarctic

are similar to those in the CCIR-322 report (1998). It is known that one year is divided into four seasons and one day into six time intervals. The global distribution for the root-mean-square (RMS) values of the atmospheric noise spectral density can then be known at given season and time. For the frequency range of 1 kHz–100 kHz, the data can be estimated by extrapolation according to the relation curve of the atmospheric noise spectral density to the operating frequency, with the spectral density at 10 kHz known. However, this book leaves no information in the range below 1 kHz.

Table 8.4 The speculated data of the atmospheric noise at $f = 10$ kHz

Year or season	Speculated data (10 kHz)		
	Values	Time interval	Location
Year	−86 (Max)	Summer 16 : 00–20 : 00	Caribbean Sea
	−110 (Min)	Autumn 08 : 00–12 : 00	Near the Antarctic
Winter	−86 (Max)	08 : 00–12 : 00	Southeast of Africa North of Australia
	−108 (Min)	04 : 00–8 : 00	Near the Antarctic
Spring	−90 (Max)	12 : 00–16 : 00	Central of Africa Gulf of Guinea
	−110 (Min)	12 : 00–16 : 00	Near the Antarctic
Summer	−86 (Max)	16 : 00–20 : 00	Caribbean Sea
	−110 (Min)	08 : 00–12 : 00	Near the Antarctic
Autumn	−86 (Max)	16 : 00–20 : 00	Central of Africa Brazil
	−108 (Min)	All time intervals	Near the Antarctic

8.4 Speculation of Global Atmospheric Noise Distributions in SLF/ELF Ranges

As a global communications system, the service of SLF/ELF radio communication system will cover almost the entire Earth's surface. Therefore, it is not enough just to know the distributions of the atmospheric noise spectral density in the individual observation points. Meanwhile, the distributions of the atmospheric noise will change significantly with the seasons and time intervals. In the view of the engineering construction, it is urgent to search for a method which is similar to that used in HF/VLF ranges (Watt 1967), to speculate on the distribution of the atmospheric noise in SLF/ELF ranges.

From the 1980s to the 1990s, the monitoring of the atmospheric noise from 10 Hz to 32 kHz in the eight ground-based sites was carried out by Stanford University. With monitoring for several years, some observation data of the four sites and analyses were summarized in the two reports (Chrissan and Fraser-Smith 1996a; Chrissan 1998). Stanford University announced that the four sites were located at Arrival Heights (77.80°S, 193.30°W) in the South Pole, Dunedin (45.80°S, 189.50°W) in New Zealand, the Sondrestromford (67.00°N, 50.10°W) in Greenland, and Stanford (37.40°N, 122.20°W) in California.

In practical engineering design, it is necessary to obtain the materials in the range of the atmospheric noise spectrum density and its maximum and minimum values, as well as those on the seasonal, hourly, and regional distributions of the atmospheric noise. The actual measured data on the four stations were given by the scientists

Table 8.5 The speculated data of the atmospheric noise at $f = 80$ Hz

Year or season	Speculated data (80 Hz)		
	Values	Time interval	Location
Year	−124 (Max)	Summer 12 : 00–16 : 00	Europe
	−144 (Min)	Spring 08 : 00–12 : 00	Near the Antarctic
Winter	−127 (Max)	20 : 00–24 : 00	North of Southern America
	−144 (Min)	04 : 00–08 : 00	Near the Antarctic
Spring	−127 (Max)	16 : 00–20 : 00	Central of Africa
	−144 (Min)	08 : 00–12 : 00	Near the Antarctic
Summer	−124 (Max)	12 : 00–16 : 00	Europe
	−143 (Min)	16 : 00–20 : 00	Near the Antarctic
Autumn	−125 (Max)	20 : 00–24 : 00	Central of Africa North of Southern America
	−143 (Min)	20 : 00–24 : 00	Near the Antarctic

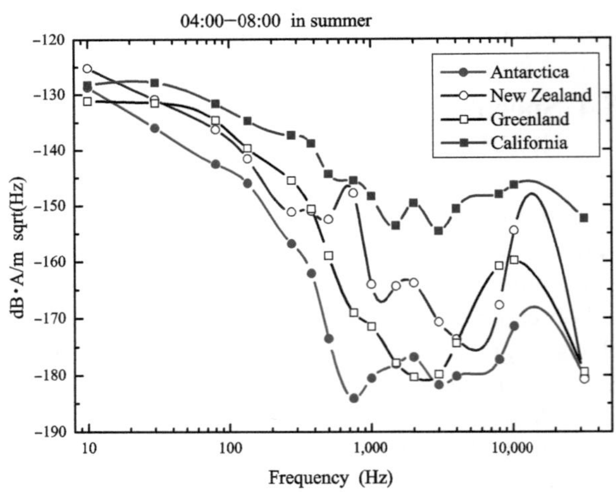

Fig. 8.1 The RMS values of the atmospheric noise spectral density versus the frequency at 4 : 00–8 : 00 in summer

in Stanford University (Chrissan and Fraser-Smith 1996a, 1996b) are listed in Tables 8.2 and 8.3. Specifically, the maximum and minimum values of the atmospheric noise spectrum density at 10 kHz and 80 Hz and the corresponding seasons and time intervals are given. The data for the electric field strength of the atmospheric noise in dB·V/m $\sqrt{\text{Hz}}$ are given in Table 8.2, while the data for the magnetic field strength in dB·A/m are given in Table 8.3. The phase difference by calculating the spectral density of the electric field strength and that of the magnetic field strength is in the range of 51–52 dB.

The global distributions of the atmospheric noise spectrum density at 10 kHz are given by Watt (1967) and listed in Table 8.4. Correspondingly, the global distribu-

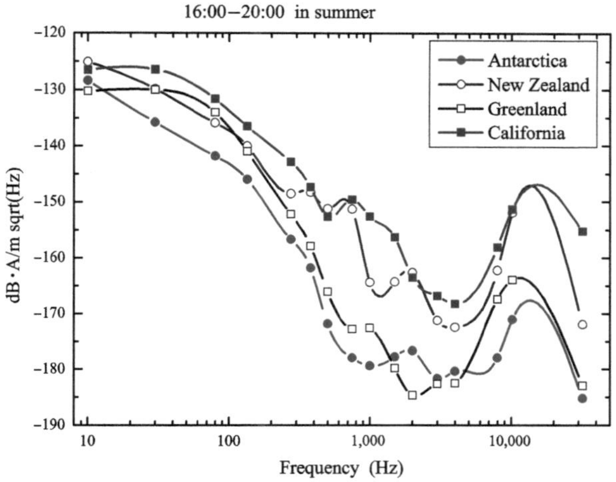

Fig. 8.2 The RMS values of the atmospheric noise spectral density versus the frequency at 16 : 00–20 : 00 in summer

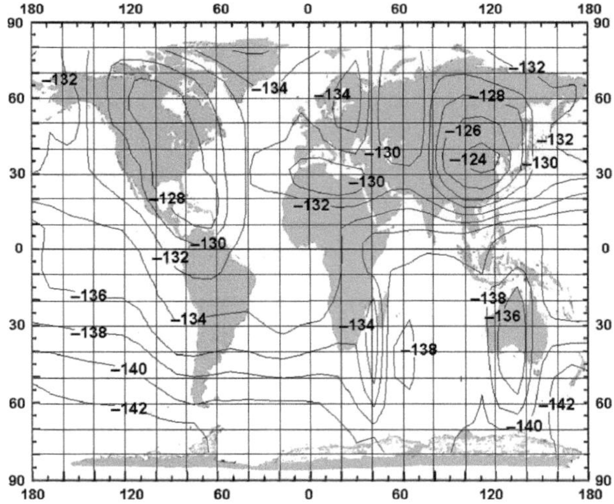

Fig. 8.3 The global contour map for the RMS values of the atmospheric noise spectral density at $f = 80$ Hz in 04 : 00–08 : 00 in summer

tions of the atmospheric noise spectral density at 80 Hz were deduced and listed in Table 8.5. In Tables 8.4 and 8.5, the maximum and minimum values of the atmospheric noise spectrum density at 10 kHz and 80 Hz and the corresponding seasons, time intervals, and regions are listed, respectively. Similarly, the data for the electric field strength of the atmospheric noise in dB·V/m $\sqrt{\text{Hz}}$ are given for 10 kHz, while the data for the magnetic field strength in dB·A/m are given for 80 Hz.

Based on the available results on the global atmospheric noise distribution in VLF range by Watt (1967), by using the curves for the changes between the atmospheric noise spectrum density and the operating frequency in VLF/SLF/ELF ranges, as shown in Figs. 8.1 and 8.2, the global distributions of the atmospheric noise spectral density in SLF/ELF ranges can be deduced readily (Wang et al. 2009). It is noted that the hourly distributions of the atmospheric noise spectral density at

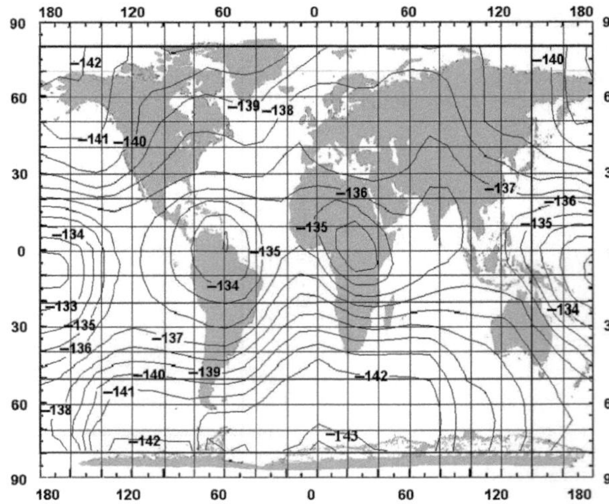

Fig. 8.4 The global contour map for the RMS values of the atmospheric noise spectral density at $f = 80$ Hz in $16:00$–$20:00$ in summer

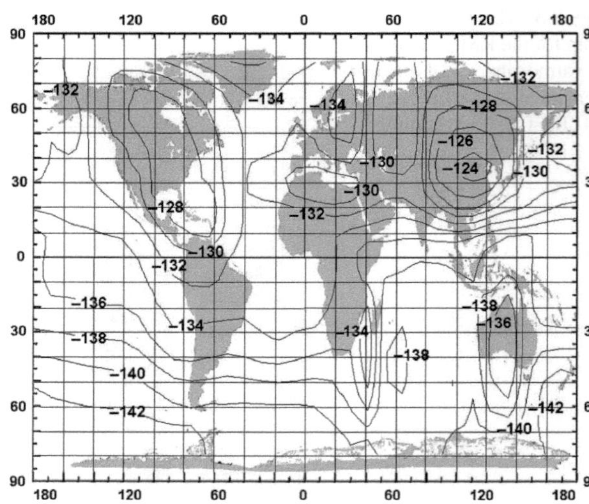

Fig. 8.5 The global distribution of the RMS values of the atmospheric noise spectral density at $f = 90$ Hz in $00:00$–$04:00$ in winter

$f = 80$ Hz are shown in Figs. 8.3–8.4. The global distributions of the RMS values of the atmospheric noise spectral density at $f = 90$ Hz in $00:00$–$04:00$ in winter are deduced readily and shown in Fig. 8.5.

8.5 Statistical Distributions of Atmospheric Noise in SLF/ELF Ranges

Besides the atmospheric noise RMS values, the statistical distributions of the atmospheric noise level are also necessary in engineering applications. It is known that the statistical distributions in SLF/ELF ranges are different from Gaussian distributions. In available works (Ginsberg 1974), the distributions of the atmospheric noise spectrum density are measured at Malta in $f = 33$ Hz, 41 Hz, 83 Hz, and 180 Hz, respectively. It is noted that the data are measured in autumn and spring, where the atmospheric noise level in autumn is significantly larger than that in spring. The distribution of the atmospheric spectrum density was measured at Guam in $f = 33$ Hz, 41 Hz, 83 Hz, and 158 Hz, respectively (Ginsberg 1974). From the data measured in Guam, it is seen that the observed phenomenon is the same as that in Malta. Namely, the atmospheric noise level in autumn was significantly larger than that in spring. Generally, the distributions of the atmospheric noise spectrum density in SLF range are close to log-normal distributions.

References

Chrissan DA (1998) Statistical Analysis and modeling of low-frequency radio noise and improvement of low-frequency communications. USA: ADA360417

Chrissan DA, Fraser-Smith AC (1996a) Seasonal variations of globally measured ELF/VLF radio noise. Radio Sci 31(5):1141–1152

Chrissan DA, Fraser-Smith AC (1996b) Seasonal variations of globally measured SLF/VLF radio noise. Technical Report D177-1, ADA358414/XAB

Enge PK, Sawate DV (1988) Spread-spectrum multiple-access performance of orthogonal codes: impulsive noise. IEEE Trans Commun 36(1):98–106

Evans J, Griffiths A (1974) Design of a Sanguine noise processor based upon world-wide extremely low frequency (ELF) recordings. IEEE Trans Commun 22(4):528–539

Feldman DA (1972) An atmosphere noise model with applications to low frequency navigation systems. PhD thesis, Massachusetts Institute of Technology, Boston, USA

Fraser-Smith AC, Bowen MM (1992) The natural background levels of 50/60 Hz radio noise. IEEE Trans Electromagn Compat 34:330–337

Fraser-Smith AC, Turtle JP (1993) ELF/VLF radio noise measurements at high latitudes during solar particle events. In: AGARD conference proceedings, vol 529, pp 161–168

Fu T, Zhou S, Zhang X (2010) Detection for signal in ELF atmospheric noise interference. In: 2010 3rd IEEE international conference on computer science and information technology (ICCSIT), pp 409–412

Galejs J (1972) Terrestrial propagation of long electromagnetic waves. Pergamon, New York

Ginsberg LH (1974) Extremely low frequency (ELF) atmospheric noise level statistics for Project Sanguine. IEEE Trans Commun 22(4):555–561

International Radio Consultative Committee (1983) Characteristics and applications of atmospheric radio noise. Technical Report, CCIR Recommendation 322-2, CCIR, Geneva, Switzerland

International Radio Consultative Committee (1988) Characteristics and applications of atmospheric radio noise. Technical Report, CCIR Recommendation 322-3, CCIR, Geneva, Switzerland

Mavromatis F et al (2009) A broadband monitoring system for electromagnetic-radiation assessment. IEEE Antennas Propag Mag 51(1):71–79

Meloni A, Palangio P, Fraser-Smith AC (1992) Some characteristics of the ELF/VLF radio noise measured near L'Aquila, Italy. IEEE Trans Antennas Propag 40:233–236

Price C, Rind D (1992) Simulating global lighting distributions from satellite cloud data. In: Proceedings of the 9th international conference on atmospheric electricity, St Petersburg, Russia, pp 327–330

Recommendation International Telecommunications Union (2001) Recommendation international telecommunications union radio noise. Technical Report, ITU Recommendation ITU-RP. 327-7, ITU, Geneva, Switzerland

Wang XQ, Pan WY, Zhang HQ, Ren Q (2009) The global distribution of SLF atmospheric noise level and the comparisons to the measured data. Technical Report, China Research Institute of Radiowave Propagation, Qingdao, China. (In Chinese)

Watt AD (1967) VLF radio engineering. Pergamon, New York

Index

W. Pan, K. Li, *Propagation of SLF/ELF Electromagnetic Waves*,
Advanced Topics in Science and Technology in China,
DOI 10.1007/978-3-642-39050-0,

MIX
Papier aus verantwortungsvollen Quellen
Paper from responsible sources
FSC® C105338

If you have any concerns about our products,
you can contact us on
ProductSafety@springernature.com

In case Publisher is established outside the EU,
the EU authorized representative is:
Springer Nature Customer Service Center GmbH
Europaplatz 3, 69115 Heidelberg, Germany

Printed by Libri Plureos GmbH
in Hamburg, Germany